Geophysical Monograph Series

Including
IUGG Volumes
Maurice Ewing Volumes
Mineral Physics Volumes

Geophysical Monograph Series

155 **The Inner Magnetosphere: Physics and Modeling** *Tuija I. Pulkkinen, Nikolai A. Tsyganenko, and Reiner H. W. Friedel (Eds.)*

156 **Particle Acceleration in Astrophysical Plasmas: Geospace and Beyond** *Dennis Gallagher, James Horwitz, Joseph Perez, Robert Preece, and John Quenby (Eds.)*

157 **Seismic Earth: Array Analysis of Broadband Seismograms** *Alan Levander and Guust Nolet (Eds.)*

158 **The Nordic Seas: An Integrated Perspective** *Helge Drange, Trond Dokken, Tore Furevik, Rüdiger Gerdes, and Wolfgang Berger (Eds.)*

159 **Inner Magnetosphere Interactions: New Perspectives From Imaging** *James Burch, Michael Schulz, and Harlan Spence (Eds.)*

160 **Earth's Deep Mantle: Structure, Composition, and Evolution** *Robert D. van der Hilst, Jay D. Bass, Jan Matas, and Jeannot Trampert (Eds.)*

161 **Circulation in the Gulf of Mexico: Observations and Models** *Wilton Sturges and Alexis Lugo-Fernandez (Eds.)*

162 **Dynamics of Fluids and Transport Through Fractured Rock** *Boris Faybishenko, Paul A. Witherspoon, and John Gale (Eds.)*

163 **Remote Sensing of Northern Hydrology: Measuring Environmental Change** *Claude R. Duguay and Alain Pietroniro (Eds.)*

164 **Archean Geodynamics and Environments** *Keith Benn, Jean-Claude Mareschal, and Kent C. Condie (Eds.)*

165 **Solar Eruptions and Energetic Particles** *Natchimuthukonar Gopalswamy, Richard Mewaldt, and Jarmo Torsti (Eds.)*

166 **Back-Arc Spreading Systems: Geological, Biological, Chemical, and Physical Interactions** *David M. Christie, Charles Fisher, Sang-Mook Lee, and Sharon Givens (Eds.)*

167 **Recurrent Magnetic Storms: Corotating Solar Wind Streams** *Bruce Tsurutani, Robert McPherron, Walter Gonzalez, Gang Lu, José H. A. Sobral, and Natchimuthukonar Gopalswamy (Eds.)*

168 **Earth's Deep Water Cycle** *Steven D. Jacobsen and Suzan van der Lee (Eds.)*

169 **Magnetospheric ULF Waves: Synthesis and New Directions** *Kazue Takahashi, Peter J. Chi, Richard E. Denton, and Robert L. Lysal (Eds.)*

170 **Earthquakes: Radiated Energy and the Physics of Faulting** *Rachel Abercrombie, Art McGarr, Hiroo Kanamori, and Giulio Di Toro (Eds.)*

171 **Subsurface Hydrology: Data Integration for Properties and Processes** *David W. Hyndman, Frederick D. Day-Lewis, and Kamini Singha (Eds.)*

172 **Volcanism and Subduction: The Kamchatka Region** *John Eichelberger, Evgenii Gordeev, Minoru Kasahara, Pavel Izbekov, and Johnathan Lees (Eds.)*

173 **Ocean Circulation: Mechanisms and Impacts—Past and Future Changes of Meridional Overturning** *Andreas Schmittner, John C. H. Chiang, and Sidney R. Hemming (Eds.)*

174 **Post-Perovskite: The Last Mantle Phase Transition** *Kei Hirose, John Brodholt, Thorne Lay, and David Yuen (Eds.)*

175 **A Continental Plate Boundary: Tectonics at South Island, New Zealand** *David Okaya, Tim Stem, and Fred Davey (Eds.)*

176 **Exploring Venus as a Terrestrial Planet** *Larry W. Esposito, Ellen R. Stofan, and Thomas E. Cravens (Eds.)*

177 **Ocean Modeling in an Eddying Regime** *Matthew Hecht and Hiroyasu Hasumi (Eds.)*

178 **Magma to Microbe: Modeling Hydrothermal Processes at Oceanic Spreading Centers** *Robert P. Lowell, Jeffrey S. Seewald, Anna Metaxas, and Michael R. Perfit (Eds.)*

179 **Active Tectonics and Seismic Potential of Alaska** *Jeffrey T. Freymueller, Peter J. Haeussler, Robert L. Wesson, and Göran Ekström (Eds.)*

180 **Arctic Sea Ice Decline: Observations, Projections, Mechanisms, and Implications** *Eric T. DeWeaver, Cecilia M. Bitz, and L.-Bruno Tremblay (Eds.)*

181 **Midlatitude Ionospheric Dynamics and Disturbances** *Paul M. Kintner, Jr., Anthea J. Coster, Tim Fuller-Rowell, Anthony J. Mannucci, Michael Mendillo, and Roderick Heelis (Eds.)*

182 **The Stromboli Volcano: An Integrated Study of the 2002–2003 Eruption** *Sonia Calvari, Salvatore Inguaggiato, Giuseppe Puglisi, Maurizio Ripepe, and Mauro Rosi (Eds.)*

183 **Carbon Sequestration and Its Role in the Global Carbon Cycle** *Brian J. McPherson and Eric T. Sundquist (Eds.)*

184 **Carbon Cycling in Northern Peatlands** *Andrew J. Baird, Lisa R. Belyea, Xavier Comas, A. S. Reeve, and Lee D. Slater (Eds.)*

185 **Indian Ocean Biogeochemical Processes and Ecological Variability** *Jerry D. Wiggert, Raleigh R. Hood, S. Wajih A. Naqvi, Kenneth H. Brink, and Sharon L. Smith (Eds.)*

186 **Amazonia and Global Change** *Michael Keller, Mercedes Bustamante, John Gash, and Pedro Silva Dias (Eds.)*

187 **Surface Ocean–Lower Atmosphere Processes** *Corinne Le Quèrè and Eric S. Saltzman (Eds.)*

188 **Diversity of Hydrothermal Systems on Slow Spreading Ocean Ridges** *Peter A. Rona, Colin W. Devey, Jérôme Dyment, and Bramley J. Murton (Eds.)*

189 **Climate Dynamics: Why Does Climate Vary?** *De-Zheng Sun and Frank Bryan (Eds.)*

Geophysical Monograph 190

The Stratosphere:
Dynamics, Transport, and Chemistry

L. M. Polvani
A. H. Sobel
D. W. Waugh
Editors

American Geophysical Union
Washington, DC

Library of Congress Cataloging-in-Publication Data

The stratosphere : dynamics, transport, and chemistry / L.M. Polvani, A.H. Sobel, D.W. Waugh, editors.
  p. cm. — (Geophysical monograph series ; 190)
 Includes bibliographical references and index.
 ISBN 978-0-87590-479-5 (alk. paper)
 1. Stratosphere. 2. Whirlwinds. 3. Dynamic meteorology. I. Polvani, L. M. (Lorenzo M.), 1961- II. Sobel, Adam H., 1967- III. Waugh, D. W. (Darryn W.)
 QC881.2.S8S875 2010
 551.51'42—dc22

                    2010043355

 ISBN: 978-0-87590-479-5
 ISSN: 0065-8448

Cover Image: A stupendous show of nacreous clouds photographed in September 2003 at McMurdo Station, Antarctica. These iridescent clouds, also known as "polar stratospheric clouds," are observed in the lower stratosphere during spring and play a crucial role in the formation of the ozone hole. Photo courtesy of Seth White.

CONTENTS

Foreword: R. Alan Plumb—A Brief Biographical Sketch and Personal Tribute
Adam H. Sobel vii

Preface
Lorenzo M. Polvani, Adam H. Sobel, and Darryn W. Waugh xiii

Introduction
Darryn W. Waugh and Lorenzo M. Polvani 1

Middle Atmosphere Research Before Alan Plumb
Marvin A. Geller 5

Planetary Waves and the Extratropical Winter Stratosphere
R. Alan Plumb 23

Stratospheric Polar Vortices
Darryn W. Waugh and Lorenzo M. Polvani 43

Annular Modes of the Troposphere and Stratosphere
Paul J. Kushner 59

Stratospheric Equatorial Dynamics
Lesley J. Gray 93

Gravity Waves in the Stratosphere
M. Joan Alexander 109

Variability and Trends in Stratospheric Temperature and Water Vapor
William J. Randel 123

Trace Gas Transport in the Stratosphere: Diagnostic Tools and Techniques
Mark R. Schoeberl and Anne R. Douglass 137

Chemistry and Dynamics of the Antarctic Ozone Hole
Paul A. Newman 157

Solar Variability and the Stratosphere
Joanna D. Haigh 173

AGU Category Index 189

Index 191

FOREWORD:
R. Alan Plumb—A Brief Biographical Sketch and Personal Tribute

Adam H. Sobel

Department of Applied Physics and Applied Mathematics and Department of Earth and Environmental Sciences, Columbia University
New York, New York, USA
Lamont-Doherty Earth Observatory, Earth Institute at Columbia University, Palisades, New York, USA

Raymond Alan Plumb was born on 30 March 1948 in Ripon, Yorkshire, United Kingdom. He is not known for talking about his childhood, but we do know that he liked to sing and was part of a group called the Avocets.

Alan did his undergraduate degree in Manchester, obtaining his BS Physics with I Honors in 1969. He was offered a fellowship to do his PhD at Cambridge, but he had a negative reaction to a visit there and decided to stay at Manchester, where he pursued his studies in Astronomy, completing his PhD in 1972. With a highly disengaged thesis advisor, Alan was largely self-taught as a graduate student. He studied planetary atmospheres. Toward the end of his studies, Alan participated in a summer school organized by Steve Thorpe in Bangor, Wales, where he came into contact with the broader international community in geophysical fluid dynamics. Raymond Hide became particularly influential and became Alan's mentor at the UK Meteorological Office (UKMO), where Alan worked for 4 years after receiving his PhD. Another key early influence whom Alan met then was Michael McIntyre. McIntyre's interest and encouragement were very important to Alan at that early time and would continue to be so in later years, including after his move to Australia.

Alan's first peer-reviewed journal article, "Momentum transport by the thermal tide in the stratosphere of Venus" [*Plumb*, 1975] was based on his PhD thesis, though it came out several years after his degree. This first paper shows that even at this early point in his career Alan was a mature scientist, with an approach that has since remained remarkably

constant. The young Dr. Plumb was already an expert practitioner of what we now know as classic geophysical fluid dynamics. His mathematics is elegant and sophisticated but never more complex than necessary and is combined with great physical insight and clarity of exposition. Certain themes from this and his other earliest papers have stayed at the forefront of his work to the present: angular momentum; wave-mean flow interaction; and the interplay of conservative and nonconservative processes (advective and diffusive transport and sources and sinks of tracers). Above all, one finds in these early papers an author seeking the most direct route from fundamental physical laws to observed behavior.

Alan's first position at the UKMO was Scientific Officer, then Senior Scientific Officer. As a member of Hide's group, Alan had great freedom to pursue his interests in fundamental geophysical fluid dynamics (GFD). UKMO policy at that time, however, commonly required anyone in Alan's position to switch groups after 3 years or so. In Alan's case, any other group he might have joined likely would have given him greater operational responsibilities, taking him away from basic research. Largely in response to this, Alan moved in 1976 to the Commonwealth Scientific and Industrial Research Organisation (CSIRO) in Aspendale, a suburb of Melbourne, Australia.

CSIRO was at that time hospitable to basic, curiosity-driven research. It was very strong in dynamical and physical meteorology with a roster of young scientists whose names are now familiar to many in our field (e.g., Webster, Stephens, Frederiksen, and Baines). Alan's contributions in stratospheric dynamics drew international attention and were proudly touted by the lab in annual reports at the time.

Alan's papers from the early CSIRO years cover a mix of explicitly middle-atmospheric topics (quasi-biennial oscillation (QBO), equatorial waves, meridional circulations, sudden warmings, and mesospheric 2 day waves) with theoretical

The Stratosphere: Dynamics, Transport, and Chemistry
Geophysical Monograph Series 190
Copyright 2010 by the American Geophysical Union.
10.1029/2010GM000998

GFD papers whose applicability was broader, though they may have been motivated by stratospheric problems. One of my favorites in the latter category is *Plumb* [1979]. In this paper, Alan shows that transport of a scalar by small-amplitude waves is diffusive in character if either the scalar is subject to damping (such as Newtonian cooling in the case of temperature, or in the case of a chemical species, reactions that can be represented as relaxation toward a chemical equilibrium state) or the waves are growing in time. At the same time, it also showed that the eddy fluxes often do not appear diffusive because when the waves are almost steady and conservative, the fluxes are dominated by the off-diagonal (i.e., advective) components of the diffusion tensor wherever the Stokes drift is nonzero, as it usually is. That was not realized at the time, and it showed how important it is to use the residual, not the Eulerian mean, velocity as the advecting velocity when trying to parameterize eddy transport. Though not one of Alan's most cited papers (as of this writing, it is ranked sixteenth, with 93 citations), this one is a contribution of the most fundamental sort. Diffusion, in the sense of Fick or Fourier (in which the local time tendency of some scalar field is proportional to its Laplacian in space), is by far the simplest and best understood transport process. It is of great value to know when nominally more complex processes lead to diffusive behavior. A. Einstein showed that Brownian motion leads to diffusive transport when viewed statistically on large scales, and G. I. Taylor showed that fluid turbulence, under some circumstances, does as well. Linear waves and turbulence are entirely different sorts of fluid flows, so Alan's explanation of the diffusive as well as advective character of linear waves deserves, in my view, to be mentioned in the same sentence as Einstein's and Taylor's papers in any historical discussion of tracer transport in fluids.

Another favorite of mine is *Plumb* [1986] in which Alan generalizes the quasi-geostrophic Eliassen-Palm flux to three dimensions. This was a great demonstration of technical mastery, but more importantly, a work of fundamental significance, building the basic toolbox our field needs to understand cause and effect in the atmosphere. Few scientists are able both to recognize when problems like this need to be solved and to solve them.

One of Alan's more dramatic achievements at CSIRO was the tank experiment demonstrating in the laboratory the mechanism for the quasi-biennial oscillation [*Plumb and McEwan*, 1978]. Figure 1 [from *Garratt et al.*, 1998] shows Alan explaining this experiment to a group of visitors to CSIRO. *Lindzen and Holton* [1968] had proposed that upward propagating gravity waves, with time scales of days or

Figure 1. Alan Plumb shows his QBO water tank experiment to Bill Priestley and other dignitaries at CSIRO [from *Garratt et al.*, 1998]. © Copyright CSIRO Australia.

less, interacted systematically with the mean flow to generate an oscillation in the stratospheric winds with a period of over 2 years. The mechanism was inherently multiscale and nonlinear, with the amplitude of the waves determining the frequency of the QBO. While this idea must have seemed exotic at the time, its essential elements were familiar to Alan from his thesis work on wave-mean flow interaction in the Venusian atmosphere. Characteristic of Alan's later work both in research and education, his essential contribution was not only in understanding the physics better than most others (as demonstrated by several classic papers from the early CSIRO period [*Plumb*, 1977; *Plumb and Bell*, 1982a, 1982b] in which Alan fleshed out the skeleton of the Holton-Lindzen theory, painting a physical picture of the QBO in three dimensions that in many respects stands unchanged today) but in recognizing what made it difficult for others to understand and how to make it easier for them.

Alan's colleagues from the CSIRO period describe him as one of the leading lights of the field in Australia at the time and as an unselfish collaborator. Robert Vincent, of Adelaide University, recounted to me regular trips Alan made to Adelaide, a relative backwater compared to Melbourne. Alan brought with him all the latest theoretical developments, but he was also profoundly interested in and knowledgeable about observations. With Vincent's group, Alan played an instrumental role in developing a technique to estimate mesospheric eddy momentum fluxes from radar measurements. Robert Bell (CSIRO) was employed as a computer programmer working with different investigators and wrote the code used to obtain the results detailed by *Plumb and Bell* [1982a, 1982b]; Bell recounted the pleasure and satisfaction of working with Alan on this project and also how it helped to establish his (Bell's) career, bringing him recognition and subsequent collaborations with other scientists.

Alan's colleagues from his Australian period also describe him with much fondness as a good friend with an active social life. He served as stage manager for a local musical theater company (though he claims that he did not sing any roles), played volleyball, and brewed a strong beer. In hearing these recollections and others, one gets hints of certain nonscientific anecdotes whose existence is acknowledged, but whose details are not divulged, at least not to Alan's students (i.e., me). It seems that Alan's reputation as the most reserved of Englishmen has been earned partly through occasional departures from that role, though the details are likely to remain unknown to those who were not near him in Melbourne at that time.

Later in Alan's time at CSIRO, during the mid and late 1980s, his scientific interests evolved toward transport problems of more direct relevance to stratospheric chemistry, more direct interaction with the comprehensive numerical models of the time, and more collaboration with American scientists. The latter may have been in part a consequence of an extended visit to NOAA's Geophysical Fluid Dynamics Laboratory in 1982.

After 1985, the discovery of the ozone hole drove excitement and growth in the study of the stratosphere. Despite the ozone hole's location in the Southern Hemisphere, much of the activity was in the United States, where F. Sherwood Rowland and Mario Molina had made the original predictions of ozone loss due to chlorofluorocarbons (CFCs). In the late 1980s, NASA began a series of aircraft experiments to better assess the chemistry and transport of ozone and the key species influencing it. Alan would play an important role in these experiments after his move to the United States in 1988, and perhaps this move was partly motivated by a desire to be closer to the center of things.

Also, however, CSIRO was changing to favor more applied work funded by short-term contracts, which made it more difficult for Alan (and other basic researchers, many of whom left around this time) to pursue his interests. Alan's international reputation earned him an offer of a faculty position at the Massachusetts Institute of Technology (MIT) in the great department that had been home to Jule Charney, Ed Lorenz, Victor Starr, and others and still was arguably the leading department in GFD. In 1988, Alan moved to the United States for reasons similar to those which had brought him to Australia: at MIT he could better pursue his interest in the basic physics controlling the circulation of the Earth's atmosphere.

At MIT, Alan's interests continued to broaden. One new direction, motivated by his participation in the NASA aircraft experiments, was in nonlinear polar vortex dynamics and transport. With Darryn Waugh, Alan used the contour advection with surgery approach to diagnosing (and even forecasting during field experiments) the generation of fine-scale filaments of polar vortex air in the midlatitude surf zone due to Rossby wave breaking events [*Waugh and Plumb*, 1994; *Waugh et al.*, 1994; *Plumb et al.*, 1994]. The discovery that the formation of such fine-scale features could be accurately predicted using only low-resolution meteorological data was a remarkable breakthrough that spawned a huge number of follow-on studies, theoretical and applied, by many other researchers.

Another new thread in Alan's portfolio was tropical tropospheric dynamics, particularly the dynamics of the Hadley circulation and monsoons [*Plumb and Hou*, 1992; *Hsu and Plumb*, 2000; *Plumb*, 2007b; *Privé and Plumb*, 2007a, 2007b; *Clift and Plumb*, 2008]. At first glance, this topic may seem disconnected from Alan's work on the stratosphere. Once one recognizes the central role played by angular momentum in this work, the connection is clear; one of the central results in the now classical axisymmetric theory developed by Edwin Schneider and Richard Lindzen [*Schneider and Lindzen*, 1977; *Schneider*, 1977], Isaac Held and Arthur Hou [*Held and Hou*, 1980], and then Alan is known as Hide's theorem, due to Alan's former mentor.

Perhaps the most broadly influential of all the work from Alan's first decade at MIT is a remarkable series of papers that grew out of Alan's study of tracer-tracer correlations in aircraft data. The series really begins with *Plumb and McConalogue* [1988], but the central ideas were established in the mind of the community by *Plumb and Ko* [1992]. This study clarified the conditions under which compact relations between simultaneous measurements of different tracers would be expected and the further conditions under which those relations would be linear, and it generally clarified the roles of transport and chemistry in creating or breaking these compact relations. It continues with *Hall and Plumb* [1994], which clearly defined the concept of age of air, continues further with *Plumb* [1996], which broadened the theory of *Plumb and Ko* [1992] to include an isolated tropics, or tropical pipe, and then has continued since with further developments [*Waugh et al.*, 1997; *Neu and Plumb*, 1999; *Plumb*, 2007a].

It is difficult to overstate the impact this work had on the field at the time. I had the good fortune to be Alan's student during this period, and he gave me the opportunity to attend a number of conferences and workshops. The roughly decade-long wave of excitement and rapid progress (and funding) in stratospheric chemistry and transport that followed the discovery of the ozone hole had not yet passed, and avalanches of results from new field experiments, satellite measurements, and numerical models of stratospheric trace gases were still pouring in at these meetings. Alan was unquestionably the most important theorist in this scene. He cast a long shadow over each meeting, even if he was not there and even though he didn't say much (apart from his own presentations) when he was. As soon as each new Plumb paper became available (often before publication), other scientists from many institutions would scramble to reorient their research, doing their best to make use of Alan's new insights or to use their own tools to try to address the new questions Alan's new conceptual framework raised.

In more recent years, Alan's work has evolved in new directions again. One of these is stratosphere-troposphere interaction, where Alan has turned his attention to the physics of annular modes and the mechanisms by which stratospheric dynamics may influence tropospheric weather. Another is physical oceanography. Here many of the ideas that evolved through the work of Alan and others in the context of the stratosphere are relevant, directly or indirectly, to the ocean; the ocean is, as is often said, more like the stratosphere than it is like the troposphere, because of the relative weakness of vertical mixing processes and internal heating and resulting strong control exerted by stratification.

Since his move to MIT, Alan has been an educator as well as a research scientist. His record as a teacher and mentor is perhaps less widely known than his research record, but it is no less stellar. Here I can speak from my own personal experience as well as that of all the other alumni I have come to know who worked with Alan or took his courses before, during, and after my time as Alan's student at MIT.

Alan's classroom courses are models of clarity. The experience of taking one of them is basically a semester-long, much more in-depth version of the experience of reading one of Alan's journal articles. One feels that one has been taken from a point of ignorance to a point of deep understanding by the shortest route. This is a very rare experience, not at all common to all classroom teachers, even those few whose research records are comparable to Alan's. His lecture notes on middle atmosphere dynamics are, in my view, better than any textbook on the subject, though it is the field's loss that he has never published them. He has, more recently, coauthored with John Marshall an outstanding textbook [*Marshall and Plumb*, 2008] based on their undergraduate course.

As a mentor (speaking again from my own experience), Alan was hands-off while still providing critical insightful guidance. Owing to the many demands on Alan's time, I could not necessarily get to see him very frequently or on short notice. When I did, the dynamic range of his reactions to the results I showed him was narrow; it took me a year or two to learn that a furrowed brow and mildly perplexed look was a pretty negative reaction even if not accompanied by any harsh words, while the phrase "that's good" was the highest praise. Once I understood that, Alan was the best of mentors. If I was doing well, he let me go my own way, allowing me to develop as a scientist without micromanagement. If I started to drift in an unproductive direction, I was redirected in a way that left me feeling wiser rather than chastised. In a discussion with Alan, no words were wasted, at least none of his. Whatever the source of my confusion, Alan grasped it quickly and saw how to move me past it.

Alan's former graduate students, postdocs, and junior collaborators on whom his influence has been formative have gone on to positions of prominence at a wide range of scientific institutions around the world; on the faculty of Columbia University alone, where the PlumbFest was held, three of us (Lorenzo Polvani, Tim Hall, and myself) consider ourselves Alan's proteges.

Alan is famous among all who have encountered him, either at MIT or in the broader scientific sphere, for the kind respect with which he treats everyone. Alan never makes one feel stupid, even when one is. This trait stands out because it is far from universal among scientists of Alan's caliber (or even much lesser ones).

At the present time, Alan continues down the path he has been on since the start of his career in Manchester: finding elegant solutions to difficult and important scientific problems and explaining them in the most effective and clear way to students and colleagues. On the occasion of his 60th birthday, some of us gathered in New York City to mark the occasion and to discuss the science of the stratosphere, to which he has contributed so much. On behalf of those of us who were present there, and those who were not but shared our feelings, I wish Alan health, happiness, and many more years in which to keep doing what he does.

Acknowledgments. Conversations with a number of people informed this piece, though I take responsibility for any errors. I thank Robert Bell, Paul Fraser, Jorgen Frederiksen, Harry Hendon, Michael McIntyre, and Robert Vincent, as well as, of course, Alan himself, for discussions and insight into R. Alan Plumb's career. Darryn Waugh provided useful feedback on the first draft.

REFERENCES

Clift, P. D., and R. A. Plumb (2008), *The Asian Monsoon: Causes, History and Effects*, Cambridge Univ. Press, Cambridge, U. K.

Garratt, J., D. Angus, and P. Holper (1998), *Winds of Change: Fifty Years of Achievements in the CSIRO Division of Atmospheric Research 1946–1996*, 1st ed., CSIRO, Collingwood, Victoria, Australia

Hall, T. M., and R. A. Plumb (1994), Age as a diagnostic of stratospheric transport, *J. Geophys. Res.*, *99*, 1059–1070.

Held, I. M., and A. Y. Hou (1980), Nonlinear axially symmetric circulations in a nearly inviscid atmosphere, *J. Atmos. Sci.*, *37*, 515–533.

Hsu, C.-H., and R. A. Plumb (2000), Nonaxisymmetric thermally driven circulations and upper-tropospheric monsoon dynamics, *J. Atmos. Sci.*, *57*, 1255–1276.

Lindzen, R. S., and J. R. Holton (1968), A theory of the quasi-biennial oscillation, *J. Atmos. Sci.*, *25*, 1095–1107.

Marshall, J., and R. A. Plumb (2008), *Atmosphere, Ocean, and Climate Dynamics: An Introductory Text*, Elsevier, New York

Neu, J. L., and R. A. Plumb (1999), Age of air in a "leaky pipe" model of stratospheric transport, *J. Geophys. Res.*, *104*, 19,243–19,255.

Plumb, R. A. (1975), Momentum transport by the thermal tide in the stratosphere of Venus, *Q. J. R. Meteorol. Soc.*, *101*, 763–776.

Plumb, R. A. (1977), The interaction of two internal gravity waves with the mean flow: Implications for the theory of the quasi-biennial oscillation, *J. Atmos. Sci.*, *34*, 1847–1858.

Plumb, R. A. (1979), Eddy fluxes of conserved quantities by small-amplitude waves, *J. Atmos. Sci.*, *36*, 1699–1704.

Plumb, R. A. (1986), Three-dimensional propagation of transient quasi-geostrophic eddies and its relationship with the eddy forcing of the time-mean flow, *J. Atmos. Sci.*, *43*, 1657–1678.

Plumb, R. A. (1996), A "tropical pipe" model of stratospheric transport, *J. Geophys. Res.*, *101*, 3957–3972.

Plumb, R. A. (2007a), Tracer interrelationships in the stratosphere, *Rev. Geophys.*, *45*, RG4005, doi:10.1029/2005RG000179.

Plumb, R. A. (2007b), Dynamical constraints on monsoon circulations, in *The Global Circulation of the Atmosphere*, edited by T. Schneider, and A. H. Sobel, Princeton Univ. Press, Princeton, N. J.

Plumb, R. A., and R. C. Bell (1982a), Equatorial waves in steady zonal shear flow, *Q. J. R. Meteorol. Soc.*, *108*, 313–334.

Plumb, R. A., and R. C. Bell (1982b), A model of the quasi-biennial oscillation on an equatorial beta-plane, *Q. J. R. Meteorol. Soc.*, *108*, 335–352.

Plumb, R. A., and A. Hou (1992), The response of a zonally-symmetric atmosphere to subtropical thermal forcing, *J. Atmos. Sci.*, *49*, 1790–1799.

Plumb, R. A., and M. K. W. Ko (1992), Interrelationships between mixing ratios of long-lived stratospheric constituents, *J. Geophys. Res.*, *97*, 10,145–10,156.

Plumb, R. A., and D. D. McConalogue (1988), On the meridional structure of long-lived tropospheric constituents, *J. Geophys. Res.*, *93*, 15,897–15,913.

Plumb, R. A., and A. D. McEwan (1978), The instability of a forced standing wave in a viscous stratified fluid: A laboratory analogue of the quasi-biennial oscillation, *J. Atmos. Sci.*, *35*, 1827–1839.

Plumb, R. A., D. W. Waugh, R. J. Atkinson, P. A. Newman, L. R. Lait, M. R. Schoeberl, E. V. Browell, A. J. Simmons, and M. Loewenstein (1994), Intrusions into the lower stratospheric Arctic vortex during the winter of 1991–1992, *J. Geophys. Res.*, *99*, 1089–1105.

Privé, N. C., and R. A. Plumb (2007a), Monsoon dynamics with interactive forcing. Part I: Axisymmetric studies, *J. Atmos. Sci.*, *64*, 1417–1430.

Privé, N. C., and R. A. Plumb (2007b), Monsoon dynamics with interactive forcing. Part II: Impact of eddies and asymmetric geometries, *J. Atmos. Sci.*, *64*, 1431–1442.

Schneider, E. K. (1977), Axially symmetric steady-state models of the basic state for instability and climate studies. Part II. Nonlinear calculations, *J. Atmos. Sci.*, *34*, 280–296.

Schneider, E. K., and R. S. Lindzen (1977), Axially symmetric steady-state models of the basic state of instability and climate studies. Part I. Linearized calculations, *J. Atmos. Sci.*, *34*, 253–279.

Waugh, D. W., and R. A. Plumb (1994), Contour advection with surgery: A technique for investigating finescale structure in atmospheric transport, *J. Atmos. Sci.*, *51*, 530–540.

Waugh, D. W., et al. (1994), Transport out of the stratospheric Arctic vortex by Rossby wave breaking, *J. Geophys. Res.*, *99*, 1071–1088.

Waugh, D. W., et al. (1997), Mixing of polar vortex air into middle latitudes as revealed by tracer-tracer correlations, *J. Geophys. Res.*, *102*, 13,119–13,134.

PREFACE

The year 2008 marked the 60th birthday of R. Alan Plumb, one of the great atmospheric scientists of our time. To celebrate this anniversary, a symposium was held at Columbia University on Friday and Saturday, 24–25 October 2008: this event was referred to, affectionately, with the nickname PlumbFest. A dozen invited speakers gave detailed presentations, reviewing the recent advances and the current understanding of the dynamics, transport, and chemistry of the stratosphere. In order to make the PlumbFest an event of lasting significance, it was decided to invite the symposium speakers to write chapter-length review articles, summarizing our present knowledge of the stratosphere: hence the present Festschrift volume. With heartfelt gratitude, it is dedicated to our mentor, colleague, and friend, Alan Plumb, *il miglior fabbro*!

Lorenzo M. Polvani
Columbia University

Adam H. Sobel
Columbia University

Darryn W. Waugh
Johns Hopkins University

The Stratosphere: Dynamics, Transport, and Chemistry
Geophysical Monograph Series 190
Copyright 2010 by the American Geophysical Union.
10.1029/2010GM001019

Introduction

Darryn W. Waugh

Department of Earth and Planetary Sciences, Johns Hopkins University, Baltimore, Maryland, USA

Lorenzo M. Polvani

Department of Applied Physics and Applied Mathematics and Department of Earth and Environmental Sciences
Columbia University, New York, New York, USA

Over the past few decades there has been intensive research into the Earth's stratosphere, which has resulted in major advances in our understanding of its dynamics, transport, and chemistry and its coupling with other parts of the atmosphere. This interest in the stratosphere was originally motivated by concerns regarding the stratospheric ozone layer, which plays a crucial role in shielding Earth's surface from harmful ultraviolet light. In the 1980s the depletion of ozone was first observed, with the Antarctic ozone hole being the most dramatic example, and then linked to increases in chlorofluorocarbons (CFCs). These findings led to the signing of the Montreal Protocol, which regulates the production of CFCs and other ozone-depleting substances. Over the subsequent decades, extensive research has led to a much better understanding of the controls on stratospheric ozone and the impact of changes in CFC abundance (including the recovery of the ozone layer as the abundance of CFCs returns to historical levels). More recently, there has been added interest in the stratosphere because of its potential impact on surface climate and weather. This surface impact involves changes in the radiative forcing, the flux of ozone and other trace constituents into the troposphere, and dynamical coupling.

The aim of this monograph is to summarize the last two decades of research in stratospheric dynamics, transport, and chemistry and to provide a concise yet comprehensive overview of the state of the field. By reviewing the recent advances this monograph will act, we hope, as a companion to the *Middle Atmosphere Dynamics* textbook by *Andrews et al.* [1987]. This is the most widely used book on the stratosphere and provides a comprehensive treatment of the fundamental dynamics of the stratosphere. However, it was published over 20 years ago, and major advances in our understanding of the stratosphere, on very many fronts, have occurred during this period. These advances are described as in this monograph.

The chapters in this monograph cover the dynamical, transport, chemical, and radiative processes occurring within the stratosphere and the coupling and feedback between these processes. The chapters also describe the structure and variability (including long-term changes) in the stratosphere and the role played by different processes. Recent advances in our understanding of the above issues have come from a combination of increased observations and the development of more sophisticated theories and models. This is reflected in the chapters, which each include discussions of observations, theory, and models.

The first chapter [*Geller*, this volume] provides a historical perspective for the material reviewed in the following chapters. It describes the status of research and understanding of stratospheric dynamics and transport before Alan Plumb's entrance into stratospheric research.

The second chapter (by Alan Plumb himself [*Plumb*, this volume]) describes recent developments in the dynamics of planetary-scale waves, which dominate the dynamics of the winter stratosphere and play a key role in stratosphere-

The Stratosphere: Dynamics, Transport, and Chemistry
Geophysical Monograph Series 190
Copyright 2010 by the American Geophysical Union
10.1029/2010GM001018

troposphere couplings. While there is a long history in understanding the propagation of these waves in the stratosphere, some very basic questions remain unsolved, the most important being the relationship between planetary-scale Rossby wave activity and the mean flow, which are discussed in chapter 2.

The chapter by *Waugh and Polvani* [this volume] covers the dynamics of stratospheric polar vortices. The observed climatological structure and variability of the vortices are reviewed, from both zonal mean and potential vorticity perspectives, and then interpreted in terms of dynamical theories for Rossby wave propagation and breaking. The role of vortices in troposphere-stratosphere coupling and possible impact of climate change of vortex dynamics are also discussed.

Kushner [this volume] provides a review of the so called "annular modes," which are the principal modes of variability of the extratropical circulation of the troposphere and stratosphere on time scales greater than a few weeks. The observed characteristics of these annular modes in each hemisphere are presented, together with a discussion of their dynamics and their role in extratropical climate variability and change.

Gray [this volume] focuses on the dynamics of the equatorial stratosphere. The characteristics of the quasi-biennial oscillation (QBO) and semiannual oscillation (SAO), which dominate the variability in zonal winds and temperatures near the equator, are summarized. The interaction of thee QBO and the SAO with the solar cycle and their impact on the extratropics and the troposphere, as well as on the transport of ozone and other chemical species, are also reviewed.

The chapter by *Alexander* [this volume] focuses on gravity waves in the stratosphere. Recent research on the direct effects of these waves in the stratosphere, including their effects on the general circulation, equatorial oscillations, and polar ozone chemistry, are highlighted. Advances in our understanding of the sources of gravity waves and in parameterizing these waves in global models are also discussed.

Randel [this volume] describes the observed interannual variability and recent trends in stratospheric temperature and water vapor. There is also a discussion of mechanisms causing these changes, including long-term increases in carbon dioxide, volcanic eruptions, the QBO, and other dynamical variability, as well as an examination of the link between variability in stratospheric water vapor and temperature anomalies near the equatorial tropopause.

Schoeberl and Douglass [this volume] provide an overview of stratospheric circulation and transport as seen through the distribution of trace gases. They also summarize the techniques used to analyze trace gas distributions and transport and the numerical methods used in models of tracer transport.

The chapter by *Newman* [this volume] deals with polar ozone and chemistry, with a focus on the Antarctic ozone hole. The chapter offers an updated overview of observed changes in polar ozone, our current understanding of polar ozone losses, the heterogeneous chemistry behind those loss processes, and a short prognosis of the future of ozone levels.

The final chapter [*Haigh*, this volume] reviews what is known about solar variability and the evidence for solar signals in the stratosphere. It discusses the relevant radiative, chemical, and dynamical processes and to what extent climate models are able to reproduce the observed signals. It also discusses the potential for a solar impact on the stratosphere to influence tropospheric climate through dynamical coupling.

REFERENCES

Alexander, M. J. (2010), Gravity waves in the stratosphere, in *The Stratosphere: Dynamics, Transport, and Chemistry, Geophys. Monogr. Ser.*, doi: 10.1029/2009GM000864, this volume.

Andrews, D. G., J. R. Holton, and C. B. Leovy (1987), *Middle Atmosphere Dynamics*, 489 pp., Academic, San Diego, Calif.

Geller, M. A. (2010), Middle atmosphere research before Alan Plumb, in *The Stratosphere: Dynamics, Transport, and Chemistry, Geophys. Monogr. Ser.*, doi: 10.1029/2009GM000871, this volume.

Gray, L. J. (2010), Stratospheric equatorial dynamics, in *The Stratosphere: Dynamics, Transport, and Chemistry, Geophys. Monogr. Ser.*, doi: 10.1029/2009GM000868, this volume.

Haigh, J. D. (2010), Solar variability and the stratosphere, in *The Stratosphere: Dynamics, Transport, and Chemistry, Geophys. Monogr. Ser.*, doi: 10.1029/2010GM000937, this volume.

Kushner, P. J. (2010), Annular modes of the troposphere and stratosphere, in *The Stratosphere: Dynamics, Transport, and Chemistry, Geophys. Monogr. Ser.*, doi: 10.1029/2009GM000924, this volume.

Newman, P. A. (2010), Chemistry and dynamics of the Antarctic ozone hole, in *The Stratosphere: Dynamics, Transport, and Chemistry, Geophys. Monogr. Ser.*, doi: 10.1029/2009 GM000873, this volume.

Plumb, R. A. (2010), Planetary waves and the extratropical winter stratosphere, in *The Stratosphere: Dynamics, Transport, and Chemistry, Geophys. Monogr. Ser.*, doi: 10.1029/2009 GM000888, this volume.

Randel, W. J. (2010), Variability and trends in stratospheric temperature and water vapor, in *The Stratosphere: Dynamics, Transport, and Chemistry, Geophys. Monogr. Ser.*, doi: 10.1029/ 2009GM000870, this volume.

Schoeberl, M. R., and A. R. Douglass (2010), Trace gas transport in the stratosphere: Diagnostic tools and techniques, in *The*

Stratosphere: Dynamics, Transport, and Chemistry, Geophys. Monogr. Ser., doi: 10.1029/2009GM000855, this volume.

Waugh, D. W., and L. M. Polvani (2010), Stratospheric polar votices, in *The Stratosphere: Dynamics, Transport, and Chemistry, Geophys. Monogr. Ser.*, doi: 10.1029/2009GM000887, this volume.

L. M. Polvani, Department of Applied Physics and Applied Mathematics, Columbia University, New York, NY 10027, USA. (lmp@columbia.edu)

D.W. Waugh, Department of Earth and Planetary Sciences, Johns Hopkins University, Baltimore, MD 21218, USA.

Middle Atmosphere Research Before Alan Plumb

School of Marine and Atmospheric Science, State University of New York at Stony Brook, Stony Brook, New York, USA

Alan Plumb received his Ph.D. in 1972. Since that time, he has made very great contributions to middle atmosphere research. This paper briefly examines the status of middle atmosphere research upon Alan's arrival on the scene and his development into one of the world's leading researchers in this area.

1. INTRODUCTION

Alan Plumb has been one of the principal contributors to research into middle atmosphere dynamics and transport for over 3 decades now, so it is difficult to imagine the field without his great contributions, but it is good to remember the famous quote from Isaac Newton's 1676 letter to Robert Hooke, "If I have seen a little further it is by standing on the shoulders of Giants." Alan's work similarly built on the work of those that came before him, just as many younger atmospheric scientists make their contributions standing on Alan's shoulders.

Alan has made significant contributions in many areas, but I will concentrate on those aspects of his work that are in the broad areas of wave–mean flow interactions and middle atmosphere transport. The following then is my version of the status of our understanding of these fields in the "before Alan Plumb" years.

2. A LITTLE HISTORY

The study of the middle atmosphere had its beginnings in the early balloon measurements of *Teisserenc De Bort* [1902], who established that above the troposphere where the temperature decreases with increasing altitude, there existed a region where the temperature became approximately isothermal (i.e., the lower stratosphere). This is nicely seen in Figure 1 of *Goody* [1954], which shows balloon mea-

The Stratosphere: Dynamics, Transport, and Chemistry
Geophysical Monograph Series 190
Copyright 2010 by the American Geophysical Union.
10.1029/2009GM000871

surements of temperature up to an altitude of about 14 km. Proceeding up in altitude, before the advent of rocket and lidar measurements of atmospheric temperature profiles, the main information on the atmospheric temperature between about 30 and 60 km was from the refraction of sound waves. It was thought curious that the guns fired at Queen Victoria's funeral were heard far to the north of London. Later, during World War I, it was found that the gunfire from the western front was frequently heard in southern England, but there was a "zone of silence" in between where the gunfire was not heard. *Whipple* [1923] explained these observations in terms of the existence of a stratosphere where the temperatures increased appreciably with increasing altitude. It is interesting to note that *Whipple* [1923, p. 87] said the following: "Further progress in our knowledge of the temperature of the outer atmosphere and of its motion would be made if Prof. Goddard could send up his rockets."

In fact, after the end of the World War II, the expansion of the radiosonde balloon network and the use of rockets provided a much better documentation of the temperature and wind structure of the middle atmosphere. *Murgatroyd* [1957] synthesized these measurements, and his Figure 4 shows the very cold polar night stratospheric temperatures (at about 30 km), the warm stratopause temperatures (at about 50 km), and the warm winter mesopause and cold summer mesopause (at about 80 km). Consistent with the thermal wind relation, the wind structure was seen to be dominated by strong winter westerly and strong summer easterly jets centered at about 60 km.

Research into stratospheric ozone can trace its beginnings to the early work of *Hartley* [1881], who correctly attributed the UV shortwave cutoff in solar radiation reaching the ground as being due to stratospheric ozone; to *Chapman*

[1930], who advanced the first set of chemical reactions for ozone formation and destruction (neglecting catalytic reactions); and to *Dobson and Harrison* [1926], who developed the ground-based instrument for measuring the ozone column that is still being used today. Ground-based measurements [*Götz*, 1931; *Götz et al.*, 1934] and in situ measurements [*Regener*, 1938, 1951] of ozone concentrations clearly indicated that ozone concentrations are highest in the stratosphere.

Early British measurements, using the techniques of *Brewer et al.* [1948], indicated that lower stratospheric water vapor water concentrations are very low (on the order of 10^{-3} times that of the troposphere. These results are summarized by *Murgatroyd et al.* [1955]. Later measurements in the United States indicated larger water vapor concentrations, and this led to some controversy [*Gutnick*, 1961], but the U.K. measurements proved to be correct. This turned out to be very important in establishing the nature of the Brewer-Dobson circulation (as will be seen later), where virtually all tropospheric air enters the stratosphere by rising through the cold tropical tropopause.

This is but a much abbreviated version of the early history of our sources of knowledge of the middle atmosphere well before Alan entered the field. In subsequent sections, we discuss in more detail some previous work in specific areas of research where Alan would be a seminal contributor.

3. WAVE–MEAN FLOW INTERACTIONS

Alan's Ph.D. dissertation in 1972 from the University of Manchester was on the "moving flame" phenomenon, with reference to the atmosphere of Venus. The problem he addressed was the following: Venus's surface rotates once every 243 Earth days, while observations of Venus's cloud tops indicate that the atmosphere at that altitude rotates once every 4–5 days. The question then is by what process does the atmosphere at that level come to rotate so much faster than Venus's surface? A nice explanation of the "moving flame" process is given in *Lindzen's* [1990] textbook. It basically involves a propagating heat source for gravity waves leading to acceleration at the altitude of this heat source. For Venus, solar heating of the cloud tops is pictured as this propagating heat source.

The *Plumb* [1975] article was largely based on this dissertation work. Among this paper's reference list was the classic paper by *Eliassen and Palm* [1961], who along with *Charney and Drazin* [1961] put forth the famous noninteraction theorem. In the following, some of the results from these classic papers will be briefly reviewed.

The *Charney and Drazin* [1961] paper is a classic. It addresses two important issues: Observations indicate that

the scales of stratospheric disturbances were much larger than those seen in the troposphere, so there must be some reason that upward propagating disturbances experience shortwave filtering. The other issue is that while monthly mean stratospheric maps in winter showed planetary-scale wave patterns, such wave patterns were absent during summer.

The first result of the *Charney and Drazin* [1961, p. 83] paper is summarized in its abstract as follows: "It is found that the effective index of refraction for the planetary waves depends primarily on the distribution of the mean zonal wind with height. Energy is trapped (reflected) in regions where the zonal winds are easterly or are large and westerly." To obtain this result, *Charney and Drazin* [1961] derived the following equation for the vertical variations of the perturbation northward velocity in the presence of a mean zonal wind u_0 for quasi-geostrophic flow on a β plane and where the time, longitude, and latitude dependence of the perturbation is $e^{i(kx+ly-kct)}$:

$$(u_0-c)\frac{\mathrm{d}}{\mathrm{d}z}\left(\frac{\rho_0}{N^2}\frac{\mathrm{d}v}{\mathrm{d}z}\right) \tag{1}$$
$$-\left[\frac{\mathrm{d}}{\mathrm{d}z}\left(\frac{\rho_0}{N^2}\frac{\mathrm{d}u_0}{\mathrm{d}z}\right)+\frac{\beta\rho_0}{f_0^2 u_c}(u_0-c-u_c)\right]v=0,$$

where z is the upward directed vertical coordinate, ρ_0 is the basic state density that only depends on z, N is the Brunt-Väisälä frequency, f is the Coriolis parameter, v is the northward directed wave velocity amplitude, and $u_c = \beta/(k^2 + l^2)$. Letting $\chi \equiv \sqrt{\frac{\rho_0}{N^2}}v$ gives the equation

$$\frac{\mathrm{d}^2\chi}{\mathrm{d}z^2}+n^2\chi=0, \tag{2}$$

where

$$n^2=-\left\{\frac{(k^2+l^2)N^2}{f_0^2}+\sqrt{\frac{N^2}{\rho_0}}\frac{\mathrm{d}^2}{\mathrm{d}z^2}\sqrt{\frac{\rho_0}{N^2}}\right\} \tag{3}$$
$$+\frac{N^2}{u_0-c}\left\{\frac{\beta}{f_0^2}-\frac{1}{\rho_0}\frac{\mathrm{d}}{\mathrm{d}z}\left(\frac{\rho_0}{N^2}\frac{\mathrm{d}u_0}{\mathrm{d}z}\right)\right\}$$

is the local index of refraction for the problem. Here k is the zonal wave number, l is the meridional wave number, x is the eastward directed coordinate, and y is the northward directed coordinate. *Charney and Drazin* [1961] consider a number of special cases, but the classic case is also the simplest case, where u_0 and $\bar{T}$, the basic state temperature, are constant. In this case, it is easily derived that

$$n^2=-\frac{1}{4H^2}-\frac{N^2}{f_0^2}\left\{(k^2+l^2)-\frac{\beta}{u_0-c}\right\}, \tag{4}$$

where H is pressure scale height. In this case, vertical wave propagation can only occur when $n^2 > 0$ or when

$$0 < u_0 - c < \frac{\beta}{(k^2 + l^2) + (f_0^2/4H^2N^2)} \equiv U_c. \quad (5)$$

This yields the following two famous results. One is that small-scale tropospheric planetary waves cannot propagate a substantial amount into the stratosphere (because $k^2 + l^2$ large implies U_c is small). Thus, vertical propagation can only occur for synoptic scales (i.e., $k^2 + l^2$ large) when $u_0 - c$ is small, implying vertical propagation can occur only in a very narrow window of phase speeds. Also, stationary ($c = 0$) planetary waves cannot propagate through easterlies ($u_0 < 0$) or through strong westerlies ($u_0 > U_c$).

A simple physical interpretation of this result can be seen with the aid of results given by *Pedlosky* [1979]. He showed that the dispersion relation for Rossby waves in a stratified atmosphere is given by the following slight modification of his equation (6.11.6):

$$u_0 - c = \frac{\beta}{k^2 + l^2 + \frac{1}{N^2}\left(m^2 + \frac{1}{4H^2}\right)} \quad (6)$$

where m is the vertical wave number. This gives the familiar result that Rossby waves must propagate westward relative to the mean zonal flow so that stationary Rossby waves cannot exist in an easterly "or westward" flow where $u_0 < 0$. Furthermore, the maximum of $u_0 - c$ occurs for $m = 0$ (infinite vertical wavelength). Thus, the famous *Charney and Drazin* [1961] result of equation (5) can be restated as follows: stationary planetary waves cannot propagate vertically through easterlies (since Rossby waves cannot exist in such a flow), nor can they propagate westward relative to the mean zonal flow at a phase velocity that exceeds the maximum phase velocity for Rossby waves in an atmosphere with constant u_0 and $\overline{T}$.

As an aside, note that the Rossby radius of deformation $L_R \equiv NH/f_0$ for a continuously stratified fluid, so that equation (6) can be rewritten as

$$c = u_0 - \frac{\beta}{(k^2 + l^2) + \frac{1}{4L_R^2}} \quad (7)$$

This is analogous to the case for free barotropic Rossby waves where the $1/4L_R^2$ would be replaced with $1/L^2 \equiv f_0^2/gH$ (where g is the acceleration due to gravity), the reciprocal of the barotropic Rossby radius of deformation squared [see *Holton*, 2004; *Rossby et al.*, 1939].

The second major result of *Charney and Drazin* [1961, p. 83] is stated as follows in their abstract: ". . . when the wave disturbance is a small stationary perturbation on a zonal flow that varies vertically but not horizontally, the second-order effect of the eddies on the zonal flow is zero." *Charney and Drazin* [1961] say that this result was first obtained by A. Eliassen, who communicated it to them. In the following, we more closely follow the discussions of *Eliassen and Palm* [1961] than those of *Charney and Drazin* [1961].

Eliassen and Palm [1961] considered the propagation of stationary ($c = 0$) mountain waves both when rotation was ignored (i.e., when $f = 0$) and also for the case when $f \neq 0$. For the $f = 0$ case, a more general form of their equation (3.2), for the case of a steady gravity wave propagating with phase velocity c in a shear flow in the absence of diabatic effects, is

$$\overline{p'w'} = -\rho_0(u_0 - c)\overline{u'w'}, \quad (8)$$

where p, u, and w are pressure and horizontal and vertical velocities, respectively, the overbars denote averaging over wave phase, and the primes indicate the wave perturbations. Equation (8) is sometimes referred to as Eliassen and Palm's first theorem. It implies that for upward wave energy flux ($\overline{p'w'} > 0$), the wave momentum flux ($\rho_0\overline{u'w'}$) is negative when the mean flow u_0 is greater than the phase velocity c and is positive when $u_0 < c$. Thus, any physical process that leads to a decrease of the wave amplitude as it propagates (e.g., dissipation) will force the mean flow toward the wave phase velocity.

For gravity waves with phase velocity $c \neq u_0$, Eliassen and Palm's second theorem, their equation (3.3), is

$$\rho_0\overline{u'w'} = \text{constant} \quad (9)$$

in the case of no wave transience and no diabatic effects. Thus, in this case, there is no gravity wave interaction with the mean flow.

The implications of Eliassen and Palm's first and second theorems are far-reaching. They indicate that unless there is dissipation, other diabatic effects, wave transience, or $u_0 = c$, atmospheric gravity waves do not interact with the mean flow. Conversely, if any of these are present, the waves do interact with the mean flow, and this interaction gives rise to a deceleration or acceleration of the mean flow toward the wave's phase velocity.

The $f \neq 0$ case is more complex. To discuss this, I will use a mixture of results from *Eliassen and Palm* [1961] and *Dickinson* [1969], which reproduce the noninteraction results from *Charney and Drazin* [1961]. Eliassen and Palm's equation (10.8) can be written as

$$\frac{\partial}{\partial y}\left[\frac{\rho_0}{N^2}\frac{\partial u_0}{\partial z}\overline{v'\frac{\partial \Phi'}{\partial z}}-\rho_0\overline{u'v'}\right]+ \qquad (10)$$

$$\frac{\partial}{\partial z}\left[\rho_0\left\{\frac{1}{N^2}\left(f-\frac{\partial u_0}{\partial y}\right)\overline{v'\frac{\partial \Phi'}{\partial z}}-\overline{u'w'}\right\}\right]=0,$$

and it holds for steady state conditions, no dissipation, and so long as $u_0 \neq 0$. This is now a familiar result, which is most often written as $\nabla \cdot \mathbf{F} = 0$, where the terms in the square brackets are the y and z components of the Eliassen and Palm flux. Now, *Charney and Drazin* [1961] show that for steady state, nondissipative conditions, and for quasi-geostrophic conditions, when $u_0(z)$ only, this implies that u_0 does not change with time. Thus, there is noninteraction between the planetary waves and the mean zonal flow. Under quasi-geostrophic conditions ($R_0 \ll 1$, where R_0 is the Rossby number U/fL, where U is the characteristic horizontal velocity scale and L is the characteristic horizontal length scale), w' is small, $f - (\partial u_0/\partial y) \rightarrow f$, and the first term in the top square brackets is much smaller in magnitude than the second term within these brackets, in which case equation (10) becomes

$$\frac{\partial}{\partial y}[-\rho_0\overline{u'v'}]+\frac{\partial}{\partial z}\left[\frac{\rho_0 f_0}{N^2}\overline{v'\frac{\partial \Phi'}{\partial z}}\right]=0. \qquad (11)$$

Given these results, the Charney-Drazin, or noninteraction, result can be easily obtained by noting that for steady state conditions and in the absence of diabatic effects,

$$\frac{\partial \bar{q}}{\partial t}=-\frac{\partial}{\partial y}\overline{v'q'}, \qquad (12)$$

where q is the quasi-geostrophic potential vorticity [see *Holton*, 2004, p. 160].

That is to say, the zonal mean quasi-geostrophic potential vorticity can only change with time if the planetary waves induce an eddy transport of the quasi-geostrophic potential vorticity [e.g., *Dickinson*, 1969], but it is easily seen that, under quasi-geostrophic conditions,

$$\overline{v'q'}=\frac{1}{\rho_0}\nabla \cdot \mathbf{F}. \qquad (13)$$

Now, in the absence of diabatic effects, in steady state, and when there are no singular lines where $u_0 = 0$,

$$\nabla \cdot \mathbf{F}=\frac{\partial}{\partial y}[-\rho_0\overline{u'v'}]+\frac{\partial}{\partial z}\left[\frac{\rho_0 f_0}{N^2}\overline{v'\frac{\partial \Phi'}{\partial z}}\right]=0. \qquad (14)$$

This generalizes the Charney-Drazin noninteraction theorem to include the case where there can be latitudinal shears in the mean zonal wind.

Of course, later work by *Boyd* [1976] and by *Andrews and McIntyre* [1978a, 1978b] further generalized this noninteraction, or nonacceleration theorem, but by this time Alan was already established as a leading middle atmosphere researcher.

4. GRAVITY WAVES, CRITICAL LEVELS, AND WAVE BREAKING

Hines' [1960] paper on internal gravity waves is also a classic. He presented observational evidence for gravity waves in the atmosphere. He developed the linear theory for these waves. He discussed some of their effects, and he predicted which waves could be observed in ionospheric regions. *Hines* [1960] showed that there were two distinct classes of waves in a compressible, gravitationally stratified atmosphere: internal gravity waves with frequencies less than the Brunt-Väisälä frequency and acoustic-gravity waves with frequencies greater than the acoustic cutoff frequency. Further, he showed that the internal gravity waves have the asymptotic behavior of internal gravity waves in an incompressible fluid for low frequencies and the acoustic-gravity waves have the asymptotic behavior of sound waves for frequencies much higher than the acoustic cutoff frequency.

One of the most fundamental results of the *Hines* [1960] paper was that the vertical component of an internal gravity wave's phase velocity is opposite to the vertical component of the internal gravity wave's group velocity, the speed at which

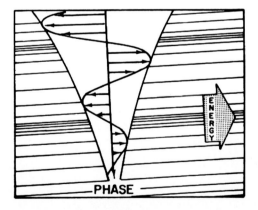

Figure 1. Pictorial representation of internal gravity waves. Instantaneous velocity vectors are shown, as are their instantaneous and overall envelopes. Density variations are depicted by a background lying in surfaces of constant phase. The vertical component of the phase velocity is downward while energy is being propagated upward. Note that gravity is directed vertically downward. From *Hines* [1960]. © NRC Canada or its licensors. Reproduced with permission.

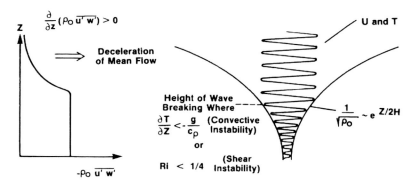

Figure 2. Schematic of wave breaking, with the resultant convergence of gravity wave momentum flux. From *Geller* [1983].

the gravity wave energy propagates. This is shown in Figure 1, which is Figure 2 in *Hines'* [1960] paper.

In the previous discussion of the noninteraction theorems, one of the conditions for noninteraction was $u_0 \neq c$; that is, the mean zonal flow is unequal to the wave phase velocity. *Bretherton* [1966] examined the case of a gravity wave in a shear flow where $u_0 = c$ (the critical level) and the Richardson number (to be defined shortly) is very large. He found that in this case the gravity wave vertical group velocity $\rightarrow 0$ as $u_0 \rightarrow c$. Thus, the gravity wave energy flux $\overline{p'w'}$ vanishes on the far side of the critical level, in which case the momentum flux $\rho_0\overline{u'w'}$ is also zero. Since there is no wave interaction below the critical level, this implies a convergence (or divergence) of the wave momentum flux at the critical level. *Booker and Bretherton* [1967] generalized this result to the case of finite Richardson number, in which case they derived the results that in passing through the critical level, the wave momentum flux is attenuated by a factor of $e^{-2\pi\sqrt{Ri-\frac{1}{4}}}$, where Ri, the Richardson number, is given by

$$Ri = \frac{N^2}{\left(\frac{\partial \mathbf{v}}{\partial z}\right)^2}.$$

Thus, at a gravity wave critical level, the absorption of the wave will tend to bring the mean flow toward the wave phase velocity (by Eliassen and Palm's first theorem).

There followed a period of very active research into the nature of gravity wave critical levels. *Hazel* [1967] showed that the *Booker and Bretherton* [1967] result was essentially correct in the case of a fluid with viscosity and heat conduction. *Breeding* [1971] suggested that nonlinear effects might lead to some wave reflection in addition to absorption, but *Geller et al.* [1975] suggested that as the wave approached a critical level, it produces turbulence that would likely lead to wave absorption before nonlinear effects would lead to wave reflection.

In an isothermal atmosphere, the density decreases exponentially with increasing altitude z as $e^{\frac{-z}{H}}$, H being the pressure scale height. Without dissipation or critical levels, the gravity wave kinetic energy per unit volume $\rho_0\mathbf{v}'^2$ should remain constant, in which case the amplitude of the wave's horizontal velocity (and as it turns out temperature) fluctuations should grow as $e^{+\frac{z}{2H}}$. This being the case, the wave eventually becomes unstable. *Hodges* [1967] was the first to point out that this will be a source of turbulence in the middle and upper atmosphere.

This provided the starting point for *Lindzen's* [1981] seminal paper that suggested a self-consistent way of parameterizing the effects of unresolved gravity waves in climate models. The principle for this parameterization is illustrated in Figure 2. On the right is illustrated a gravity wave whose wind and temperature amplitude are exponentially increasing with height, and as pictured, the wave momentum flux $\rho_0\overline{u'w'}$ is constant with height. Since the vertical wavelength of this wave is fixed, $\partial v'/\partial z$ and $\partial T'/\partial z$ also increase with height exponentially, as illustrated by the outer envelope. Eventually, the wave becomes either convectively unstable or shear unstable and breaks down. *Lindzen* [1981] made the assumption that above the level where the wave breaks down, it loses just enough energy to turbulence to keep the wave amplitude constant above that level, as illustrated. This means that $\rho_0\overline{u'w'}$ decreases with height above the breaking level so that there is a divergence of wave momentum flux above the level where the gravity wave begins to break, also as pictured in Figure 2. Of course, one could make different assumptions of what occurs above the breaking level. For instance, *Alexander and Dunkerton* [1999] assume that gravity waves deposit all of their momentum at the breaking level.

This gravity wave breaking and the subsequent drag on mesospheric winds (by Eliassen and Palm's first theorem) gave physical justification to the Rayleigh drag used by *Leovy*

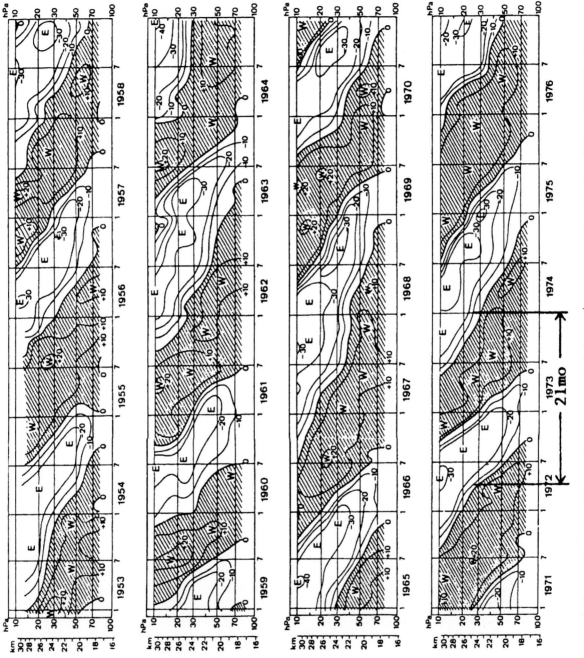

Figure 3. Time-height section of the monthly mean zonal winds (in m s^{-1}) over equatorial stations from *Geller et al.* [1997], which was an update from *Naujokat* [1986]. Copyright American Meteorological Society.

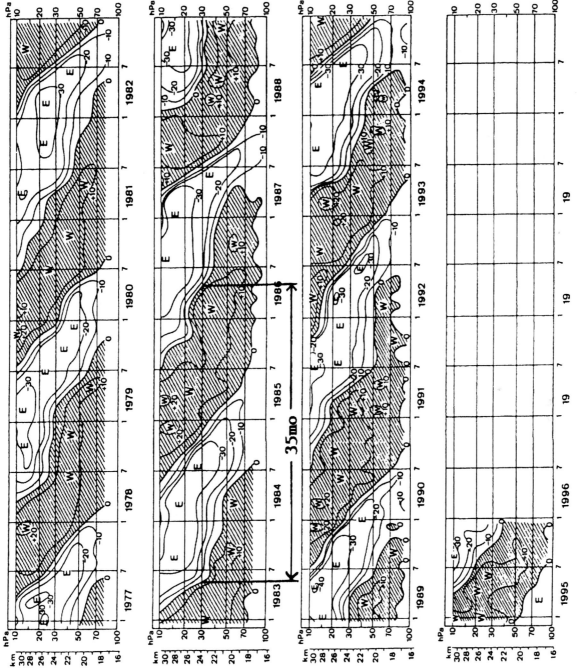

Figure 3. (continued)

[1964] in his modeling of the mesospheric wind structure, since many gravity waves have their source in the troposphere where their source phase velocity is small. Developing and implementing ways of parameterizing the effects of unre-solved gravity waves in climate models is a research topic of great current interest, but one might say that this had its intellectual roots in the papers of *Eliassen and Palm* [1961], *Hodges* [1967], and *Booker and Bretherton* [1967], since critical levels are also of great importance in this.

5. QUASI-BIENNIAL OSCILLATION

The quasi-biennial oscillation (QBO) was discovered independently by *Reed et al.* [1961] and by *Veryard and Ebdon* [1961]. Figure 3 shows its structure over the equator. A quasiperiodic pattern of descending easterlies (unshaded) followed by descending westerlies (shaded) is evident. The average period of a complete cycle is about 28 months, but the period varies considerably, being about 21 months in 1972–1974 and about 35 months in 1983–1986. Moreover, the westerlies descend more quickly than the easterlies. The maximum amplitude of the QBO is about 20 m s^{-1}, occurring in the middle stratosphere.

Following the discovery of the QBO, there were many attempts to explain why this phenomenon occurred, but the key papers that led to today's generally accepted explanation for the QBO were those of *Wallace and Holton* [1968], *Lindzen and Holton* [1968], and *Holton and Lindzen* [1972].

There were many efforts that tried to explain the QBO in terms of a hypothesized periodic radiative forcing, but *Wallace and Holton* [1968] constructed a diagnostic model to

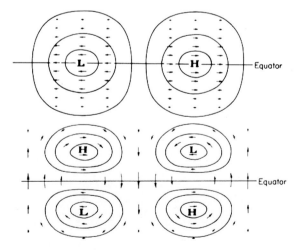

Figure 4. Schematic illustration of the geopotential and wind fields for the equatorial trapped (top) Kelvin and (bottom) mixed Rossby-gravity waves. Adapted from *Andrews et al.* [1987], who, in turn, adapted it from *Matsuno* [1966].

Table 1. Characteristics of the Dominant Observed Planetary-Scale Waves in Equatorial Lower Stratosphere[a]

Theoretical Description	Kelvin Wave	Rossby-Gravity Wave
Discovery	*Wallace and Kousky* [1968]	*Yanai and Maruyama* [1966]
Period (ground-based)	15 days	4–5 days
Zonal wave number	1–2	4
Vertical wavelength	6–10 km	4–5 km
Average phase speed relative to ground	+25 m s^{-1}	−23 m s^{-1}
Observed when mean zonal flow is	easterly (maximum ≈ −25 m s^{-1}	westerly (maximum ≈ +7 m s^{-1})
Average phase speed relative to maximum zonal flow	+50 m s^{-1}	−30 m s^{-1}
Average phase speed relative to maximum zonal flow		
w'	8 m s^{-1}	2–3 m s^{-1}
v'	0	2–3 m s^{-1}
T'	2–3°K	1°K
Approximate inferred amplitudes		
ϕ'/g	30 m	4 m
w'	1.5 × 10^{-3} m s^{-1}	1.5 × 10^{-3} m s^{-1}
Approximate meridional scales $\left(\frac{2N}{\beta|m|}\right)^{1/2}$	1300–1700 km	1000–1500 km

[a]From *Andrews et al.* [1987].

see what kind of radiative and momentum forcings would be necessary to explain the observed characteristics of the QBO. They found that only an unrealistic radiative forcing could explain the observed features. On the other hand, they found that momentum forcings could explain the observations but only if the momentum forcing itself had a downward pro-pagation. *Lindzen and Holton* [1968], noting the results of *Wallace and Holton* [1968], published their famous paper that gave essentially today's accepted explanation for the QBO only 8 months after the appearance of the *Wallace and Holton* [1968] paper. They noted that there were reasons to believe that there were strong gravity waves in the equatorial region. They noted that *Matsuno* [1966] had predicted the existence of equatorially trapped eastward propagating Kelvin waves and westward propagating mixed Rossby-gravity waves (see

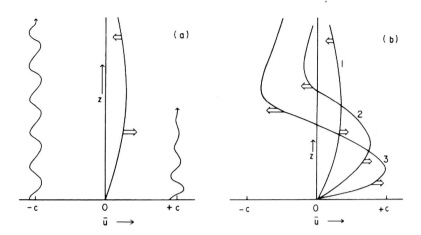

Figure 5. Schematic representation of the *Lindzen and Holton* [1968]/*Holton and Lindzen* [1972] theory for the QBO: (a) initial state and (b) initial state (curve 1) and evolutionary progression. Curves 2 and 3 show successive stages of evolution, as explained in the text. After *Plumb* [1984].

Figure 4). These waves had subsequently been observed by *Yanai and Maruyama* [1966] and *Wallace and Kousky* [1968] (see Table 1). They noted that these equatorial gravity waves would encounter critical levels and that the theory of *Booker and Bretherton* [1967] implied the needed downward propagating momentum flux to explain the QBO. Their theory was updated by *Holton and Lindzen* [1972] so that the gravity wave momentum absorption now occurred through radiative damping together with critical levels to produce the QBO.

Since the pioneering work of J. M. Wallace, J. R. Holton, and R. S. Lindzen, much more work on the theory of the QBO has taken place, and Alan's work on this topic has been seminal. An interesting laboratory analogue to the QBO was demonstrated by *Plumb and McEwan* [1978]: a standing wave pattern was forced by pistons oscillating a membrane at the bottom of a cylinder filled with a stratified fluid. A descending pattern of alternating angular velocities was observed to result. This has been nicely interpreted by *Plumb* [1977], as is

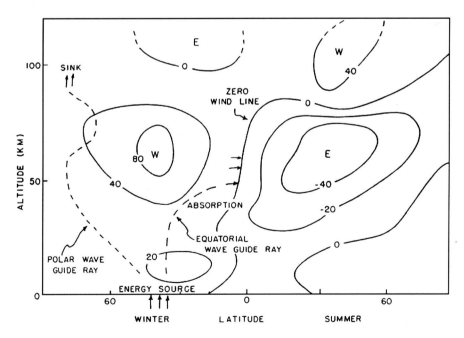

Figure 6. Schematic sketch of the winter and summer mean zonal wind patterns. Also shown are the planetary wave raypaths for the weak westerly wind waveguides in the winter hemisphere. From *Dickinson* [1968]. Copyright American Meteorological Society.

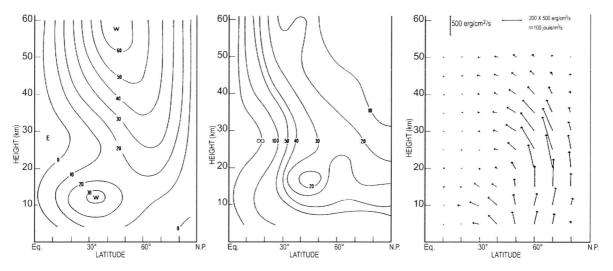

Figure 7. (left) Mean zonal wind state, (middle) $m = 0$ refractive index, (right) $m = 1$ stationary planetary wave energy propagation. From *Matsuno* [1970]. Copyright American Meteorological Society.

illustrated in Figure 5. This clearly showed that the essence of the mechanism for the QBO was to have both eastward and westward momentum fluxes that would be preferentially absorbed in regions of small Doppler-shifted intrinsic wave frequencies. Thus, in Figure 5a, positive phase speed waves are preferentially absorbed, leading to a downward propagating westerly shear zone as shown in curve 1 of Figure 5b. The negative phase speed waves, having high intrinsic frequencies, propagated to higher altitudes, but they were ultimately absorbed as indicated by the arrows at the top of Figure 5a and of curve 1 in Figure 5b. As time passes, the absorption of the two waves leads to curve 2 and then to curve 3 in Figure 5b. Ultimately, the bottom shear zone gets so extreme, it is subject to diffusive smoothing, which effectively leads to the mirror image of Figure 5a, so that the oscillation continues. While equatorially trapped waves no doubt play a role in forcing the QBO, *Haynes* [1998] has demonstrated that a geographically uniform source of gravity waves gives rise to the QBO through the different manner in which the equatorial atmosphere reacts to momentum and heat fluxes, which is distinct from the situation in the extratropics. Thus, both "garden variety" gravity waves and equatorially trapped waves play an important role in forcing the QBO.

6. PLANETARY WAVES AND STRATOSPHERIC SUDDEN WARMINGS

The importance of the *Charney-Drazin* [1961] results on planetary wave propagation was quickly appreciated. For instance, *Dickinson* [1968] considered stationary planetary wave propagation through a basic state that had its mean zonal winds varying both in altitude and latitude. He

concluded that the strong mean zonal winds of the polar night jet would be a "barrier" to planetary wave propagation, resulting in the picture shown in Figure 6. In Figure 6, note that the planetary wave rays refracted toward the equator are hypothesized to be absorbed at the $u_0 = 0$ critical line. (Later work showed that there is also reflection.) Also pictured is the polar waveguide ray where the waves are refracted poleward by the strongest winds of the polar night jet and reflected by the polar geometry. Some of the waves in this polar waveguide are pictured as propagating to very high altitudes.

Matsuno [1970] advanced an alternative picture for the propagation of stationary planetary waves in winter. His formulation differed from that of *Dickinson* [1968] in that *Matsuno* [1970] formulated his quasi-geostrophic equations on a sphere so that they conserved energy. His picture is shown in Figure 7. Figure 7 (left) shows Matsuno's winter mean zonal wind state. Figure 7 (middle) shows the effective refractive index for the $m = 0$ planetary wave (m here being the zonal wave number). The refractive index minimum results from the minimum in the latitudinal gradient of the zonally averaged quasi-geostrophic potential vorticity (not shown here). Figure 7 (right) shows the energy propagation vectors. Note that, unlike in the work of *Dickinson* [1968], the minimum in the refractive index acts as a "barrier" to planetary wave propagation rather than the "barrier" necessarily being the region of strongest westerly winds. The planetary wave propagation shows a bifurcation around this "barrier" with propagation toward the equator but also upward propagation through the lower portion of the strong polar night jet. *Matsuno's* [1970] picture is confirmed by observations [e.g., *Geller*, 1993].

Stationary planetary waves propagate both vertically and latitudinally, so they encounter a critical line where $u_0 = 0$.

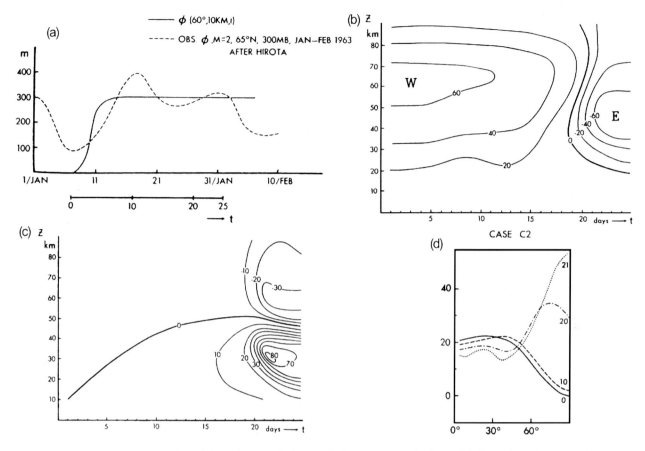

Figure 8. *Matsuno*'s [1971] model results. (a) His imposed planetary wave forcing at his lower boundary at 10 km compared to an observation by I. Hirota. (b) Time-height section of the evolution of his modeled wind at 60°N. (c) Time-height section of his modeled changes in temperature at 60°N. (d) Changes in the zonal mean temperature as a function of latitude at t = 0, 10, 20, and 21 days at an altitude of 30 km. Figures 8b–8d are from *Matsuno* [1971] for his experiment C2. Copyright American Meteorological Society.

This situation was analyzed by *Dickinson* [1970]. His time-dependent linear analysis indicated that the planetary wave perturbation zonal velocities $u' \to \infty$ as the planetary wave energy approaches the critical level but that the time scale for this process is long compared with the times scale on which u_0 varies. He concluded from his analysis that in realistic situations, planetary wave–mean flow interactions take place over a region hundreds of kilometers in width rather than at a singular line. Later nonlinear analyses by *Stewartson* [1977], *Warn and Warn* [1976], *Killworth and McIntyre* [1985], and *Haynes* [1985] indicate that it is likely that planetary waves are partially reflected by these critical regions. Observations indicate that these critical regions now correspond to what we now call the subtropical "surf zone" [e.g., *McIntyre and Palmer*, 1983, 1984]. These later works were published while Alan was already a leading researcher in middle atmosphere dynamics and transport, and Alan would go on to clarify many aspects of this "surf zone" on stratospheric transport.

A sudden stratospheric warming is an event in which lower stratospheric temperatures increase dramatically by several tens of degrees Celsius, and the winter westerly vortex actually reverses to easterlies in a period of only a few days. The sudden stratospheric warming was discovered by *Scherhag* [1952], and what is now the accepted explanation for these warmings was given by *Matsuno* [1971]. The basis for *Matsuno*'s [1971] treatment is found in results derived by *Eliassen and Palm* [1961]. Their equations (10.11) and (10.12) give the following results for quasi-geostrophic planetary waves:

$$\overline{v'\varphi'} = -u_0\overline{u'v'} \tag{15}$$

$$\overline{\omega'\varphi'} = -u_0\frac{fR}{\sigma p}\overline{v'T'} \tag{16}$$

where ϕ' is the wave geopotential perturbation, ω' is the pressure vertical velocity perturbation, the overbars represent

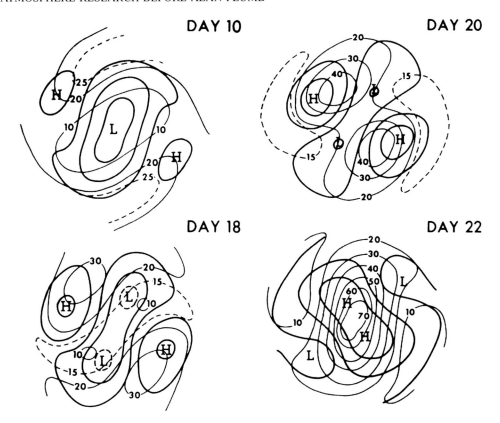

Figure 9. Evolution of *Matsuno*'s [1971] modeled stratospheric warming at 30 km (about 10 hPa) for the case shown in Figure 8. Thick lines show the isobaric height (with 500 m contours), and the thin lines show temperature deviations (in °C) from its value at the pole before the warming. Modeled from *Matsuno* [1971]. Copyright American Meteorological Society.

zonal averaging, R is the gas constant, and σ is the static stability. Equation (16) implies that extratropical planetary waves with upward energy flux must have an associated northward heat flux. Furthermore, one can show from this that such waves must have their phase lines sloping to the west with height. *Matsuno* [1971] noted that an intensifying planetary wave propagating energy upward in winter will be accompanied by a northward heat flux and a thermally indirect circulation with ascending motion at the cold pole and descending motion at low latitudes. This transient wave intensification implies that not all of the polar heating caused by the convergence of the meridional heat flux is canceled by the ascending motion there. This acts to diminish the meridional temperature gradient, thus leading to a decreasing mean zonal westerly flow. In time, a critical level can develop, in which case, there is even a greater convergence of the planetary wave heat flux so that very rapid heating occurs. Eventually, radiation reestablishes the cold winter stratospheric pole, and the easterly polar vortex is also reestablished.

This sequence of events is illustrated in Figure 8 from *Matsuno* [1971]. *Matsuno* [1971] imposed a sharp increase in the amplitude of the planetary waves at his lower boundary at 10 km altitude. The results shown in Figure 8 are for a spherical domain with wave number 2 forcing. Note that a very large increase in planetary wave forcing occurs between day 7 and 11. A corresponding decrease in the strength of the westerlies is seen between about day 7 and day 18, after which time a critical level exists where $u_0 = 0$ (Figure 8b). Large temperature increases are then seen below about 50 km, with somewhat smaller temperature decreases above this altitude (Figure 8c). The polar temperature rises by some 50°K over a period of 10 days. Notice that although the transient rise in planetary wave forcing causes significant changes, much more rapid changes occur after the establishment of the critical layer. What we see then in the work of *Matsuno* [1971] is that the planetary wave transience initially breaks the noninteraction, leading to changes in the mean zonal state, but the much larger zonal mean changes occur after the $u_0 = 0$ condition occurs.

Sudden stratospheric warmings can occur because of planetary wave number 1 or 2 forcing. The results above are shown for *Matsuno*'s [1971] wave number 2 case. The change in the polar vortex is seen in the 30 km maps in Figure 9, corresponding to the case shown in Figure 8. This wave number

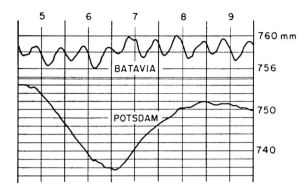

Figure 10. Barometric variations (on two different scales) at Batavia (the present Jakarta at 6°S) and Potsdam (52°N) in November 1919. From *Chapman and Lindzen* [1970], who, in turn, took this image from *Bartels* [1928]. Reprinted with kind permission of Springer Science and Business Media.

2 warming proceeds by splitting the polar westerly vortex (day 10–20) after which a polar easterly vortex is set up (day 22). A number of subsequent papers followed *Matsuno*'s [1971] work, modeling stratospheric warmings, but they all followed his basic prescription, with relatively minor variations.

7. ATMOSPHERIC TIDES

The theory for atmospheric tides goes back to the work of *Laplace* [1799], who derived the idealized equations for the free and forced oscillations for a thin atmosphere on a spherical planet. While the Moon's gravitation forces the oceanic tide, which is semidiurnal, it is the Sun's heating that forces the Earth's atmospheric tides. Under simplifying assumptions, Laplace used the traditional method of separation of variables to obtain an ordinary differential equation for the latitudinal structure of the tide as well as one for its vertical structure. The equation for the latitude structure is called Laplace's tidal equation. Its eigenvalues are often referred to in terms of equivalent depths (an analogy to the ocean), and its eigenfunctions are called Hough functions. The equivalent depth then occurs as a parameter in the vertical structure equation. One can then expand the latitudinal structure of the solar heating in terms of these Hough functions.

Early observations contradicted simple intuition, in that the surface pressure showed tidal variations that were predominantly semidiurnal, while the solar forcing is predominantly diurnal. Figure 10, from *Chapman and Lindzen* [1970], illustrates two things. One is that surface pressure variations in connection with extratropical systems are much greater than surface pressure variations in the tropics (except when tropical cyclones occur). The other is that tidal variations are larger in the tropics and are predominantly semidiurnal rather than diurnal, as expected.

Initial suggestions to explain the dominance of the semidiurnal tide in surface pressure over the expected diurnal tide were that the semidiurnal solar forcing excited a resonance in the atmosphere, while the diurnal forcing was off resonance. As more and more was learned about atmospheric tides and their forcing, this did not prove to be the case. It was not until *Kato* [1966] and *Lindzen* [1966] independently discovered the existence of negative equivalent depth eigenmode solutions to Laplace's tidal equation that this problem was solved. The solution is best explained with the aid of Figure 11, from *Chapman and Lindzen* [1970]. In Figure 11, *V* the vertical variation of the solar heating of atmospheric water vapor (*V*1) and ozone (*V*2) are shown, along with the latitudinal variations of these heating functions. Note the different scales for the diurnal and semidiurnal heating functions. When these latitudinal heating functions are expanded in the diurnal and semidiurnal Hough function solutions of Laplace's tidal equation, the semidiurnal heating only gives rise to positive equivalent depth modes, the principal one of which has a very long vertical wavelength of about 100 km, while the diurnal heating gives rise to negative equivalent depth modes, which cannot propagate vertically, and a principal positive equivalent depth mode of rather short vertical wavelength (about 25 km). For the diurnal tide, the ozone heating forces a negative equivalent depth mode that cannot propagate down to the surface plus a short vertical wavelength mode. In contrast, the semidiurnal ozone heating principally forces a long wavelength mode that reaches the surface. For the semidiurnal solar tide, the vertical wavelength is longer than the depth of the ozone heating, and it propagates down to the surface, so the tidal pressure variations forced by ozone and water vapor add to produce a sizable response in surface

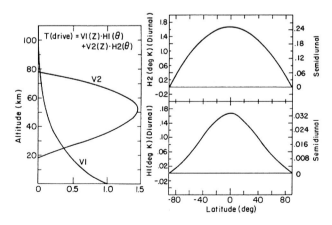

Figure 11. Vertical and horizontal variations of solar heating of atmospheric water vapor and ozone. From *Chapman and Lindzen* [1970], who, in turn, took this image from *Lindzen* [1968]. Reprinted with kind permission of Springer Science and Business Media.

pressure. For the diurnal ozone heating, not only does the negative equivalent depth mode not propagate down to the surface, but there is also destructive interference over the depth of the ozone heating for the short vertical wavelength solution. Therefore, the sizable semidiurnal variation in surface pressure is a consequence of the additive solutions from water vapor and ozone heating, while for the diurnal tide there is very little surface response to the ozone heating.

Not only did *Lindzen*'s [1966] and *Kato*'s [1966] papers lead to a resolution of the long-standing questions about the solar atmospheric tides, their work also led to an appreciation of the completeness of the Hough functions. This, in turn, led to *Longuet-Higgins* [1968] paper that examined the family of oscillations on a fluid envelope on a rotating sphere, which gave a theoretical framework for understanding global atmospheric wave motions.

8. STRATOSPHERIC OZONE CHEMISTRY

Alan Plumb's entrance on the middle atmosphere scene occurred just before a fundamental change occurred in the field with the famous publication by *Molina and Rowland* [1974]. While the study of stratospheric ozone already had a rich history, it was the *Molina and Rowland* [1974] publication that thrust stratospheric ozone research into the prominent position it occupies today.

The *Chapman* [1930] reactions

$$O_2 + h\nu(\lambda < 240 \text{ nm}) \rightarrow O + O, \tag{17}$$

$$O + O_2 + M \rightarrow O_3 + M, \tag{18}$$

$$O_3 + h\nu(\lambda < 320 \text{ nm}) \rightarrow O + O_2, \tag{19}$$

$$O + O_3 \rightarrow 2O_2 \tag{20}$$

are the simplest set of chemical reactions to account for the production and destruction of stratospheric ozone. In these reactions, ozone is produced by the dissociation of molecular oxygen by ultraviolet radiation (reaction (17)) followed by the attachment of one of the freed oxygen atoms to an oxygen molecule (reaction (18)). It is then destroyed by reaction (20), where an oxygen atom combines with an ozone molecule to make two stable oxygen molecules. Note that reaction (19) does not really destroy

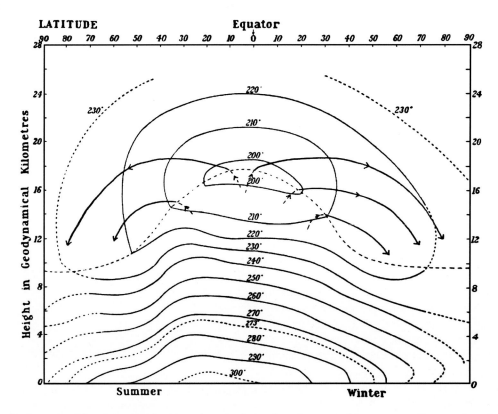

Figure 12. A supply of dry air is maintained by a slow circulation from the equatorial tropopause. Figure and caption from *Brewer* [1949]. Reprinted with permission.

ozone since the resulting oxygen atom quickly combines with an oxygen molecule to remake ozone.

A problem arose with these simple *Chapman* [1930] reactions when the rate of reaction (20) was measured by *Benson and Axworthy* [1957] and was found to proceed too slowly to account for measured ozone concentrations. This led *Hampson* [1964] to suggest the following reactions as being important. Note that the net result of reactions

$$OH + O_3 \rightarrow HO_2 + O_2 \qquad (21)$$

$$HO_2 + O \rightarrow OH + O_2 \qquad (22)$$

is reaction (20) since the OH radical acts as a catalyst. *Hunt* [1966] suggested a set of rate constants for reactions (21) and (22) that accounted for observed ozone concentrations. This was followed by papers by *Crutzen* [1970] and *Johnston* [1971] that pointed out the importance of the catalytic cycle involving nitrogen oxides,

$$NO + O_3 \rightarrow NO_2 + O_2 \qquad (23)$$

$$NO_2 + O \rightarrow NO + O_2. \qquad (24)$$

Crutzen [1970] suggested that reactions (23) and (24) dominate ozone loss at altitudes between 25 and 40 km, while reactions (21) and (22) increase in importance at higher altitudes. *Johnston* [1971] suggested that reactions (23) and (24) would lead to decreased stratospheric ozone loss if a large fleet of supersonic transport planes were to be implemented, emitting large amounts of nitrogen oxides.

In 1974, just as Alan was entering the scene, the paper by *Molina and Rowland* [1974] appeared, suggesting that manmade chlorofluorocarbon atmospheric concentrations were rapidly increasing because of their increasing use as aerosol propellants in refrigeration and other industrial sources. These chlorofluorocarbons are very stable in the troposphere but are dissociated in the stratosphere where they encounter the energetic solar ultraviolet radiation. This released chlorine to the stratosphere that participates in the very fast catalytic reactions

$$Cl + O_3 \rightarrow ClO + O_2 \qquad (25)$$

$$ClO + O \rightarrow Cl + O_2. \qquad (26)$$

The *Molina and Rowland* [1974] paper changed the face of stratospheric research, as did the discovery of the Antarctic ozone hole by *Farman et al.* [1985]. Large national and international programs were implemented to study stratospheric ozone. These included new satellite programs, extensive international assessments of stratospheric ozone, the implementation of the Montreal Protocol, and the award of the 1995 Nobel Prize in Chemistry to P. J. Crutzen, M. J. Molina, and F. S. Rowland.

9. STRATOSPHERIC TRANSPORT

Given the fact that ozone is mainly produced in the tropical stratosphere, where the UV radiation is sufficiently intense,

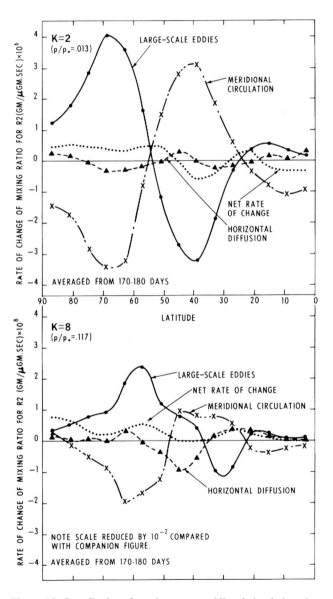

Figure 13. Contributions from the mean meridional circulation, the large-scale eddies, and horizontal diffusion to the continuity equation for an ozone-like tracer at two levels. The net rate of change of this ozone-like tracer is also shown. From *Hunt and Manabe* [1968]. Copyright American Meteorological Society.

and the fact that the measured ozone column amounts are largest in winter high latitudes, where the UV radiation is either zero or very weak, it was already clear to *Dobson* [1956] that there had to exist a global circulation that transported ozone from its source in the tropics to high latitudes.

The clearest case for such a circulation, though, was *Brewer*'s [1949] classic paper where he used the facts that stratospheric water vapor concentrations were very small and helium concentrations were very similar in the troposphere and stratosphere to infer that this circulation must consist of rising motion through the cold tropical tropopause and poleward and downward motion in the extratropical stratosphere. This is what we now refer to as the Brewer-Dobson circulation, which is shown in Figure 12, from *Brewer* [1949].

The injection of radioactive isotopes from the testing of nuclear weapons into the atmosphere gave additional valuable information on stratospheric transport that was very useful in early two-dimensional ozone models. Particularly notable in this regard was the paper by *Reed and German* [1965]. They noted that when mixing took place on sloping surfaces, such as that produced by the action of the a mean meridional circulation, the eddy transports could be written in terms of Fickian diffusion as

$$\overline{\chi'v'} = -K_{yy}\frac{\partial\overline{\chi}}{\partial y} - K_{yz}\frac{\partial\overline{\chi}}{\partial z} \qquad (27)$$

$$\overline{\chi'w'} = -K_{zy}\frac{\partial\overline{\chi}}{\partial y} - K_{zz}\frac{\partial\overline{\chi}}{\partial z}, \qquad (28)$$

where v (w) are northward (upward) velocities, χ is a conservative tracer, overbars denote zonal averaging, primes denote eddy departures from the zonal average, and K represent diffusion coefficients. *Reed and German* [1965] went on to estimate the K from heat flux data. These *Reed and German* [1965] diffusion coefficients were used in early two-dimensional ozone models. *Plumb and Mahlman* [1987] cleverly used this concept to later estimate these Ks based on general circulation modeling.

In section 2 of this paper, there was extensive discussion of the noninteraction theorem, but there is an analogous nontransport theorem that became clear when examining early troposphere-stratosphere general circulation model results. This is illustrated in Figure 13 from *Hunt and Manabe* [1968].

Note the near cancellation of the flux divergence terms from the mean meridional circulation and the large-scale eddies. The nontransport theorem says that for steady state conditions, no dissipation, no critical levels, and for a conserved tracer, there is no net transport by the planetary-scale eddies since these eddies give rise to canceling transport

effects by the mean meridional circulation. Incomplete cancellation occurs because these conditions are not met in either the general circulation model or in the atmosphere. More complete discussion on this point is provided by *Mahlman et al.* [1984]. Alan has gone on to be a primary contributor to the evolution of our present-day picture of stratospheric transport [e.g., see *Plumb*, 2002].

10. CONCLUDING REMARKS

The purpose of this review is to portray the research landscape that Alan found upon leaving graduate school and starting his illustrious research career studying the middle atmosphere. Over the next 30 plus years, he has greatly advanced these subjects by his personal research, that of his students, and his collaborative research with many others. Later chapters will elaborate on these themes. Certainly, Alan has left a very large footprint on our field, and . . . he is not finished.

REFERENCES

Alexander, M. J., and T. J. Dunkerton (1999), A spectral parameterization of mean-flow forcing due to breaking gravity waves, *J. Atmos. Sci.*, *56*, 4167–4182.

Andrews, D. G., and M. E. McIntyre (1978a), Generalized Eliassen-Palm and Charney-Drazin theorems for waves on axismmetric mean flows in compressible atmospheres, *J. Atmos. Sci.*, *35*, 175–185.

Andrews, D. G., and M. E. McIntyre (1978b), An exact theory of nonlinear waves on a Lagrangian-mean flow, *J. Fluid Mech.*, *89*, 609–646.

Andrews, D. G., J. R. Holton, and C. B. Leovy (1987), in *Middle Atmosphere Dynamics*, 489 pp., Academic, San Diego, Calif.

Bartels, J. (1928), Gezeitenschwingungen der Atmosphäre, *Handb. Experimentalphys.*, *25*, 163–210.

Benson, S. W., and A. E. Axworthy (1957), Mechanism of the gas phase, thermal decomposition of ozone, *J. Chem. Phys.*, *26*, 1718–1726.

Booker, J. R., and F. P. Bretherton (1967), The critical layer for internal gravity waves in shear flow, *J. Fluid Mech.*, *27*, 513–539.

Boyd, J. P. (1976), The noninteraction of waves with the zonally averaged flow on a spherical Earth and the interrelationships on eddy fluxes of energy, heat and momentum, *J. Atmos. Sci.*, *33*, 2285–2291.

Breeding, R. J. (1971), A non-linear investigation of critical levels for internal gravity waves, *J. Fluid Mech.*, *50*, 545–563.

Bretherton, F. P. (1966), The propagation of groups of internal gravity waves in a shear flow, *Q. J. R. Meteorol. Soc.*, *92*, 466–480.

Brewer, A. W. (1949), Evidence for a world circulation provided by the measurements of helium and water vapour distribution in the stratosphere, *Q. J. R. Meteorol. Soc.*, *75*, 351–363.

Brewer, A. W., B. Cwilong, and G. M. B. Dobson (1948), Measurement of absolute humidity in extremely dry air, *Proc. Phys. Soc. London*, *60*, 52–70.

Chapman, S. (1930), A theory of upper atmospheric ozone, *R. Meteorol. Soc. Mem.*, *3*, 103–125.

Chapman, S., and R. S. Lindzen (1970), in *Atmospheric Tides*, 200 pp., D. Reidel, Dordrecht, Netherlands.

Charney, J. G., and P. G. Drazin (1961), Propagation of planetary-scale disturbances from the lower into the upper atmosphere, *J. Geophys. Res.*, *66*, 83–109.

Crutzen, P. J. (1970), The influence of nitrogen oxides on the atmospheric ozone content, *Q. J. R. Meteorol. Soc.*, *96*, 320–325.

Dickinson, R. E. (1968), Planetary Rossby waves propagating vertically through weak westerly wind wave guides, *J. Atmos. Sci.*, *25*, 984–1002.

Dickinson, R. E. (1969), Theory of planetary wave-mean flow interaction, *J. Atmos. Sci.*, *26*, 73–81.

Dickinson, R. E. (1970), Development of a Rossby wave critical level, *J. Atmos. Sci.*, *27*, 627–633.

Dobson, G. M. B. (1956), Origin and distribution of the polyatomic molecules in the atmosphere, *Proc. R. Soc. London, Ser. A*, *236*, 187–193.

Dobson, G. M. B., and D. N. Harrison (1926), Measurements of the amount of ozone in the Earth's atmosphere and its relation to other geophysical conditions, *Proc. R. Soc. London, Ser. A*, *110*, 660–693.

Eliassen, A., and E. Palm (1961), On the transfer of energy in stationary mountain waves, *Geofys. Publ.*, *22*, 1–23.

Farman, J. C., B. G. Gardiner, and J. D. Shanklin (1985), Large losses of total ozone in Antarctica reveal seasonal ClO_x/NO_x interaction, *Nature*, *315*, 207–210.

Geller, M. A. (1983), Dynamics of the middle atmosphere, *Space Sci. Rev.*, *34*, 359–375.

Geller, M. A. (1993), Planetary wave coupling—Observations and theory, in *Coupling Processes in the Lower and Middle Atmosphere, NATO ASI Ser., Ser. C*, vol. 387, edited by E. V. Thrane, T. A. Blix, and D. C. Fritts, pp. 95–123, Kluwer Acad., Dordrecht, Netherlands.

Geller, M. A., H. Tanaka, and D. C. Fritts (1975), Production of turbulence in the vicinity of critical levels for internal gravity waves, *J. Atmos. Sci.*, *32*, 2125–2135.

Geller, M. A., W. Shen, and W. Wu (1997), Calculations of the stratospheric QBO for time-varying wave forcing, *J. Atmos. Sci.*, *54*, 883–894.

Goody, R. M. (1954), in *The Physics of the Stratosphere*, 187 pp., Cambridge Univ. Press, London.

Götz, F. W. P. (1931), Zum Strahlungsklima des Spitzbergensommers, Strahlungs- und Ozonmessungen in der Königsbucht 1929, *Gerlands Beitr. Geophys.*, *31*, 119–154.

Götz, F. W. P., A. R. Meetham, and G. M. B. Dobson (1934), The vertical distribution of ozone in the atmosphere, *Proc. R. Soc. London, Ser. A*, *145*, 416–446.

Gutnick, M. (1961), How dry is the sky?, *J. Geophys. Res.*, *66*, 2867–2871.

Hampson, J. (1964), C. A. R. D. E. (Canadian Armament Research and Development Establishment), *Tech. Note1627/64*, 280 pp., Valvartier, Quebec, Canada.

Hartley, W. N. (1881), On the absorption of solar rays by atmospheric ozone, *J. Chem. Soc.*, *39*, 111–128.

Haynes, P. H. (1985), Nonlinear instability of a Rossby-wave critical layer, *J. Fluid Mech.*, *161*, 493–511.

Haynes, P. H. (1998), The latitudinal structure of the quasi-biennial oscillation, *Q. J. R. Meteorol. Soc.*, *124*, 2645–2670.

Hazel, P. (1967), The effect of viscosity and heat conduction on internal gravity waves at a critical level, *J. Fluid Mech.*, *30*, 775–784.

Hines, C. O. (1960), Internal atmospheric gravity waves at ionospheric heights, *Can. J. Phys.*, *38*, 1441–1481.

Hodges, R. R., Jr. (1967), Generation of turbulence in the upper atmosphere by internal gravity waves, *J. Geophys. Res.*, *72*, 3455–3458.

Holton, J. R. (2004), in *An Introduction to Dynamic Meteorology*, 4th ed., 535 pp., Elsevier, New York.

Holton, J. R., and R. S. Lindzen (1972), An updated theory for the quasi-biennial cycle of the tropical stratosphere, *J. Atmos. Sci.*, *29*, 1076–1080.

Hunt, B. G. (1966), Photochemistry of ozone in a moist atmosphere, *J. Geophys. Res.*, *71*, 1385–1398.

Hunt, B. G., and S. Manabe (1968), Experiments with a stratospheric general circulation model: II. Large-scale diffusion of tracers in the stratosphere, *Mon. Weather Rev.*, *96*, 503–539.

Johnston, H. (1971), Reduction of stratospheric ozone by nitrogen oxide catalysts from supersonic transport exhaust, *Science*, *173*, 517–522.

Kato, S. (1966), Diurnal atmospheric oscillation: 1. Eigenvalues and Hough functions, *J. Geophys. Res.*, *71*, 3201–3209.

Killworth, P. D., and M. E. McIntyre (1985), Do Rossby-wave critical levels absorb, reflect, or over-reflect?, *J. Fluid Mech.*, *161*, 449–492.

Laplace, P. S. (1799), in *Mécanique Céleste*, Paris.

Leovy, C. B. (1964), Simple models of thermally driven mesospheric circulation, *J. Atmos. Sci.*, *21*, 327–341.

Lindzen, R. S. (1966), On the theory of the diurnal tide, *Mon. Weather Rev.*, *94*, 295–301.

Lindzen, R. S. (1968), The application of classical atmospheric tidal theory, *Proc. R. Soc. London, Ser. A*, *303*, 299–316.

Lindzen, R. S. (1981), Turbulence and stress owing to gravity wave and tidal breakdown, *J. Geophys. Res.*, *86*, 9707–9714.

Lindzen, R. S. (1990), in *Dynamics in Atmospheric Physics*, 310 pp., Cambridge Univ. Press, New York.

Lindzen, R. S., and J. R. Holton (1968), A theory of the quasi-biennial oscillation, *J. Atmos. Sci.*, *25*, 1095–1107.

Longuet-Higgins, M. S. (1968), The eigenfunctions of Laplace's tidal equations over a sphere, *Philos. Trans. R. Soc. London, Ser. A, 262*, 511–607.

Mahlman, J. D., D. G. Andrews, T. Hartmann, T. Matsuno, and R. G. Murgatroyd (1984), Transport of trace constituents in the stratosphere, in *Dynamics of the Middle Atmosphere*, edited by J. R. Holton, and T. Matsuno, pp. 387–416, Terra Sci., Tokyo.

Matsuno, T. (1966), Quasi-geostrophic motions in the equatorial area, *J. Meteorol. Soc. Jpn., 44*, 25–43.

Matsuno, T. (1970), Vertical propagation of stationary planetary waves in the winter Northern Hemisphere, *J. Atmos. Sci., 27*, 871–883.

Matsuno, T. (1971), A dynamical model of the stratospheric sudden warming, *J. Atmos. Sci., 28*, 1479–1494.

McIntyre, M. E., and T. N. Palmer (1983), Breaking planetary waves in the stratosphere, *Nature, 305*, 593–600.

McIntyre, M. E., and T. N. Palmer (1984), The "surf zone" in the stratosphere, *J. Atmos. Terr. Phys., 46*, 825–850.

Molina, M. J., and F. S. Rowland (1974), Stratospheric sink for chlorofluoromethanes: Chlorine atom-catalyzed destruction of ozone, *Nature, 249*, 810–812.

Murgatroyd, R. J. (1957), Winds and temperatures between 20 km and 100 km—A review, *Q. J. R. Meteorol. Soc., 83*, 417–458.

Murgatroyd, R. J., P. Goldsmith, and W. E. H. Hollings (1955), Some recent measurements of humidity from aircraft up to heights of about 50,000 ft over southern England, *Q. J. R. Meteorol. Soc., 81*, 533–537.

Naujokat, B. (1986), An update of the observed quasi-biennial oscillation of the stratospheric winds over the tropics, *J. Atmos. Sci., 43*, 1873–1877.

Pedlosky, J. (1979), in *Geophysical Fluid Dynamics*, 624 pp., Springer, New York.

Plumb, R. A. (1975), Momentum transport by the thermal tide in the stratosphere of Venus, *Q. J. R. Meteorol. Soc., 101*, 763–776.

Plumb, R. A. (1977), The interaction of two internal waves with the mean flow: Implications for the theory of the quasi-biennial oscillation, *J. Atmos. Sci., 34*, 1847–1858.

Plumb, R. A. (1984), The quasi-biennial oscillation, in *Dynamics of the Middle Atmosphere*, edited by J. R. Holton, and T. Matsuno, pp. 217–251, Terra Sci., Tokyo.

Plumb, R. A. (2002), Stratospheric transport, *J. Meteorol. Soc. Jpn., 80*, 793–809.

Plumb, R. A., and J. D. Mahlman (1987), The zonally averaged transport characteristics of the GFDL general circulation/transport model, *J. Atmos. Sci., 44*, 298–327.

Plumb, R. A., and A. D. McEwan (1978), The instability of a forced stationary wave in a viscous stratified fluid: A laboratory analogue of the quasi-biennial oscillation, *J. Atmos. Sci., 35*, 1827–1839.

Reed, R. J., and K. E. German (1965), A contribution to the problem of stratospheric diffusion by large-scale mixing, *Mon. Weather Rev., 93*, 313–321.

Reed, R. J., W. J. Campbell, L. A. Rasmussen, and D. G. Rogers (1961), Evidence of downward-propagating, annual wind reversal in the equatorial stratosphere, *J. Geophys. Res., 66*, 813–818.

Regener, V. H. (1938), Neue Messungen der vertikalen Verteilung des Ozons in der Atmosphäre, *Z. Phys., 109*, 642–670.

Regener, V. H. (1951), Vertical distribution of atmospheric ozone, *Nature, 167*, 276–277.

Rossby, C.-G., et al. (1939), Relation between variations in the intensity of the zonal circulation of the atmosphere and the displacements of the semi-permanent centers of action, *J. Mar. Res., 2*, 38–55.

Scherhag, R. (1952), Die explosionsartigen Stratosphärenerwärmungen des Spätwinters 1951/52, *Ber. Dtsch. Wetterdienst. U. S. Zone, 38*, 51.

Stewartson, K. (1977), The evolution of a critical level of a Rossby wave, *Geophys. Astrophys. Fluid Dyn., 9*, 185–200.

Teisserenc de Bort, L. P. (1902), Variations de la temperature del'air libre dans la zone comprise 8 km et 13 km d'altitude, *C. R. Hebd. Seances Acad. Sci., 24*, 987–989.

Veryard, R. G., and R. A. Ebdon (1961), Fluctuations is tropical stratospheric winds, *Meteorol. Mag., 90*, 125–143.

Wallace, J. M., and J. R. Holton (1968), A diagnostic numerical model of the quasi-biennial oscillation, *J. Atmos. Sci., 25*, 280–292.

Wallace, J. M., and V. E. Kousky (1968), Observational evidence of Kelvin waves in the tropical stratosphere, *J. Atmos. Sci., 25*, 900–907.

Warn, T., and H. Warn (1976), On the development of a Rossby wave critical level, *J. Atmos. Sci., 33*, 2021–2024.

Whipple, F. J. W. (1923), The high temperature of the upper atmosphere as an explanation of zones of audibility, *Nature, 111*, 187.

Yanai, M., and T. Maruyama (1966), Stratospheric wave disturbances propagating over the equatorial Pacific, *J. Meteorol. Soc. Jpn., 44*, 291–294.

M. A. Geller, School of Marine and Atmospheric Science, State University of New York at Stony Brook, Stony Brook, NY 11794-5000, USA. (Marvin.Geller@stonybrook.edu)

Planetary Waves and the Extratropical Winter Stratosphere

R. Alan Plumb

Department of Earth, Atmospheric, and Planetary Sciences, Massachusetts Institute of Technology, Cambridge, Massachusetts, USA

Planetary-scale Rossby waves dominate the dynamics of the winter stratosphere. In their classic analysis of the propagation of such waves on a mean state that varied only with height, Charney and Drazin (1961) concluded that deep vertical propagation is permitted only around the equinoxes, when the mean winds are westerly and sufficiently weak. Subsequent developments, especially by Matsuno, incorporated spherical geometry and latitudinal variations of the mean state into the analysis. A refractive index for the waves can be determined from the mean state; it has been widely used to diagnose wave propagation characteristics, and usually leads one to conclude that the winter westerlies, even in southern winter when the westerlies are strongest, are transparent to such waves. However, such conclusions rest on Wentzel-Kramers-Brillouin (WKB) assumptions, which are often inappropriate in the presence of realistic latitudinal variations of the mean state. It is here argued that the original conclusions of Charney and Drazin are qualitatively correct and that in its undisturbed, radiative equilibrium state, the winter stratosphere does not permit deep wave propagation. Such propagation requires the westerlies to be weakened by the waves themselves; it is argued that the consequent feedback between waves and the mean flow is at the heart of the strong variability of the stratospheric circulation, including the occurrence of major warmings, and may be central to stratosphere-troposphere interactions.

1. INTRODUCTION

Stratospheric meteorology blossomed once reasonably frequent observational data became available around half a century ago. It did not take long for the essential features of the extratropical northern winter stratosphere, the dominance of planetary scales in the wavefield, and the dramatic Arctic "sudden warmings," to be recognized (see, e.g., the remarks of *Schoeberl* [1978] and *Labitzke* [1981] about the early history of the subject). It also soon became evident that the waves are responsible for the warming events, partly on the basis of the empirical evidence that warmings are invariably

associated with amplification of the waves and partly on the theoretical grounds that there appeared to be no other plausible explanation. It is now recognized that the planetary-scale waves, together with contributions from smaller-scale internal gravity waves, control almost all aspects of the extratropical stratospheric circulation. Since the planetary waves have a large quasi-stationary component, it is clear that they must be forced, rather than arising from in situ instability of the flow, and that their structure (specifically, a westward phase tilt with height) is consistent with a tropospheric source. Sources of quasi-stationary waves have been identified as large-scale topography [*Charney and Eliassen*, 1949], planetary-scale heat sources and sinks [*Smagorinsky*, 1953] and, somewhat more recently, the statistically averaged effects of synoptic-scale eddies [*Scinocca and Haynes*, 1998].

In the ensuing decades, much effort has been aimed at documenting and understanding the dynamics of the extratropical stratosphere, and much progress has been made. We

The Stratosphere: Dynamics, Transport, and Chemistry
Geophysical Monograph Series 190
Copyright 2010 by the American Geophysical Union.
10.1029/2009GM000888

have learned not only how the mean state of the stratosphere determines the wave structure and wave breaking, but also how the waves, in turn, influence the mean state. Moreover, we have also learned that, rather than the stratosphere being simply a slave to the troposphere, internal stratospheric dynamics plays a major role in effecting the very large variability of the wave, mean flow system, including the wave flux out of the troposphere. We are even beginning to understand what impact such variability has on the troposphere and thereby on surface climate.

Despite all the progress, significant gaps remain. On the one hand, the seemingly difficult and nonlinear problem of the impact of the waves on the mean state is actually quite straightforward and well understood, especially outside the tropics, thanks largely to the development of "transformed Eulerian mean" theory [Andrews and McIntyre, 1976]. The role of gravity waves is still a matter of some uncertainty, but that will not be our focus here. (It is discussed in detail by Alexander [this volume].) It is the main contention of this review that our understanding of how the characteristics of even linear waves depend on the mean state remains inadequate and is the chief roadblock to a complete theory of stratospheric meteorology. To be sure, a body of linear theory exists in principle: one can write down a set of linear equations (although to do so requires that nonlinear wave breaking be represented by some linear damping process) and solve them numerically, but a simple characterization of the relationship between the wave structure and the mean state remains incomplete. The development of the linear theory of planetary waves is outlined in section 2 of this paper, along with some remarks of caution about the dangers of overly simplistic interpretations about the nature of wave propagation. Wave breaking is briefly described in section 3, and in section 4, the theory of how the waves impact the mean state is reviewed, rather briefly since this is not the focus of this review and because the topic has been thoroughly reviewed elsewhere [e.g., Holton et al., 1995].

Observations and models of the variability of the circulation, the topic of section 5, have yielded a good deal of insight into the importance of wave, mean flow, feedback processes within the stratosphere. Even the seasonal cycle is of interest, as seasonal variations are large in the stratosphere, and the two hemispheres behave rather differently. More remarkable is the occurrence of persistent vacillations found in stratospheric models of various degrees of complexity and realism; these vacillations are unambiguously indicative of internal dynamical feedback, as the planetary waves occasionally amplify and the mean flow weakens, followed by wave collapse and recovery of the mean state. Consideration of why the waves amplify leads us to return to linear theory. In section 6, it is shown that, in a one-dimensional (1-D) model

of forced, damped waves, such as might be appropriate for waves on a vortex with a sharp edge, large wave amplitudes at altitude, and westward phase tilt with height, are not necessarily indicative of upward propagation and that evanescent or quasi-modal structures may be a better paradigm for waves in midwinter westerlies of realistic magnitude. This will lead us in section 7 to revisit resonance, including nonlinear self-tuning, as an explanation of events of wave amplification and high-latitude warmings. In the concluding remarks of section 8, we will note the implications that quasi-modal behavior might have on our understanding of stratosphere-troposphere interactions.

2. LINEAR THEORY OF WAVE PROPAGATION

The theory of planetary-scale Rossby waves developed hand in hand with observations. To some extent, in fact, the theory preceded the observations. When *Charney and Drazin* [1961] published their seminal paper on the vertical propagation of large-scale quasi-geostrophic waves, their stated motivation was not to explain observed stratospheric waves (observations of which were few at the time) but to explain why, given the highly energetic eddies of the troposphere, the upper atmosphere is not heated to extreme temperatures when this energy is transferred via wave propagation into the high atmosphere and dissipated as heat, in the manner of the solar corona. To this end, they considered the vertical structure of waves on an extratropical beta plane, in a medium whose background properties (mean density ρ, zonal wind U, and buoyancy frequency N) vary only with height, to show that a zonally propagating wave whose geopotential perturbation is of the form (we depart from their notation here)

$$\phi'(x,y,z,t) = \frac{N}{\sqrt{\rho}} Re\{\Phi(z) exp[i(kx + ly - kct)]\} \quad (1)$$

satisfies a wave equation

$$\frac{d^2\Phi}{dz^2} + n^2\Phi = 0, \quad (2)$$

where n is a dimensional refractive index, defined by

$$n^2 = -\frac{N^2}{f^2}(k^2 + l^2) - \frac{N}{\sqrt{\rho}}\frac{d^2}{dz^2}\left(\frac{\sqrt{\rho}}{N}\right) + \frac{N^2}{f^2}\frac{dQ/dy}{U-c} \quad (3)$$

and where

$$\frac{dQ}{dy} = \beta - \frac{f^2}{\rho}\frac{d}{dz}\left(\frac{\rho}{N^2}\frac{dU}{dz}\right)$$

is the gradient of quasi-geostrophic potential vorticity (QGPV), f and β being constant values of the Coriolis parameter and its latitudinal gradient, respectively.

The refractive index becomes constant when U and N are constant and ρ decreases exponentially with height with constant scale height H; then equation (2) has wavelike solutions $\Phi \sim e^{imz}$, where the vertical wave number m is

$$m^2 = n^2 = \frac{N^2}{f^2}\left[\frac{\beta}{U-c}-(k^2+l^2)\right]-\frac{1}{4H^2}. \quad (4)$$

As has been discussed many times [e.g., *Andrews et al.*, 1987], equation (4) leads to the celebrated Charney-Drazin propagation condition that vertically propagating waves are possible, i.e., n is real, only under certain circumstances. Only the term $\beta/(U-c)$ in equation (4) is not negative definite, so propagation can occur only by virtue of the term involving the potential vorticity gradient: the waves are Rossby waves. Specifically, propagation requires

$$0 < U-c < U_c, \quad (5)$$

where

$$U_c = \frac{\beta}{k^2+l^2+f^2/4N^2H^2} \quad (6)$$

is the "Rossby critical velocity." In particular, stationary waves of the kind that dominate the winter stratosphere can propagate only when the mean winds are westerly and less than U_c. Since U_c is a decreasing function of wave number, in fact, being a fraction of 1 m s^{-1} for typical synoptic-scale waves, only the planetary-scale waves are capable of vertical propagation. Charney and Drazin estimated that $U_c \simeq 38$ m s^{-1} for a disturbance of zonal wave number 2 (which they identified as the peak wave number of Northern Hemisphere (NH) topography); for zonal wave number 1, one obtains a slightly larger $U_c \simeq 57$ m s^{-1}. Since the peak climatological westerlies exceed these values in midwinter, they concluded that, even for the planetary scales, deep vertical propagation is permitted only around the equinoxes, when the mean winds are relatively weak and westerly.

The importance of making proper allowance for the spatial variation of U was emphasized by the work of *Dickinson* [1968] and *Matsuno* [1970]. Like Charney and Drazin, Dickinson noted the need for weak westerlies to permit propagation, but proposed the existence of a "polar cap wave guide" in the weaker winds poleward of the wintertime polar jet, within which the waves are confined by the strength of the jet. (He also proposed a midlatitude wave guide bounded by strong winds both poleward and equatorward but, given what is now known of stratospheric wind climatology, this does not now appear realistic.) He also pointed out that observed large wave amplitudes through the winter stratosphere may not necessarily indicate wave propagation: amplitudes may be large at altitude even when the waves are evanescent, unless

the wave activity decreases sufficiently rapidly with height to overcome the effects of decreasing density.

Further, *Dickinson* [1968] drew attention to the singularity of equation (2) when n^2 becomes infinite, as it does at the subtropical "critical line" where $U=0$ (for stationary waves), arguing that waves propagating through the westerlies toward the critical line will be absorbed there and showing, in fact, that all wave packets in the region of westerlies, whatever their initial orientation, will be refracted toward the critical line. (This conclusion was later confirmed by *Karoly and Hoskins* [1982] on the basis of explicit ray tracing calculations.)

A similar analysis by *Matsuno* [1970] led, for a stationary wave in an atmosphere of uniform N^2, with

$$\phi' = e^{z/2H}Re\{\psi(\varphi,z)e^{is\lambda}\},$$

to the wave equation

$$\mathcal{L}(\psi) + v_s\psi = 0, \quad (7)$$

where

$$\mathcal{L}(\psi) \equiv \frac{\sin^2\varphi}{\cos\varphi}\frac{\partial}{\partial\varphi}\left(\frac{\cos\varphi}{\sin^2\varphi}\frac{\partial}{\partial\varphi}\right) + \frac{4\Omega^2a^2}{N^2}\sin^2\varphi\frac{\partial^2}{\partial z^2} \quad (8)$$

$$v_s = a\frac{\partial Q/\partial\varphi}{U} - \frac{\Omega^2a^2}{N^2H^2}\sin^2\varphi - \frac{s^2}{\cos^2\varphi}. \quad (9)$$

Here λ and φ are longitude and latitude, Ω is the planetary rotation rate, and a the planetary radius. The significance of v_s is that it represents the refractive index squared in spherical geometry and is obviously analogous to n^2 in the expression (3) of *Charney and Drazin* [1961] (but in this case, v_s is dimensionless: v_s should be multiplied by $(N/2\Omega a\sin\varphi)^2$ to compare with n^2 in expression (3)).

Matsuno's calculation of v_0, the value for zero zonal wave number, for a representative distribution of zonal winds in the winter NH, is illustrated in Figure 1. The contribution from the second term in equation (9) is small, so the values of v_0 are positive everywhere poleward of the zero wind line, increasing almost monotonically equatorward from the pole, becoming infinitely large as the line is approached. Note that the shape of the jet is hardly evident in the structure of v_0, despite the appearance of U in the first term of equation (9): the distribution of the potential vorticity gradient, shown in the middle frame of Figure 1, is similar to that of U, so much so that the ratio $(dQ/dy)/U$ does not reflect the jet structure, except for a broad tendency for v_0 to decrease upward in concert with the increasing strength of the jet. The single feature embedded within the large-scale distribution of Q_0 is a pronounced localized minimum in the lower stratosphere, between the two jets where the QGPV gradient is weak.

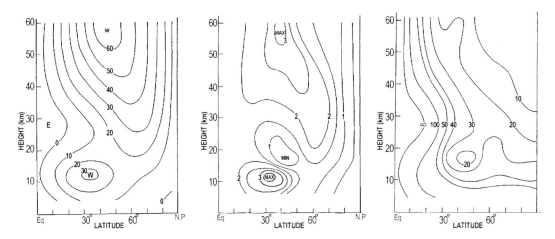

Figure 1. The mean state considered by *Matsuno* [1970] in his analysis of planetary wave propagation in the winter stratosphere, plotted versus latitude and height. (left) Zonal mean zonal wind U (m s^{-1}); (center) mean latitudinal gradient $\partial Q/\partial \phi$ of potential vorticity (in units of the Earth's rotation rate); (right) calculated refractive index squared ν_0 for a wave of zonal wave number $s = 0$. After *Matsuno* [1970]. Copyright American Meteorological Society.

Now, the operator $\mathcal{L}$ in equations (7) and (8) has the form of a modified Laplacian in the meridional plane; in fact, when the latitudinal length scale of the waves is much less than the equator-pole distance (so that the trigonometric factors in equation (8) can be regarded as slowly varying),

$$\mathcal{L} \cong a^2 \left(\frac{\partial^2}{\partial y^2} + \frac{f^2}{N^2} \frac{\partial^2}{\partial z^2} \right),$$

where $y = a\varphi$. Hence, if ν_0 can also be regarded as slowly varying in space, equation (7) has wavelike solutions $\psi = Re\psi_0\exp[i(kx + ly + mz)]$, where $x = a\lambda\cos\varphi$ and $k = s/(a\cos\varphi)$, with

$$k^2 + l^2 + \frac{f^2}{N^2} m^2 = \frac{1}{a^2}\nu_0. \qquad (10)$$

Thus, a positive ν_0 implies that the total wave number squared (in the stretched-coordinate sense implied by equation (10)) is positive, and in that sense, propagation is permitted. For a given zonal wave number s, the total meridional wave number is determined by

$$l^2 + \frac{f^2}{N^2} m^2 = \frac{1}{a^2}\nu_s, \qquad (11)$$

where

$$\nu_s = \nu_0 - \frac{s^2}{\cos^2\varphi}. \qquad (12)$$

For planetary waves $s = 1$ and 2, ν_s differs relatively little from ν_0 except near the pole.

The significance of the refractive index goes beyond its sign; Wentzel-Kramers-Brillouin (WKB) theory predicts that waves propagating in regions of positive ν_s will be refracted toward increasing ν_s. For a typical stratospheric climatology,

therefore, including the one considered by Matsuno and shown in Figure 1, upward propagating waves in extratropical latitudes will be refracted equatorward, toward the zero wind line, as *Dickinson* [1968] showed, to be absorbed there, provided the QGPV gradient is positive (as it usually is). Hence, the gross picture suggested by the distribution of ν_0 seen in Figure 1 is one in which large-scale Rossby waves propagate upward and refract equatorward to become dissipated in the subtropics. In the calculations of *Dickinson* [1968], most wave trajectories encounter the critical layer within a scale height or so, thus severely inhibiting vertical propagation. However, *Matsuno* [1970] noted the significance of the minimum in refractive index in the midlatitude lower stratosphere and argued that refraction around this feature creates a partial wave guide for vertical propagation on its poleward side. As evidence for the real impact of this feature, the wave propagation vectors (Figure 2) of a solution he determined in this flow, in response to a stationary low-level forcing of a wave, show a clear tendency to propagate around the ν_s minimum. Of course, Matsuno's calculation is directly applicable only to the particular climatology of Figure 1 assumed in his calculation, but on the basis of many similar analyses, these results appear to be robust; a lower stratospheric minimum in QGPV gradient and the consequent ν_s minimum above and slightly poleward of the tropospheric jet seem to occur whenever the stratospheric and tropospheric jets are separated, as is usually the case in practice.

Within this wave guide, waves are partly trapped by the weak QGPV gradient equatorward of the jet, at least in the lower and middle stratosphere, rather than by the strong winds of the jet as had been suggested by Dickinson, whose assumed background structure did not include this region of weak

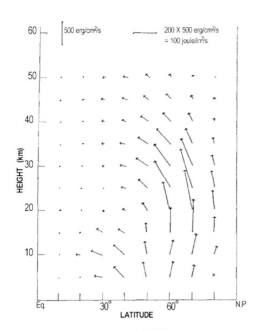

Figure 2. Calculated energy flux $\overline{\rho v'\phi'}$ for a stationary wave of zonal wave number 1 for the background state shown in Figure 1. For this stationary, quasi-geostrophic, inviscid wave, this flux equals $\bar{u}\mathbf{F}$, where $\mathbf{F}$ is the Eliassen-Palm flux. After *Matsuno* [1970]. Copyright American Meteorological Society.

gradient. The obvious importance of the wave guide is that refraction of wave activity toward the zero wind line and wave absorption is inhibited, thus permitting greater vertical penetration than Dickinson's calculations implied. It also raises the possibility of cavity-like behavior and that (to quote Matsuno) there "may be an approximately resonant state for a suitable wave number, although resonance in the rigorous sense is impossible because the walls are so leaky." A cavity within which resonance might occur would require at least partial reflection at its top, as well as its sides. Like Charney and Drazin and Dickinson, Matsuno suggested that the strong upper level winds cap the cavity, although there is little to suggest this in the refractive index distribution shown in Figure 1, nor in the wave propagation vectors of Figure 2. Nevertheless, by comparing solutions at different zonal wave numbers, he found a peak in response (around $s = 1.25$) suggestive of the quasi-resonant behavior of a damped system.

It is worth reiterating the key differences between the analyses just described. Charney and Drazin allowed only vertical structure in their basic state within which they considered wave propagation on a beta plane; in fact, most of their discussion was based on the assumption of uniform flow. The analyses of Dickinson and of Matsuno incorporated spherical geometry and, most importantly, latitudinal

structure in the background state. In fact, the key difference between the two definitions (4) and (12) for "refractive index squared" is that the QGPV gradient in the latter includes the terms describing barotropic curvature of the background flow. (The major differences between the conclusions of Dickinson and Matsuno derived from different assumptions they made about the background state.) Recall that the only term allowing propagation (i.e., the only possibly positive contribution to refractive index squared) is that involving the ratio $(\partial Q/\partial y)/U$. So, while in the simplest description of the Charney-Drazin analysis, $\partial Q/\partial y = \beta$, and propagation depends solely on the strength of the background zonal wind (as expressed in equation (5)), the later analyses emphasized the importance of the full structure of the QGPV gradient, especially the potential role of trapping by weak gradients equatorward of stratospheric jet. Moreover, the role of the wind speed is hardly evident in the distribution of v_s, simply because of the dependence of $\partial Q/\partial y$ on U itself: the QGPV gradient is increased at a jet maximum, such that the ratio becomes less sensitive to U.

There are, however, several caveats that, with the benefit of hindsight, caution against a simple interpretation of wave propagation characteristics on the basis of calculations of refractive index. The first point to be made is that equation (12) determines not the vertical wave number, but the total meridional wave number. Hence the condition for vertical propagation (real m) is not simply $v_s > 0$ but rather $v_s > l^2 a^2$. Meridional boundedness constrains the meridional wave number l. At the very least, the finite size of the Earth restricts l: if one-half wavelength is confined to the winter hemisphere, $l = 2a^{-1}$, so the criterion becomes $v_s > 4$, not a significant modification. More likely, however, the wave structure will be influenced by the structure of the background state, which varies on length scales much smaller than $\pi a/2$, and then l will be much greater than $2a^{-1}$, at least in places. If, for example, the wave were confined locally to the refractive index ridge poleward of the minimum in Matsuno's figure, the constraint on vertical propagation may be quite restrictive. In fact, *Simmons* [1974] argued that the latitudinal structure of the wave mirrors that of the mean flow, in which case the term l^2 in equation (12) approximately cancels the contribution of the barotropic flow curvature to the term involving $(\partial Q/\partial y)/U$, and the criterion for vertical propagation reverts essentially to the Charney-Drazin form. After the fact (if one knows the wave structure from observations or from calculation), one can allow for finite l^2 in equation (12), as done by *Harnik and Lindzen* [2001], who thereby diagnosed the presence of barriers to propagation, which were not at all evident from a simple inspection of the distribution of v_s, and demonstrated a clear case of downward reflection of a planetary wave packet in the winter Southern Hemisphere (SH).

A second caveat is that a climatology of the mean state of the stratosphere, such as that used by Matsuno (and by most subsequent analyses of the same type) may misrepresent its actual state in such a way as to give a misleading impression of wave propagation characteristics. In reality, a typical snapshot of the winter stratosphere, except during highly disturbed periods, shows a polar vortex bounded by an undulating jet marked by a sharp gradient of PV, with only a weak PV gradient equatorward of the jet. (We shall discuss this in more detail in section 3.) The smoother distributions evident in the climatological picture result from zonal and temporal averaging of such states. (Note from equation (9) that v_s is a nonlinear function of the background state, and so v_s calculated from the climatological average is not the same as a climatological average of v_s calculated daily.)

An equally tractable and perhaps more realistic basic state can be constructed in which all the PV gradient at a given altitude is concentrated at the edge of a circular vortex. Wave perturbations are then trapped at the vortex edge, and any meridional propagation must occur vertically along the edge. *Esler and Scott* [2005], following similar but more restricted calculations by *Waugh and Dritschel* [1999] and *Wang and Fyfe* [2000], showed that the dispersion relation one obtains for waves on a barotropic vortex of radius R has the form

$$\hat{\omega}_s - \Omega_e = -\Delta Q \; \mathcal{A}_s \left(\frac{fR}{N} \sqrt{m^2 + \frac{1}{4H^2}} \right), \qquad (13)$$

where $\hat{\omega}_s$ is the angular phase speed of a wave of zonal wave number s, Ω_e is the angular velocity of the flow at the vortex edge, ΔQ is the jump in QGPV at the vortex edge (both Ω_e and ΔQ are independent of height), and $\mathcal{A}_s(x) = K_s(x)I_s(x)$, K_s and I_s being modified Bessel functions of order s. Equation (13) is very much like equation (4) obtained by *Charney and Drazin* [1961]; as in that case, if we write $U = R\Omega_e$ and $c = R\hat{\omega}$, respectively, the flow and wave velocities at the vortex edge, the condition for propagation remains of the Charney-Drazin form (equation (5)), where now the Rossby critical velocity is

$$U_c = R \; \Delta Q \; \mathcal{A}_s \left(\frac{fR}{2NH} \right). \qquad (14)$$

Using parameter values chosen by *Waugh and Dritschel* [1999] to represent a reasonably realistic midwinter vortex with $U = 58.9$ m s^{-1}, *Esler and Scott* [2005] obtained $U_c = 42.8$ m s^{-1} for $s = 1$ (when U_c attains its maximum value), in which case, stationary waves are vertically evanescent.

These two caveats alone caution that refractive index diagnostics are to be interpreted with care and illustrate how far, almost half a century after *Charney and Drazin's* [1961] paper, we remain from a clear understanding of linear planetary waves in realistic stratospheric states. A third caveat

concerns the extent to which linear theory is applicable at all, given the large amplitude that planetary waves frequently attain in the wintertime middle and upper stratosphere. In fact, consideration of nonlinear effects becomes unavoidable in the vicinity of the critical line where, for stationary waves, $U = 0$. This question will be the focus of the next section.

3. CRITICAL LAYERS AND BREAKING

The importance of the subtropical zero wind line, the "singular" or "critical" line for stationary waves, was recognized by *Dickinson* [1968] who predicted, within the context of linear theory, that waves will suffer absorption there. *Matsuno's* [1970] analysis, by highlighting the tendency of waves to be refracted toward the zero wind line (in the absence of an intervening barrier to propagation), makes it unreasonable to ignore its effects, as emphasized by *McIntyre* [1982]. In fact, the validity of linear theory for stationary waves requires, among other things, that the perturbation flow (u',v') be weak compared to the mean flow $\bar{u}$, an assumption that obviously breaks down as the critical line is approached. What really happens there was revealed by the observational analysis of *McIntyre and Palmer* [1983] and, in idealized situations, by the barotropic model integrations of *Juckes and McIntyre* [1987] and *Haynes* [1989]. For a wave whose stream function amplitude is of typical magnitude ψ, say, propagating through westerlies toward a region within which the mean zonal winds change sign, what linear theory sees as a "critical line," in fact, is a critical layer, with finite latitudinal width proportional to $\psi^{1/2}$ and which therefore collapses to a infinitesimally narrow region around $\bar{u} = 0$ in the linear limit. The key structures that define the critical layer are closed anticyclonic eddies, the so-called "Kelvin cat's eyes." Thus, while outside the critical layer, material lines (such as PV contours, for sufficiently conservative flow) are simply wavy; within the critical layer, material lines are irreversibly stretched and wrapped up within the closed eddies. This redistribution of PV is responsible for absorption of the waves within the critical layer, a fact that can be seen directly from the now well-known [e.g., *Andrews et al.*, 1987] relationship between the eddy flux of QGPV, $\rho\overline{v'q'}$, and the divergence of the Eliassen-Palm (EP) flux $\mathbf{F}$:

$$\rho\overline{v'q'} = \nabla \cdot \mathbf{F},$$

where

$$\mathbf{F} = \begin{pmatrix} F_y \\ F_z \end{pmatrix} = \begin{pmatrix} -\rho\overline{u'v'} \\ \rho f \left(\frac{\partial \bar{\theta}}{\partial z} \right)^{-1} \overline{v'\theta'} \end{pmatrix}.$$

In the usual situation when the background PV gradient has the same sign as β (i.e., is positive), a downgradient flux of

PV caused by the critical layer stirring has $\overline{v'q'} < 0$, whence **F** is convergent, corresponding to a dissipation of wave activity within the layer. However (*Killworth and McIntyre*, 1985; and see the discussion in the work of *Andrews et al.* [1987]), if PV is truly conserved, absorption can occur only for a finite time, since once the mean PV gradient has been stirred away within the layer, the PV flux will vanish: only a finite amount of wave activity can be absorbed. Sustained absorption within the critical layer requires restoration of the PV gradient by nonconservative effects at a sufficient rate to compete with the wave-induced stirring.

McIntyre and Palmer [1983] showed from midstratospheric PV analyses during a moderately disturbed period of northern winter that, far from being a narrow zone around the zero wind line as linear theory would assume, the layer encompasses a large fraction of the hemisphere. The Kelvin cat's eye that defines the layer manifests itself as the Aleutian anticyclone, which typically grows to large amplitude during disturbed periods. These observations, complemented by the high-resolution, one-layer numerical simulations of *Juckes and McIntyre* [1987], led to a picture of the winter stratosphere as a region of high PV (the polar vortex) bounded by a sharp edge at the location of the polar night jet, from which filaments are episodically stripped and entrained into, and stirred within, the "surf zone" of middle latitudes. The stripping of these filaments off the vortex tightens the PV gradients at the vortex edge, while surf zone stirring weakens the PV gradients there. Similar processes occur at the equatorward edge of the surf zone (as is to be expected, given the finite latitudinal width of the critical layer), thereby producing a second, subtropical region of high PV gradients [*Norton*, 1994]; however, these subtropical gradients appear to be substantially weaker than those at the vortex edge, which might be a simple geometric effect [*McIntyre*, 1982]. The resulting picture of horizontal (actually, quasi-isentropic) stratospheric transport is one of vigorous stirring within a midlatitude surf zone, bounded by partial transport barriers at the edge of a relatively impermeable vortex and in the subtropics. The separation into these three regions (vortex, surf zone, and tropics) has a major impact on the distribution of stratospheric trace gases (e.g., see *Plumb* [2002, 2007], *Shepherd* [2007], and references therein).

4. THE WAVES' IMPACT ON STRATOSPHERIC STRUCTURE

With the availability of sufficient stratospheric data to provide broadscale synoptic analyses, it quickly became evident that the dominant planetary-scale waves have a profound impact on the overall state of the stratosphere. In particular, the dramatic events that have become known as major warmings (and which we shall discuss further in what follows) are clearly associated with unusually large wave amplitudes. Even from the less-disturbed climatological perspective, the winter polar stratosphere (especially in the NH) is much warmer than radiative equilibrium calculations would predict, and this can only be explained as a consequence of angular momentum transport by eddy motions [e.g., *Haynes et al.*, 1991]. The argument is both simple and elementary. First, write the equations of zonal mean angular momentum balance, continuity, and entropy as

$$\frac{\partial M}{\partial t} + \overline{\mathbf{v}}_* \cdot \nabla M = \frac{a\,\cos\varphi}{\rho}\,\nabla\cdot\mathbf{F},$$

$$\frac{\partial \rho}{\partial t} + \nabla\cdot\rho\overline{\mathbf{v}}* = 0, \qquad\qquad (15)$$

$$\frac{\partial \overline{\theta}}{\partial t} + \overline{\mathbf{v}}* \cdot \nabla\overline{\theta} = \frac{1}{\rho c_p}\left(\frac{p_0}{p}\right)^{\kappa} J.$$

Here $M = \Omega a^2\cos^2\varphi + \overline{u}a\cos\varphi$ is the specific mean absolute angular momentum, $\overline{\mathbf{v}}_* = (\overline{v}_*, \overline{w}_*)$ the meridional residual circulation, ρ density, θ potential temperature, p pressure, $p_0 = 1000$ hPa, c_p the specific heat of air at constant pressure, $\kappa = 2/7$, and J is the diabatic heating rate. Viscosity is neglected, since above the boundary layer and below the upper mesosphere, its effects on these large scales are completely negligible. We have also neglected an additional term in the thermodynamic equation, involving the divergence of the diabatic eddy flux of heat. This term vanishes for adiabatic waves; in other circumstances, it is formally negligible under quasi-geostrophic scaling and is also negligible, at leading order, in a WKB analysis of waves on a slowly varying background flow [*Andrews and McIntyre*, 1976].

If, for simplicity, we fix attention on the steady case (such as a climatological midwinter state), it follows from the first of equation (15) that if there is no eddy forcing ($\nabla \cdot \mathbf{F} = 0$), there can be no mean meridional flow across the mean angular momentum contours. Outside the tropics, contours of M are nearly vertical [*Haynes et al.*, 1991], and so, no mean circulation is possible. Then the third of equation (15) requires $J = 0$: there is no diabatic heating; hence (by definition), temperatures are in radiative equilibrium.

What happens when eddy forcing is present was described in some detail by *Haynes et al.* [1991] and *Holton et al.* [1995]. The essence is depicted in Figure 3. On the whole, one expects that $\nabla \cdot \mathbf{F} < 0$ wherever the waves are dissipated, which for the sake of argument is assumed in Figure 3 to occur within a localized region in middle latitudes. Thus, dissipating waves act as a drag on westerly mean flow. Since M decreases

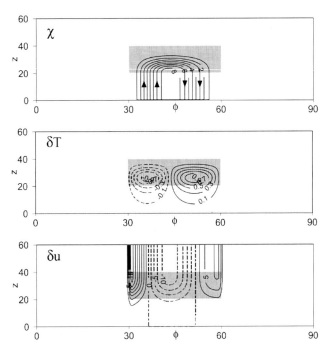

Figure 3. An example of the steady, quasi-geostrophic, response to stratospheric wave drag. The drag is applied entirely within the gray rectangle (it has a cosine-squared profile in both directions, has a maximum value of -2×10^{-6} m s^{-2} in the center, and vanishes at the edges). (top) Mass stream function (10^7 kg s^{-1}); (middle) temperature response (K); (bottom) zonal wind response (m s^{-1}). Dashed contours show zero or negative values. The calculation assumes an isothermal background atmosphere with a density scale height of 7 km, Newtonian cooling with a time constant of 20 days, and an infinitely large drag coefficient at the ground. See text for discussion.

poleward in the extratropical stratosphere, the wave drag drives a mean flow poleward across the angular momentum contours (the extratropical "Rossby wave pump") [*Holton et al.*, 1995]. Continuity of mass then requires corresponding vertical motion; at the poleward side of the wave drag, for example, the flow must turn upward or downward (or both). In fact, it cannot go upward there, since it would then, at some higher altitude, have to turn back equatorward again, and there is no "reversed wave drag" at higher altitude to allow it to do so. (This might not, however, be entirely true in models that allow a net source or sink of angular momentum in a "sponge layer" near the upper boundary [*Shepherd et al.*, 1996].) Instead, it must turn downward; the flow is able to return at lower altitude by virtue either of frictional or topographic form drag at the surface or of a region of divergent EP flux if the waves are forced internally (such as by diabatic heating). (In practice, because of the much greater density at low altitudes, the required torque is rather weak, and it is a moot point as to whether a true steady state needs to be established

at low altitudes over the course of a winter.) *Haynes et al.* [1991] referred to this as "downward control" of the extratropical meridional circulation.

Thus, the steady meridional residual circulation (except in the tropics, which we shall address below) can be deduced simply from consideration of the angular momentum budget and mass continuity alone: we have not yet needed to invoke the thermodynamic equation. This illustrates the power of the angular momentum constraint on the problem. Wave drag must be present to permit flow across angular momentum contours; flow across isentropic surfaces is not so constrained, since radiative relaxation permits potential temperature to adjust as necessary. In fact, rather than seeing radiation as a driver of the circulation, we can, in fact, calculate diabatic heating as a consequence of the vertical motion induced by the wave drag. In the downwelling region below and poleward of the region of wave drag in Figure 3, high-latitude air must be warmed sufficiently above radiative equilibrium such that diabatic cooling balances the adiabatic warming associated with the circulation. Similarly, in the low-latitude rising branch, the air is cooled to produce diabatic warming. Thus, the pattern of low-latitude diabatic heating/high-latitude diabatic cooling is not a straightforward consequence of greater solar input at low latitudes (since shortwave heating would, in radiative equilibrium, simply be balanced by longwave cooling) but is rather a result of the eddy-driven circulation forcing the stratosphere out of radiative equilibrium. In short, the thermal effect of the wave drag is to reduce the latitudinal temperature gradient below the altitude of the drag. Thermal wind balance then dictates reduced westerly wind shear below the drag and a barotropic reduction of the zonal flow above [*Haynes et al.*, 1991]. Thus, dissipation of planetary waves in the winter stratosphere warms the high latitudes, cools the tropics (to a lesser degree, for simple geometric reasons), and reduces the strength of the polar vortex.

In its strict form, the argument just outlined does not apply to unsteady situations. A localized wave drag of the kind exemplified in Figure 3, impulsively applied, will initially induce circulation cells both above and below the forcing (the "Eliassen response") [*Eliassen*, 1951]. Nevertheless, because of the decrease of density with altitude, when the wave drag acts on large horizontal scales, the mass circulation is, in practice, dominated by the lower branch, and the meridional circulation and its consequences are not qualitatively very different from the steady case, although they are broader in horizontal extent [*Haynes et al.*, 1991].

Calculations of diabatic heating rates in the stratosphere [*Rosenfield et al.*, 1994; *Rosenlof*, 1995; *Eluszkiewicz et al.*, 1996] are broadly consistent with this picture of the eddy-driven diabatic circulation, with upwelling in the tropics, somewhat on the summer side of the equator, and downwelling in the extratropics. However, *Plumb and Eluszkiewicz* [1999] questioned how the circulation could extend so deep into the tropics if the wave drag is essentially confined to the extratropical surf zone. According to the first of equation (15), the wave drag is required to extend latitudinally as far as the circulation does. Note, however, that the angular momentum gradient (which is proportional to the absolute vorticity) becomes very small in the tropics, so the wave drag needed to explain the observed circulation there is weak. In fact, Plumb and Eluszkiewicz found that weak friction in their 2-D model was sufficient to permit the circulation driven by subtropical wave drag to extend across the equator. In reality, any drag must be provided by waves [*Kerr-Munslow and Norton*, 2006; *Randel et al.*, 2008]. In theory, in the absence of tropical wave drag and friction, the tropical circulation could extend nonlinearly across a finite range of latitudes, by virtue of the elimination of the angular momentum gradient (and consequent elimination of inertial rigidity associated with a nonzero gradient) by the circulation itself, just as described by the theory of the nonlinear, inviscid, tropospheric Hadley circulation [*Held and Hou*, 1980]. Straightforward calculations show that a qualitatively realistic circulation can thus be driven by wave drag that terminates in the subtropics, but the subtropical zonal winds produced by the unopposed poleward advection of equatorial angular momentum are then far greater than those observed. In practice, it may be a moot point as to whether one regards tropical waves as responsible for driving an essentially linear response or for preventing the buildup of strong subtropical westerlies in a nonlinear circulation.

The possibility of a nonlinear, angular momentum-conserving circulation in the stratosphere, in fact, removes the angular momentum constraint that leads to the argument following equation (15). Such a circulation, if it exists, could be driven thermally, at least in the tropics, where the background absolute vorticity is weak enough to be overcome by the effects of the circulation itself. The calculations of *Dunkerton* [1989] and of *Semeniuk and Shepherd* [2001] have shown that a substantial, and qualitatively realistic, tropical circulation can be driven by radiative forcing alone. While the calculated circulation appears weaker than the estimated circulation in the lower stratosphere (including the important rate of upwelling at the tropical tropopause), the thermally driven component of the circulation may make a substantial contribution to the whole in the middle and upper tropical stratosphere.

5. VARIATIONS OF THE STRATOSPHERE-TROPOSPHERE SYSTEM

The most obvious component of stratospheric variability is, of course, its marked seasonal cycle, but even this is not as straightforward as it might seem. Despite the almost identical seasonal march of the incoming solar fluxes in the two hemispheres (except, of course, for a 6-month phase shift), their planetary wave climatologies are quite different [*Randel*, 1988; *Randel and Newman*, 1998]. This is illustrated in Figure 4: while all waves are weak in summer, in the NH, the quasi-stationary planetary waves are active throughout the winter until the "final warming" in the spring, whereas in the SH, they are usually strongest in autumn and (especially) spring and weaker during midwinter (but not every year, most notably during the highly disturbed winter of 2002 [e.g., *Newman and Nash*, 2005]). This north-south asymmetry is evident in the waves' impact on the mean state: there is much greater year-to-year variability of temperature at the North Pole during winter, contrasted with the weaker variability at the South Pole, with the latter becoming marked only during spring [*Labitzke*, 1977; *Taguchi and Yoden*, 2002]. Since, as expected from the arguments in section 4, anomalously warm polar temperatures are indicative of strong wave activity [*Newman and Nash*, 2000], the occurrence of high variability during certain months indicates high levels of wave activity during those months in some years.

The hemispheric asymmetry evident in the wave climatology appears to be a consequence of feedback between the waves and the mean flow. Recall that *Charney and Drazin* [1961] predicted that waves will propagate through the stratosphere only around the equinoxes and not during midwinter when the mean flow exceeds the Rossby critical velocity U_c. In a highly truncated wave, mean flow model like that of *Holton and Mass* [1976], to be discussed further in what follows, *Plumb* [1989] found Charney and Drazin's prediction to hold when the undisturbed mean flow in

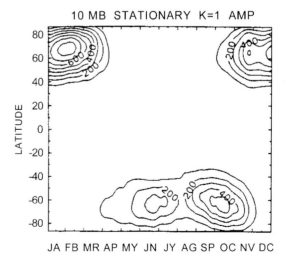

Figure 4. Seasonal variation of the amplitude (geopotential, m) of stationary zonal wave number 1 at 10 hPa. From *Randel* [1988]. Copyright Royal Meteorological Society, reprinted with permission.

midwinter exceeds U_c, and the wave amplitudes are sufficiently weak. With stronger wave amplitudes, however, the waves weaken the mean flow, thereby permitting their own propagation and remaining strong all winter. There is a positive feedback here, since the more readily the waves propagate, the more they reduce the mean flow (other things being equal). In a more complete general circulation model with simple topography of zonal wave number 1 and various heights, *Taguchi and Yoden* [2002] found similar results. For weak wave forcing, maximum wave amplitudes and variability were found in spring, with a weaker maximum in autumn, while wave amplitudes were strong throughout the winter with sufficiently strong forcing. The agreement of the strong/weak forcing cases with the observed behavior in the SH/NH is striking. *Scott and Haynes* [2002] obtained similar results, but reached a somewhat different interpretation: they argued that the waves continue to propagate vertically throughout the southern winter and that the springtime peak, and final warming, occurs as the mean state passes through resonance in its seasonal evolution. As we shall address in section 6, it is, however, difficult to establish from wave structures whether or not the waves are propagating in the usual sense (and the point may be moot). The possible role of resonance in warmings is the focus of section 7. By either interpretation, it appears that the southern stratosphere is responding in an essentially linear fashion to a relatively weak forcing of planetary-scale waves, whereas nonlinear feedback between the waves and mean flow is responsible for the NH climatology.

On subseasonal time scales, stratospheric wave amplitudes may fluctuate markedly on time scales of a week or two, with corresponding fluctuations in the mean state. Periods of strong waves are usually manifested as strong breaking events of the kind described by *McIntyre and Palmer* [1983] and very strong events as total breakdowns of the vortex (i.e., "major warmings"). In fact, *McIntyre and Palmer* [1983] (see also *McIntyre* [1982]) described a sequence of events in 1979 in which midlatitude breaking events eventually led to vortex breakdown. Such fluctuations must, in part, reflect events in the tropospheric forcing of the waves, but the degree of stratospheric variability is so large that it is difficult to ascribe it to tropospheric variability alone, in which case one looks to internal dynamics for an explanation. On week-to-week time scales, superposition of stationary and free, traveling Rossby waves will produce such fluctuations; indeed, in the barotropic simulations of the interaction between a vortex and a forced stationary wave by *Polvani and Plumb* [1992], breaking events generated such transient waves and led to subsequent, quasi-periodic occurrences of breaking. There are, however, indications of other influences which may act on both these and longer time scales.

The potential for internal dynamics, rather than unspecified variations in tropospheric forcing, to be responsible for dramatic fluctuations of the state of the winter stratosphere was raised by the seminal study of *Holton and Mass* [1976]. They used what is perhaps the simplest model of stratospheric wave, mean flow interaction (in fact, a modification of that introduced by *Geisler* [1974]): a truncated baroclinic quasi-geostrophic model in a "beta channel," in which the wave is of a specified zonal wave number, and both wave and mean flow are constrained to the gravest meridional structure, a half sine across the channel. Thus, wave-wave interactions, which would generate higher-order latitudinal structure of both wave and mean flow, are ignored. Nevertheless, the essentials of wave, mean flow interaction are captured: the wave responds to the vertical structure of the mean state, while the mean state responds to dynamical transport by the wave. Thus, the model is "quasi-linear": with the given mean flow at any instant, the wave calculation is a linear problem, the nonlinearity arising solely through the action of the wave on the zonal mean state. The mean flow is forced by Newtonian relaxation toward a specified radiative equilibrium state (that which would exist in the absence of the waves), while the wave is forced simply by specifying its amplitude on the bottom boundary and dissipated by the Newtonian cooling.

This simplified model has rich behavior. In particular, as Holton and Mass demonstrated, there is a critical wave forcing amplitude at which the system's response changes from a steady state to one exhibiting sustained, large-amplitude, quasi-periodic vacillations. With subcritical forcing, the steady state structure of the wave is almost "equivalent barotropic," in the sense that the wave exhibits almost no phase change with height, and consequently, the vertical component of the Eliassen-Palm flux is essentially zero. (The horizontal component is zero, as a consequence of the truncation.) Accordingly, the wave has little impact on the mean state, which is indistinguishable from the specified radiative equilibrium state. With supercritical forcing, the time-averaged wave magnitude is not substantially altered, but its phase structure is quite different, tilting westward with height indicating upward Eliassen-Palm flux and a consequent drag on the mean flow, whose effects are manifested in a profound reduction in the average strength of the mean flow. Even more dramatically, the supercritical case is marked by fluctuations (which may be periodic or irregular, depending on parameters) in which the wave amplifies, and the mean westerlies are weakened or reversed. Thus, this simple model displays characteristics reminiscent of those of the observed stratosphere, suggesting that it captures the essentials of the mechanisms that determine why the stratosphere behaves as it does.

Following the Holton and Mass [1976] paper, many subsequent studies have been directed at further exploration

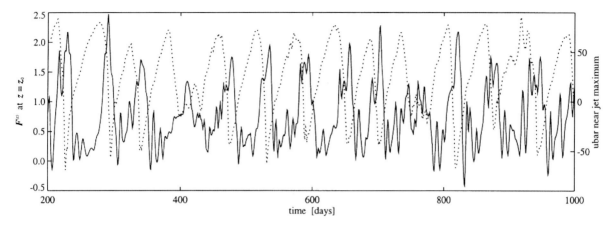

Figure 5. Vertical EP flux (solid line) at the lower boundary of a stratosphere-only model, averaged over latitude, and u at 60° latitude and 41 km altitude (dashed line). The model was forced by a specified geopotential amplitude ($\phi_0 = 600$ m) of zonal wave number 2. From *Scott and Polvani* [2006]. Copyright American Meteorological Society.

of stratospheric vacillations. Some work (most notably that of Yoden [1987a, 1987b]) has dug deeper into the properties of the Holton-Mass model. Others [e.g., *Yoden et al.*, 1996, 2002; *Christiansen*, 1999; *Scott and Haynes*, 2000; *Scott and Polvani*, 2004, 2006] have shown that such behavior is found in more realistic general circulation models. An example, from the work of *Scott and Polvani* [2006], is shown in Figure 5. In this 3-D, stratosphere-only model, stationary waves of zonal wave number 2 were forced by specifying its geopotential amplitude at the lower boundary. Once again, despite the constancy of the wave forcing, the mean winds underwent quasi-periodic cycles of marked weakening, each of which was associated with similarly marked amplification of the upward Eliassen-Palm flux into the stratosphere from below, just as one sees in observations [e.g., *Polvani and Waugh*, 2004]. These results provide very clear evidence that, in such models, the flux of wave activity out of the troposphere is under stratospheric control and imply that, to some degree, the same may be true in the real atmosphere.

6. LINEAR THEORY REVISITED: UPWARD PROPAGATION OR MODES?

We have learned, as outlined in the preceding sections, that the behavior of wintertime planetary-scale waves in the stratosphere is complex. While they are clearly generated within the troposphere, the relative contributions of the various generation processes there are poorly characterized. Within the stratosphere, the waves suffer dissipation not only by the relatively linear radiative damping of their temperature anomalies but by the highly nonlinear processes associated with wave breaking. As we have seen, the critical layer, which in linear theory is an infinitesimally thin band located at the

zero wind line, can, at times, occupy a large fraction of the hemisphere from the tropics to the polar cap. Furthermore, there is a strong nonlinear feedback between the waves and the mean flow that manifests itself in such a large degree of variability that, e.g., in a major warming event, the polar vortex is completely torn apart and the mean westerlies replaced by mean easterlies throughout much of the stratosphere. One might well ask, therefore, what value linear theory has as a guide to understanding the waves' behavior. The defeatist answer might be that one simply has to accept that the winter stratosphere is a highly complex nonlinear system and that we must, therefore, content ourselves with using models and observations to document the various classes of behavior, but this is hardly a satisfactory response. One reason for continuing to rely on linear theory as a guide to the wave behavior is that, at present, it is the only tool available that allows us to understand anything about how wave characteristics are determined by the extant conditions. A second and probably more compelling reason is that things may not be quite as bad as they appear for linear theory. For one thing, experience with simplified models such as the Holton-Mass model, or more realistic but zonally truncated models like that of *Scott and Haynes* [2000], has taught us that, at least conceptually, the nonlinear interaction between the waves and the mean state can be treated (with an important caveat) quasi-linearly, i.e., by coupling a linear model for the waves with a model for the mean state, which responds to the nonlinear wave flux of PV. (The linear wave model, in turn, responds to the evolving mean flow.) The caveat is, of course, that within such a wave model, one must somehow account for dissipation, including the effects of breaking. Since the latter is nonlinear, it is clearly impossible to represent it accurately (even if we knew how) within linear theory.

Nevertheless, since there are good dynamical reasons to regard the waves, which after all are Rossby waves, as propagating primarily up the band of strong PV gradient at the vortex edge, it may be satisfactory, at a useful level, to regard the loss of wave activity associated with leakage into the surf zone and dissipation there through stirring against the local PV gradient as a simple damping of the wave, at some appropriate rate. Of course, this will not do during very disturbed periods when the wave breaking makes major inroads into the vortex, but such events are intermittent, and in any case, one of the chief puzzles to be explained is how the wave amplitudes become large enough to do this in the first place, in which case focusing on the precursors to such events is a natural thing to do, and linear theory may be a satisfactory tool to use.

To this end, we will now revisit the linear arguments, which were, for the most part, developed more than 30 years ago, before the body of knowledge outlined in the preceding three sections was accumulated. In section 2, we raised some caveats about the overly simplistic interpretation of refractive index calculations, noting in particular that they may give an exaggerated impression of the potential for vertical propagation. In the simplest view, can we regard the typical wintertime waves as simply propagating upward from the troposphere, and dissipating in the stratosphere, in a WKB-like sense? Indeed, one might argue that one can actually detect such propagation, in the form of the typical westward tilt of the waves' phase with altitude. To be sure, for quasi-geostrophic waves, the upward component of the Eliassen-Palm flux

$$F_z = \rho f \frac{\overline{v'\theta'}}{\partial\bar{\theta}/\partial z} = \frac{\rho}{N^2} \overline{\frac{\partial\phi'}{\partial x}\frac{\partial\phi'}{\partial z}} \qquad (16)$$

is positive whenever the phase tilts westward with height, and vice versa. However, the flux will be upward in any reasonable situation where the waves are forced from below: the phase tilt may indicate an upward propagating wave or just the effects of dissipation on an evanescent or quasi-modal wave structure.

If we accept that the explanation for the usual relative lull in stationary wave activity during southern winter is that the mean westerlies are too strong to allow propagation then, since the southern stratosphere is then not too far from radiative equilibrium, and the radiative equilibrium of the northern winter stratosphere is not too different (the radiative conditions are little different), one has to conclude that the undisturbed state of the midwinter stratosphere is reflective, i.e., it does not permit wave propagation deep into the stratosphere. Deep propagation requires weakening of the westerlies, either through radiative weakening of the latitudinal temperature gradients (such as happens around the equinoxes) or through the action of the waves themselves.

Even if the waves do propagate, they may not be propagating uniquely upward; we have noted evidence for internal reflection [e.g., *Harnik and Lindzen* [2001]]. To illustrate this, we will look at results from a simple linear calculation of waves in a beta channel of width 50° latitude centered on 60°N, with a specified half-sine structure across the channel, propagating through a specified mean flow. This is essentially the wave component of the Holton-Mass model, with a mean flow that is constant in time. At the model top (at $z = 100$ km), a condition of upward radiation, or boundedness of evanescent solutions, is applied. At the surface, a wave of zonal wave number 2 and zonal phase speed c is forced at the lower boundary, either by specifying geopotential height or by applying a linearized topographic boundary condition. In dissipative cases, the dissipation occurs through Newtonian cooling and Rayleigh friction, with equal rate coefficients of $(10 \text{ day})^{-1}$.

The equilibrium response, as a function of c, is illustrated for a few cases in the following figures. The plots shown in Figure 6 are for a case with uniform flow $U = 25$ m s^{-1}, independent of z. In this case, the Charney-Drazin condition (5) has to be modified to allow for barotropic flow curvature; for these parameters, the result is $U_c = 30.2$ m s^{-1}. The propagation window $0 < U - c < U_c$, within which the vertical wave number is real at all heights, is shaded.

When the surface geopotential amplitude is specified, in the absence of dissipation (Figure 6d), the response at the stratopause ($z = 50$ km) is flat across the propagation window, since when the vertical wave number is real, the ratio $|\phi'(z = 50 \text{ km})|/|\phi'(0)| = \sqrt{\rho(0)/\rho(50 \text{ km})}$. (The slight wiggles on the plot are numerical artifacts.) The upward EP flux, F_z, is not flat, however: as *Yoden* [1987b] noted, fixing ϕ' at the lower boundary does not fix the surface value of F_z, which also depends on the vertical wave number at the surface (cf. equation (16)). Thus, F_z increases as a function of c from $F_z = 0$ at $c = U - U_c \simeq -5$ m s^{-1}, to become infinite as $c \to U = 25$ m s^{-1}, where the vertical wave number itself becomes infinite. The flux vanishes identically outside the propagation window. In the presence of the 10-day dissipation (Figure 6b), the characteristics are broadly similar. Upper level amplitudes are of course reduced, especially as c approaches U; then, the vertical wave number becomes large, the vertical group velocity becomes small, and dissipation becomes more effective. F_z remains greatest near (but no longer at) $c = U$, but its maximum magnitude is now, of course, finite. Outside the propagation window, the flux is weak but no longer zero.

Contrast this behavior with that of the topographically forced cases. In the presence of dissipation (Figure 6a), both

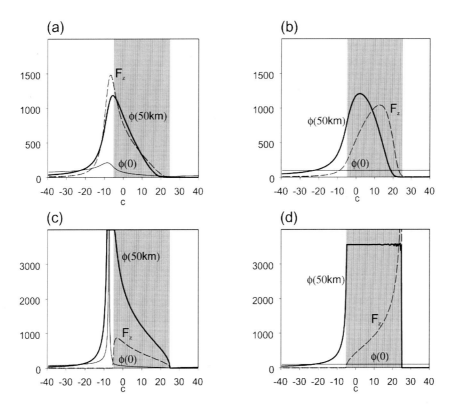

Figure 6. Solutions to the truncated channel model discussed in the text for the case of a uniform background zonal flow of 25 m s^{-1}. Shown for the channel center are $\phi(0)$, the wave geopotential amplitude at the surface; $\phi(50$ km$)$, the amplitude at 50 km altitude (both in meters, scale at left); and F_z, the Eliassen-Palm flux at the surface (which has been scaled to fit the plot axes). The abscissa is the phase speed, c, in m s^{-1}, and the shaded region denotes the "propagation window," that range of c for which the calculated wave number is real. The wave is forced (b) and (d) with a fixed geopotential amplitude $\phi(0) = 100$ m and (a) and (c) by surface topography of amplitude 100 m. The damping time scale is 10 days in Figures 6a and 6b; Figures 6c and 6d show cases with zero damping.

the wave amplitude aloft and the surface EP flux maximize outside the propagation window! The internal wave characteristics, including the ratio $|\phi'(z = 50$ km$)|/|\phi'(0)|$, are independent of the lower boundary condition, but the surface amplitude maximizes at $c \simeq -8$ m s^{-1}. Why this happens becomes evident from the undissipated results (Figure 6c): there is a resonance at $c = -7.4$ m s^{-1}, the speed of the free external mode (the only mode this system possesses when U is constant). The external mode is suppressed when the geopotential amplitude of the wave is specified at the surface, which is why no such behavior was evident in those cases. Thus, in this topographically forced case, the largest wave amplitudes, and the largest EP fluxes and hence the greatest potential to modify the mean state, are obtained when the waves are actually evanescent, when $U - c$ just exceeds U_c.

When the mean state is not uniform with height, internal reflections can occur, in which case additional baroclinic modes may exist, even when the wave's geopotential amplitude is specified at the lower boundary. The two cases shown in

Figure 7, identical to the topographically forced cases of Figure 6 with and without the 10-day dissipation, but with a piecewise linear mean zonal flow that increases from 25 m s^{-1} at 20 km to 75 m s^{-1} at 70 km, and constant above and below those altitudes, illustrate this. (With this mean flow profile, there is no propagation window to illustrate on the plots, since a wave of the chosen wave number cannot have real vertical wave number at all heights, for any c.) The undissipated case (Figure 7b) shows, in its response curves of $|\phi'|$ versus c, a sequence of resonances, beginning (with increasing c) with the external mode at $c \simeq -7$ m s^{-1}, the first internal (baroclinic) mode near $c = 5$ m s^{-1}, to modes of successively higher order with increasing c. Note that the upward EP flux is zero in all cases, consistent with the waves' inability to propagate right through the jet and the absence of dissipation. The identification of the first two modes as external and first baroclinic is confirmed by the vertical profiles on the right of the figure. At the external mode resonance, the wave structure is greatest at the surface, decreasing monotonically with

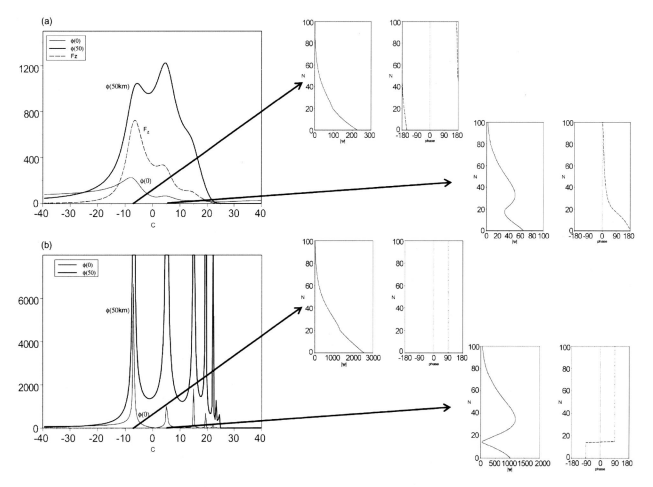

Figure 7. (left) (a) and (b) As in Figure 6 but for topographically forced waves on a basic state of nonuniform background flow, as described in the text. The dissipation rate is $(10 \text{ days})^{-1}$ in Figure 7a and zero in Figure 7b. In the case shown in Figure 7b, the Eliassen-Palm flux is zero for all c. The arrows point to plots of wave amplitude, multiplied by $\sqrt{p(z)/p(0)}$, and phase versus height for near-resonant values of c indicated by the base of the arrow. See text for discussion.

altitude, and with constant phase, while at the second resonance, the wave has a single node near 17 km altitude. When 10-day dissipation is added (Figure 7a), the resonances are of course subdued, but the presence of the first two is evident in clear peaks in the EP flux and the upper level amplitudes, and a hint of the third remains. The vertical structures of the wave near the first two peaks are similar to the corresponding peaks in the undissipated case. Note that the net upward propagation evident in the EP fluxes and in the phase tilt is entirely due to the dissipation: these are damped modes and not upward propagating waves. Nevertheless, for the same topographic forcing and dissipation, larger upper stratospheric wave amplitudes are produced in this case (Figure 7a) for which the height-dependent zonal winds do not permit deep propagation than in the case (Figure 6a) of uniform winds, which may allow deep propagation.

Of course, this kind of model, with a single wave mode in latitude, imposes an unrealistic cavity bounded by the channel walls. What in reality is leakage away from the jet into the dissipative surf zone is here represented loosely by the imposed dissipation, and the extent to which these quasi-modal structures remain classifiable as such requires that this dissipation (in the model and in the real world) be weak enough to allow sufficient internal reflection to build the structure. In the model, as dissipation is increased, the resonances weaken, and in the case of the nonuniform mean flow, the separate peaks become less distinct. The maximum of the response seen in Figure 7a around $-10 \text{ m s}^{-1} \lesssim c \lesssim 0$ remains clear, and two peaks remain barely separate when the dissipation rate is increased to $(5 \text{ days})^{-1}$, but the features are lost if it is increased still further. (By this stage, the dissipation time is comparable with or less than the time for group propagation across the domain, and any meaningful classification as propagating or modal is lost.)

While simple models such as this have serious and obvious shortcomings as analogs of the real stratosphere, it is on such models that our paradigms for understanding planetary wave behavior rest. The basic idea that the waves propagate through winter westerlies itself comes from such models, and as we have just seen, there is no reason to depart from *Charney and Drazin*'s [1961] original conclusion that typical midwinter westerlies are too strong to allow deep propagation. Nevertheless, large wave amplitudes can be reached in the upper stratosphere, even in strong westerlies, simply because, as *Dickinson* [1968] noted, even an evanescent wave may grow in amplitude with height simply because of the decreasing atmospheric density. That is not to say that the distinction, difficult though it may be to make in realistic, dissipative situations, is unimportant. For one thing, as will be addressed further in section 8, the way in which stratospheric conditions can influence tropospheric behavior may depend on whether or not upward propagating waves suffer any significant reflection. For another, the possibility that the most basic characteristics of the waves can be altered to a major degree by reductions in the mean westerlies, when those reductions are, in turn, effected by the waves themselves, implies that the entire wave, mean flow system of the winter stratosphere is susceptible to positive feedbacks, which are at the heart of the high degree of variability that the system exhibits during those times of the season when wave amplitudes are strong.

7. WAVE AMPLIFICATIONS AND WARMINGS: QUASI-RESONANCES?

From the earliest observations of sudden warmings in the Arctic stratosphere, it was recognized that such events are associated with rapid amplifications of planetary waves [e.g., *Finger and Teweles*, 1964; *Julian and Labitzke*, 1965; *Labitzke*, 1977]. We now know that the winter stratosphere displays a high degree of variability, those events classified as "major warmings" just being the most dramatic [*Coughlin and Gray*, 2009]; generally, weakenings of the vortex are preceded by bursts of planetary wave EP fluxes into the stratosphere from below [*Labitzke*, 1981; *Polvani and Waugh*, 2004].

Such situations, in which waves grow on an initially almost circular vortex to such an extent that the vortex may ultimately break apart, naturally raised the question of the stability characteristics of the vortex itself. However, for parallel flow, instability requires a reversal in sign of the PV gradient [*Charney and Stern*, 1962], which seemed not to be generally present, a conclusion that is now held more firmly, with the benefit of modern analyses of the stratospheric flow. Moreover, as *Wexler* [1959] pointed out, if the Arctic vortex

were unstable, one might expect the stronger Antarctic midwinter vortex to be even less stable, contrary to what is observed. However, the PV stability constraint is lost if the flow is not axisymmetric, and *Matsuno and Hirota* [1966] and *Hirota* [1967] suggested that a more realistic vortex, deformed by the presence of planetary waves, might be barotropically unstable. This idea was not pursued immediately, though a variant on the same theme appeared some time later.

The first mechanistic model of sudden warmings was provided by *Matsuno* [1971]. Matsuno took, as a starting point, the amplification of a planetary wave near the tropopause and showed that, given that presumption, the observed sequence of events follows. As the amplified wave front propagates upward, wave drag acting at the front produces the kind of response that was illustrated in Figure 3, generating polar warming below the front, cooling above, and deceleration of the westerlies. The impact increases with height as the front propagates into regions of progressively lower density, at some level changing the sign of the zonal flow. Subsequently, the interaction of the wave with this wave-generated critical line leads to descent of the easterlies through the stratosphere. While there may have been differences of interpretation and of emphasis in subsequent descriptions of the dynamics of warmings, Matsuno's description still serves as the foundation of our dynamical understanding, in particular that, given a strong amplification of the planetary waves, high-latitude warming and all that goes with it seems inevitable. What remains, however, is an explanation of that amplification. For some time, it was widely supposed that the flux of wave activity into the stratosphere is under tropospheric control, so that any amplification might be ascribed to the chaotic nature of the tropospheric flow. While tropospheric variations must indeed be a factor, we now understand, following the results of *Holton and Mass* [1976] and other subsequent studies, for example, those of Yoden [1987a, 1987b], *Scott and Haynes* [2000], *Yoden et al.* [2002], and most explicitly those of *Scott and Polvani* [2004, 2006], that bursts of strong wave flux into the stratosphere, and consequent warming events, can occur even without such tropospheric variations (Figure 5). At least, in part, therefore, such variations are intrinsic to the internal dynamics of the stratosphere itself, probably involving the kind of positive feedbacks alluded to in section 6.

The possibility that large wave amplitudes may occur through resonance was, as already remarked, noted as early as *Matsuno* [1970]. Resonance, of course, needs a partial cavity within which reflections can occur without too much leakage out of, or dissipation within, the cavity. Matsuno argued that such a cavity may be created by weak PV gradients equatorward of the main jet and strong westerlies aloft. *Tung and Lindzen* [1979] explored the properties in a variety of realistic (but still 1-D) flow profiles and argued that the zonally long waves could be brought into internal mode resonances without the need to invoke unreasonably large zonal winds. Similarly, in the calculations of section 6, we saw that westerlies of realistic magnitude are indeed sufficient to cause internal reflections and consequent resonances, although leakage into low latitudes was prevented by the model design (and the possible effects of such leakage represented but crudely by dissipation).

Even assuming such resonances can exist, it might seem that finding a mean flow sufficiently close to resonance, for a sufficiently long time to permit a mode to develop, must rely on serendipity. However, the nonlinear interaction with the mean flow can, in fact, drag the flow into a near-resonant state by a process of nonlinear self-tuning [*Plumb*, 1981]. If the state of the atmosphere is such that modes can exist, and one of those modes is moderately close to resonance, that mode's phase velocity c_0 is dependent on the basic state wind and static stability profiles. Introducing a forced stationary wave into such a system will induce changes to the mean state and hence to c_0. Other things being equal, there is a finite probability that the sense of that change will be to reduce $|c_0|$, thereby bringing the wave closer to resonance, increasing the wave amplitude and reinforcing the original changes. This self-tuning mechanism thus leads to unstable breakdown of the jet, in concert with amplification of the waves, and appears to be responsible for the transition from steady to vacillating solutions in the Holton-Mass model through resonant self-tuning of the first baroclinic mode [*Plumb*, 1981]. (One can also regard this process as one of the instability of a baroclinically deformed vortex and, thus, as an extension of the barotropic arguments of *Matsuno and Hirota* [1966] and *Hirota* [1967].)

Whatever the possible shortcomings of the model results just noted are, there is evidence both for the existence of the reflections that are a prerequisite for a resonant state and for the evolution into resonance, in association with warming events in full 3-D models. In an early mechanistic study of a model warming in which the specified geopotential amplitude of zonal wave 2 was turned on at the lower boundary over the course of about 10 days and then held fixed, *Dunkerton et al.* [1981] noted the continued growth of the boundary EP fluxes, leading to peak wave amplitudes within the stratosphere around day 20. As we have already noted from Figure 5, similar behavior (of strongly varying EP fluxes with fixed geopotential amplitude) is seen in the vacillating regimes of model studies such as that of *Scott and Polvani* [2006]. From equation (16), the vertical component of the EP flux for an upward propagating wave of zonal and vertical wave numbers k, m is

$$F_z = \rho \frac{1}{N^2} \overline{\frac{\partial \phi'}{\partial x} \frac{\partial \phi'}{\partial z}} = \frac{\rho}{N^2} km \, \overline{\phi'^2}.$$

Now, if the background state can be regarded as slowly varying in space, m is just a function of the local mean state; since the latter changes little near the lower boundary in the results of either *Dunkerton et al.* [1981] or *Scott and Polvani* [2006], and $\overline{\phi'^2}$ is fixed on the boundary, one cannot explain the growth of F_z there on the basis of a single, upward propagating wave. The only reasonable explanation for the growth seems to be the reinforcement of the directly forced wave by constructive interference with a component that has been reflected back down to the boundary, a necessary ingredient of resonance. More directly, *Smith* [1989] ran a series of quasi-linear simulations of the Arctic warming of 1979 and indeed found resonant behavior, including one of the key characteristics of self-tuning, namely, that the phase speed of free, transient planetary wave, slowed to match the speed of the forced wave (which in these experiments was nonzero, based on upper tropospheric observations), as the system evolves into resonance. Similar behavior is evident in observed warmings [e.g., *Labitzke*, 1981].

Taking a different approach, *Esler et al.* [2006] simulated the Antarctic major warming of September 2002 in a baroclinic "vortex patch" model of the vortex, in which at any given altitude, PV is stepwise uniform, with a single discontinuity at the vortex edge. As they argue (and as discussed in section 2), this might, in fact, be just as good a model of the actual stratosphere as the smoother states based on zonal and time averages of the observations. Using this model, they showed that vortex breakdown occurred in the model as a result of the self-tuning of the model's external mode.

8. CONCLUDING REMARKS

The extratropical stratosphere is controlled by a relatively small number of external influences, and yet, its rich behavior, though well understood in principle, depends on aspects of the problem that remain poorly clarified. One of the remaining uncertainties is of course the gravity wave climatology and its tropospheric sources; the status of our understanding of stratospheric gravity waves is addressed by *Alexander* [this volume]. The dominant motions, the planetary-scale waves, which have been the focus of this discussion, are the primary agents for driving the winter stratosphere out of radiative equilibrium and consequently for moderating the strength of (and, on occasion, destroying) the westerlies. The way in which they influence the mean state through wave drag, and the way in which the extratropical mean state responds to that drag, is now well understood, following about three decades of research effort. Progress has also been made in understanding the driving of the tropical component of the circulation, but this is a more complex issue, and much remains to be done. In turn, the mean state exerts a strong influence on the planetary-scale waves, leading to a feedback loop of such intensity that it is manifested in the vacillation cycles of stratospheric models and in observed major warming events. This feedback may be the reason for the sensitivity of the modeled stratospheric flow to relatively modest gravity wave drag [*Boville*, 1995] and for the influence of the tropical quasi-biennial oscillation of the extratropical circulation, as first suggested by *Dickinson* [1968] and documented by *Holton and Tan* [1980].

The main assertion of this review is that what might appear to be the simplest piece of the whole puzzle, and certainly the piece for which the theory is the oldest, is where the greatest conceptual uncertainty lies. There remains no clear, simple understanding of just how the changes in the wintertime mean flow influence the gross characteristics of the waves. Moreover, at any given time, what is the paradigm that best describes the waves? Under background conditions of midwinter, are they simple, upward propagating Rossby waves modified by dissipation, as is frequently supposed; are they damped evanescent structures, as Charney and Drazin originally suggested; or are they damped quasi-modes of the kind illustrated here in Figure 7? One might suppose that observations should be able to discriminate between these possibilities, but in practice, the three cases will not look qualitatively different whenever the characteristic vertical wavelengths are large, as appears to be the case: any damped disturbance forced from below will exhibit westward phase tilt and upward, decaying EP fluxes. One might, in fact, wonder whether the difference really matters, but it probably does, in at least two important respects. First, the wave, mean flow feedback that produces vacillation could depend on transitions between wave propagation and nonpropagation, or on self-tuned resonance, which requires quasi-modal structures. Second, deciding which paradigm is appropriate impacts the way we think about dynamical stratosphere-troposphere interactions, of the kind that appears to be manifested in the behavior of the extratropical "annular modes" [*Thompson and Wallace*, 1998, 2000; *Baldwin and Dunkerton*, 1999]. If upward-propagating waves suffer no significant reflection, then downward coupling would probably have to depend on the meridional circulation, whereas if the waves are quasi-modal, the waves themselves couple the two regions together. Some model studies [*Kushner and Polvani*, 2004; *Song and Robinson*, 2004] do implicate the waves in the coupling. Indeed, if the latter interpretation is correct, then thinking about the troposphere and stratosphere as separate, coupled systems may not be the most logical, nor the most transparent, approach. In fact, it may be more sensible to make the separation based on the system's dynamics, rather than on physical location. The two coupled systems are thus (1) the synoptic-scale baroclinic eddies of the troposphere and (2) the deep, planetary-scale waves. These two systems interact both directly, and indirectly, through their mutual interaction with the mean flow.

Acknowledgments. I would like to take this opportunity to thank Lorenzo Polvani, Adam Sobel, and Darryn Waugh for their considerable efforts in organizing the meeting at which this paper was presented and for editing this volume. I also thank Peter Haynes for comments on an earlier version of the paper and colleagues too numerous to list for their intellectual input. This work was supported by the National Science Foundation, through grant ATM-0808831.

REFERENCES

Alexander, M. J. (2010), Gravity waves in the stratosphere, in *The Stratosphere: Dynamics, Transport, and Chemistry, Geophys. Monogr. Ser.*, doi:10.1029/2009GM000864, this volume.

Andrews, D. G., and M. E. McIntyre (1976), Planetary waves in horizontal and vertical shear: The generalized Eliassen-Palm relation and the mean zonal acceleration, *J. Atmos. Sci., 33,* 2031–2048.

Andrews, D. G., J. R. Holton, and C. B. Leovy (1987), *Middle Atmosphere Dynamics*, 489 pp., Academic, San Diego, Calif.

Baldwin, M. P., and T. Dunkerton (1999), Propagation of the Arctic Oscillation from the stratosphere to the troposphere, *J. Geophys. Res., 104,* 30,937–30,946.

Boville, B. A. (1995), Middle atmosphere version of CCM2 (MACCM2): Annual cycle and interannual variability, *J. Geophys. Res., 100,* 9017–9039.

Charney, J. G., and P. G. Drazin (1961), Propagation of planetary-scale disturbances from the lower into the upper atmosphere, *J. Geophys. Res., 66,* 83–109.

Charney, J. G., and A. Eliassen (1949), A numerical method for predicting the perturbations of the middle latitude westerlies, *Tellus, 1*(2), 38–54.

Charney, J. G., and M. E. Stern (1962), On the stability of internal baroclinic jets in a rotating atmosphere, *J. Atmos. Sci., 19,* 159–172.

Christiansen, B. (1999), Stratospheric vacillations in a general circulation model, *J. Atmos. Sci., 56,* 1858–1872.

Coughlin, K., and L. J. Gray (2009), A continuum of stratospheric warmings, *J. Atmos. Sci., 66,* 532–540.

Dickinson, R. E. (1968), Planetary Rossby waves propagating through weak westerly wind wave guides, *J. Atmos. Sci., 25,* 984–1002.

Dunkerton, T. (1989), Nonlinear Hadley circulation driven by asymmetric differential heating, *J. Atmos. Sci., 46,* 956–974.

Dunkerton, T., C.-P. F. Hsu, and M. E. McIntyre (1981), Some Eulerian and Lagrangian diagnostics for a model stratospheric warming, *J. Atmos. Sci., 38,* 819–843.

Eliassen, A. (1951), Slow thermally or frictionally controlled meridional circulation in a circular vortex, *Astrophys. Norv.*, *5*, 19–60.

Eluszkiewicz, J. E., et al. (1996), Residual circulation in the stratosphere and lower mesosphere as diagnosed from Microwave Limb Sounder data, *J. Atmos. Sci.*, *53*, 217–240.

Esler, J. G., and R. K. Scott (2005), Excitation of transient Rossby waves on the stratospheric polar vortex and the barotropic sudden warming, *J. Atmos. Sci.*, *62*, 3661–3682.

Esler, J. G., L. M. Polvani, and R. K. Scott (2006), The Antarctic stratospheric sudden warming of 2002: A self-tuned resonance?, *Geophys. Res. Lett.*, *33*, L12804, doi:10.1029/2006GL026034.

Finger, F. G., and S. Teweles (1964), The mid-winter 1963 stratospheric warming and circulation change, *J. Appl. Meteorol.*, *3*, 1–15.

Geisler, J. E. (1974), A numerical model of the sudden stratospheric warming mechanism, *J. Geophys. Res.*, *79*, 4989–4999.

Harnik, N., and R. S. Lindzen (2001), The effect of reflecting surfaces on the vertical structure and variability of stratospheric planetary waves, *J. Atmos. Sci.*, *58*, 2872–2894.

Haynes, P. H. (1989), The effect of barotropic instability on the nonlinear evolution of a Rossby wave critical layer, *J. Fluid Mech.*, *207*, 231–266.

Haynes, P. H., C. J. Marks, M. E. McIntyre, T. G. Shepherd, and K. P. Shine (1991), On the "downward control" of extratropical diabatic circulations by eddy-induced mean zonal forces, *J. Atmos. Sci.*, *48*, 651–678.

Held, I. M., and A. Y. Hou (1980), Nonlinear axially symmetric circulations in a nearly inviscid atmosphere, *J. Atmos. Sci.*, *37*, 515–533.

Hirota, I. (1967), Dynamical stability of the stratospheric polar vortex, *J. Meteorol. Soc. Jpn.*, *45*, 409–421.

Holton, J. R., and C. Mass (1976), Stratospheric vacillation cycles, *J. Atmos. Sci.*, *33*, 2218–2225.

Holton, J. R., and H.-C. Tan (1980), The Influence of the equatorial quasi-biennial oscillation on the global circulation at 50 mb, *J. Atmos. Sci.*, *37*, 2200–2208.

Holton, J. R., P. H. Haynes, M. E. McIntyre, A. R. Douglass, R. B. Rood, and L. Pfister (1995), Stratosphere-troposphere exchange, *Rev. Geophys.*, *33*, 403–439.

Julian, P. R., and K. B. Labitzke (1965), A study of atmospheric energetics during the January–February 1963 stratospheric warming, *J. Atmos. Sci.*, *22*, 597–610.

Juckes, M. N., and M. E. McIntyre (1987), A high-resolution, one-layer model of breaking planetary waves in the winter stratosphere, *Nature*, *328*, 590–596.

Karoly, D. J., and B. J. Hoskins (1982), Three-dimensional propagation of stationary waves, *J. Meteorol. Soc. Jpn.*, *60*, 109–123.

Kerr-Munslow, A. M., and W. A. Norton (2006), Tropical wave driving of the annual cycle in tropical tropopause temperatures. Part I: ECMWF analyses, *J. Atmos. Sci.*, *63*, 1410–1419.

Killworth, P. D., and M. E. McIntyre (1985), Do Rossby-wave critical layers absorb, reflect, or over-reflect?, *J. Fluid Mech.*, *161*, 449–462.

Kushner, P. J., and L. M. Polvani (2004), Stratosphere-troposphere coupling in a relatively simple AGCM: the role of eddies, *J. Clim.*, *17*, 629–639.

Labitzke, K. (1977), Interannual variability of the winter stratosphere in the northern hemisphere, *Mon. Weather Rev.*, *105*, 762–770.

Labitzke, K. (1981), Stratospheric-mesospheric midwinter disturbances: A summary of observed characteristics, *J. Geophys. Res.*, *86*, 9665–9678.

Matsuno, T. (1970), Vertical propagation of stationary planetary waves in the winter northern hemisphere, *J. Atmos. Sci.*, *27*, 871–883.

Matsuno, T. (1971), A dynamical model of the stratospheric sudden warming, *J. Atmos. Sci.*, *28*, 1479–1494.

Matsuno, T., and I. Hirota (1966), On the dynamical stability of polar vortex in wintertime, *J. Meteorol. Soc. Jpn.*, *44*, 122–128.

McIntyre, M. E. (1982), How well do we understand the dynamics of stratospheric warmings? *J. Meteorol. Soc. Jpn.*, *60*, 37–65.

McIntyre, M. E., and T. N. Palmer (1983), Breaking planetary waves in the stratosphere, *Nature*, *305*, 593–594.

Newman, P. A., and E. R. Nash (2000), Quantifying the wave driving of the stratosphere, *J. Geophys. Res.*, *105*, 12,485–12,497.

Newman, P. A., and E. R. Nash (2005), The unusual Southern Hemisphere stratosphere winter of 2002, *J. Atmos. Sci.*, *62*, 614–628.

Norton, W. A. (1994), Breaking Rossby waves in a model stratosphere diagnosed by a vortex-following coordinate system and a technique for advecting material contours, *J. Atmos. Sci.*, *51*, 654–673.

Plumb, R. A. (1981), Instability of the distorted polar night vortex: A theory of stratospheric warmings, *J. Atmos. Sci.*, *38*, 2514–2531.

Plumb, R. A. (1989), On the seasonal cycle of stratospheric planetary waves, *Pure Appl. Geophys.*, *130*, 233–242.

Plumb, R. A. (2002), Stratospheric transport, *J. Meteorol. Soc. Jpn.*, *80*, 793–809.

Plumb, R. A. (2007), Tracer interrelationships in the stratosphere, *Rev. Geophys.*, *45*, RG4005, doi:10.1029/2005RG000179.

Plumb, R. A., and J. Eluszkiewicz (1999), The Brewer-Dobson circulation: Dynamics of the tropical upwelling, *J. Atmos. Sci.*, *56*, 868–890.

Polvani, L. M., and R. A. Plumb (1992), Rossby wave breaking, filamentation and secondary vortex formation: The dynamics of a perturbed vortex, *J. Atmos. Sci.*, *49*, 462–476.

Polvani, L. M., and D. W. Waugh (2004), Upwelling wave activity flux as a precursor to extreme stratospheric wave events and subsequent anomalous surface weather regimes, *J. Clim.*, *17*, 3548–3554.

Randel, W. J. (1988), The seasonal evolution of planetary waves in the Southern Hemisphere stratosphere and troposphere, *Q. J. R. Meteorol. Soc.*, *114*, 1385–1409.

Randel, W. J., and P. A. Newman (1998), The stratosphere in the southern hemisphere, in *Meteorology of the Southern Hemisphere*, edited by D. J. Karoly and D. G. Vincent, *Meteorol. Monogr.*, *27*, 243–282.

Randel, W. J., R. Garcia, and F. Wu (2008), Dynamical balances and tropical stratospheric upwelling, *J. Atmos. Sci.*, *65*, 3584–3595.

Rosenfield, J. E., P. A. Newman, and M. R. Schoeberl (1994), Computations of diabatic descent in the stratospheric polar vortex, *J. Geophys. Res.*, *99*, 16,677–16,689.

Rosenlof, K. H. (1995), Seasonal cycle of the residual mean meridional circulation in the stratosphere, *J. Geophys. Res.*, *100*, 5173–5191.

Schoeberl, M. R. (1978), Stratospheric warmings: Observations and theory, *Rev. Geophys.*, *16*, 521–538.

Scinocca, J. F., and P. H. Haynes (1998), Dynamical forcing of stratospheric planetary waves by tropospheric baroclinic eddies, *J. Atmos. Sci.*, *55*, 2361–2392.

Scott, R. K., and P. H. Haynes (2000), Internal vacillations in stratosphere-only models, *J. Atmos. Sci.*, *57*, 3233–3250.

Scott, R. K., and P. H. Haynes (2002), The seasonal cycle of planetary waves in the winter stratosphere, *J. Atmos. Sci.*, *59*, 803–822.

Scott, R. K., and L. M. Polvani (2004), Stratospheric control of upward wave flux near the tropopause, *Geophys. Res. Lett.*, *31*, L02115, doi:10.1029/2003GL017965.

Scott, R. K., and L. M. Polvani (2006), Internal variability of the winter stratosphere. Part I: Time-independent forcing, *J. Atmos. Sci.*, *63*, 2758–2776.

Semeniuk, K., and T. G. Shepherd (2001), Mechanisms for tropical upwelling in the stratosphere, *J. Atmos. Sci.*, *58*, 3097–3115.

Shepherd, T. G. (2007), Transport in the middle atmosphere, *J. Meteorol. Soc. Jpn.*, *85*, 165–191.

Shepherd, T. G., K. Semeniuk, and J. N. Koshyk (1996), Sponge layer feedbacks in middle-atmosphere models, *J. Geophys. Res.*, *101*, 23,447–23,464.

Simmons, A. J. (1974), Planetary-scale disturbances in the polar winter stratosphere, *Q. J. R. Meteorol. Soc.*, *100*, 76–108.

Smagorinsky, J. (1953), The dynamical influence of large-scale heat sources and sinks on the quasi-stationary mean motions of the atmosphere, *Q. J. R. Meteorol. Soc.*, *79*, 342–366.

Smith, A. K. (1989), An investigation of resonant waves in a numerical model of an observed sudden stratospheric warming, *J. Atmos. Sci.*, *46*, 3038–3054.

Song, Y., and W. A. Robinson (2004), Dynamical mechanisms for stratospheric influences on the troposphere, *J. Atmos. Sci.*, *61*, 1711–1725.

Taguchi, M., and S. Yoden (2002), Internal interannual variability of the troposphere-stratosphere coupled system in a simple global circulation model. Part II: Millenium integrations, *J. Atmos. Sci.*, *59*, 3037–3050.

Thompson, D. W. J., and J. M. Wallace (1998), The Arctic Oscillation signature in the wintertime geopotential height and temperature fields, *Geophys. Res. Lett.*, *25*, 1297–1300.

Thompson, D. W. J., and J. M. Wallace (2000), Annular modes in the extratropical circulation. Part I: Month-to-month variability, *J. Clim.*, *13*, 1000–1016.

Tung, K. K., and R. S. Lindzen (1979), A theory of stationary long waves. Part II: Resonant Rossby waves in the presence of realistic vertical shears, *Mon. Weather Rev.*, *107*, 735–750.

Wang, X., and J. Fyfe (2000), Onset of edge wave breaking in an idealized model of the polar stratospheric vortex, *J. Atmos. Sci.*, *57*, 956–966.

Waugh, D. W., and D. G. Dritschel (1999), The dependence of Rossby wave breaking on the vertical structure of the polar vortex, *J. Atmos. Sci.*, *56*, 2359–2375.

Wexler, H. (1959), Seasonal and other temperature changes in the Antarctic stratosphere, *Q. J. R. Meteorol. Soc.*, *85*, 196–208.

Yoden, S. (1987a), Bifurcation properties of a stratospheric vacillation model, *J. Atmos. Sci.*, *44*, 1723–1733.

Yoden, S. (1987b), Dynamical aspects of stratospheric vacillations in a highly truncated model, *J. Atmos. Sci.*, *44*, 3683–3695.

Yoden, S., Y. Naito, and S. Pawson (1996), A further analysis of internal variability in a perpetual January integration of a troposphere-stratosphere-mesosphere GCM, *J. Meteorol. Soc. Jpn.*, *74*, 175–188.

Yoden, S., M. Taguchi, and Y. Naito (2002), Numerical studies on time variations of the troposphere-stratosphere coupled system, *J. Meteorol. Soc. Jpn.*, *80*, 811–830.

R. A. Plumb, Department of Earth, Atmospheric, and Planetary Sciences, Massachusetts Institute of Technology, Cambridge, MA 02139, USA. (rap@rossby.mit.edu)

Stratospheric Polar Vortices

Darryn W. Waugh

Department of Earth and Planetary Sciences, Johns Hopkins University, Baltimore, Maryland, USA

Lorenzo M. Polvani

Department of Applied Physics and Applied Mathematics and Department of Earth and Environmental Sciences
Columbia University, New York, New York, USA

The intense cyclonic vortices that form over the winter pole are one of the most prominent features of the stratospheric circulation. The structure and dynamics of these "polar vortices" play a dominant role in the winter and spring stratospheric circulation and are key to determining distribution of trace gases, in particular ozone, and the couplings between the stratosphere and troposphere. In this chapter, we review the observed structure, dynamical theories, and modeling of these polar vortices. We consider both the zonal mean and three-dimensional potential vorticity perspective and examine the occurrence of extreme events and long-term trends.

1. INTRODUCTION

The most prominent feature of the stratospheric circulation is the seasonal formation and decay of an intense cyclonic vortex over the winter pole. The strong circumpolar westerly winds at the edge of this "polar vortex" are in stark contrast to the weak easterlies that occur in the summer hemisphere. In both hemispheres, a polar vortex forms in the fall, reaches maximum strength in midwinter, and decays in later winter to spring. The structure and dynamics of these polar vortices play a dominant role in the winter and spring stratospheric circulation and are key to determining distributions of trace gases, in particular ozone, and the couplings between the stratosphere and troposphere.

There has been much interest in the structure and dynamics of stratospheric polar vortices ever since the discovery in the 1950s of this stratospheric "monsoon" circulation (westerlies in the winter and easterlies in the summer) and the recording of rapid warming events in the polar stratosphere (so-called stratospheric sudden warmings); see *Hamilton* [1999] and *Labiztke and van Loon* [1999] for historical reviews. However, there was a dramatic increase in interest in the stratospheric vortices in the 1980s with the discovery of the Antarctic ozone hole: Because polar vortices act as containment vessels and allow for the occurrence of extremely low temperatures, they play a critical role in polar ozone depletion and the annual formation of the Antarctic ozone hole [e.g., *Newman*, this volume; *Solomon*, 1999]. As a result, there has been a rapid growth in the last 2 decades of observational and modeling studies to better understand the structure and dynamics of polar vortices. Interest in the vortices has further intensified in recent years as numerous studies have shown that the polar vortices can influence tropospheric weather and climate. In particular, vortices are an important component of the dynamical stratosphere-troposphere couplings and so-called "annular modes" [e.g., *Kushner*, this volume].

In this chapter, we review the observed structure of polar vortices and briefly summarize recent advances in our understanding of their dynamics. We briefly touch upon aspects relevant to polar ozone depletion and stratosphere-troposphere coupling but leave detailed discussions of these issues to *Newman* [this volume] and *Kushner* [this volume], respectively. For earlier reviews of polar vortices, see *Schoeberl and Hartmann* [1991] and *Newman and Schoeberl* [2003].

The Stratosphere: Dynamics, Transport, and Chemistry
Geophysical Monograph Series 190
Copyright 2010 by the American Geophysical Union.
10.1029/2009GM000887

The observed climatological structure and variability of the polar vortices is first summarized in section 2, focusing on zonal mean aspects. In section 3, polar vortices are examined from a potential vorticity (PV) perspective, followed by a discussion of dynamical theories and modeling based on PV, including Rossby wave propagation and "breaking" and formation of a "surf zone" surrounding the vortices. In section 4, we discuss the observations and theories of extreme vortex events, including so-called "stratospheric sudden warmings." The coupling with the troposphere is discussed in section 5, including examination of the possible impacts of stratospheric polar vortices on tropospheric weather and climate. In section 6, we review observed trends over the past 4 decades and model projections of the possible impact of climate change on stratospheric polar vortex dynamics. Concluding remarks are given in the final section.

2. CLIMATOLOGICAL STRUCTURE

The general characteristics of stratospheric polar vortices can be seen in plots of zonal mean zonal winds. For example, Figure 1a shows the latitude-height variations of climatological zonal winds for July (left plot) in the Southern Hemisphere (SH) and January (right plot) in the Northern Hemisphere (NH). (See, for example, *Andrews et al.* [1987] and *Randel and Newman* [1998] for similar plots for other months and of zonal mean temperatures.) For both hemispheres, there is a strong westerly jet, the center of which corresponds roughly to the edge of the polar vortex. The westerly jets shown in Figure 1a form because of strong pole to equator temperature gradients, and there are very low temperatures over the winter polar regions (see below).

Stratospheric polar vortices form in fall when solar heating of polar regions is cut off, reach maximum strength in midwinter, and then decay in later winter to spring as sunlight returns to polar regions. This is illustrated in Figure 1b, which shows the latitude-seasonal variations of the zonal winds in the middle stratosphere (10 hPa). In both hemispheres, there are weak easterlies during summer months (June–August in the NH and December–February in the SH), which are replaced by westerlies in fall that grow in strength until there is a strong zonal flow in midwinter. These strong westerlies flow then decay through spring, and the flow returns to easterlies in the summer.

Although radiative processes (e.g., heating by absorption of solar radiation by ozone and cooling by thermal emission by carbon dioxide) play the forcing role in setting up the large-scale latitudinal temperature gradients and resulting zonal flow, the winter stratosphere is not in radiative equilibrium. Waves excited in the troposphere (e.g., by topography, land-sea heating contrasts, or tropospheric eddies) propagate up

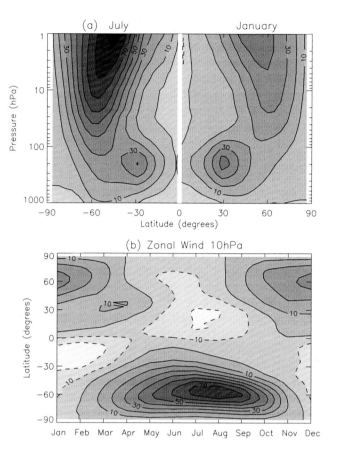

Figure 1. (a) Latitude-height variation of climatological mean zonal mean zonal winds for (left) SH in July and (right) NH in January. (b) Latitude-month variation of climatological mean zonal mean zonal winds at 10 hPa.

into the stratosphere and perturb it away from radiative equilibrium, and the zonal winds shown in Figure 1 are weaker than predicted by radiative equilibrium [see *Andrews et al.*, 1987]. Moreover, the propagation of such waves into the stratosphere varies with conditions in the stratosphere itself. *Charney and Drazin* [1961] showed that Rossby waves propagate upward only if their horizontal scale is large and if the flow is weakly eastward relative to their phase speed; that is, stationary waves only propagate through weak westerlies [see *Andrews et al.*, 1987]. As a result, stationary Rossby waves propagate up into the stratosphere in the winter (when westerlies are prevalent) and not in the summer (when easterlies are prevalent), and the stratospheric flow is more disturbed in the winter than in the summer.

Large hemispheric differences in the polar vortices can be seen in Figure 1: The Antarctic vortex is larger, stronger (more rapid westerlies), and has a longer lifespan than its Arctic counterpart. These differences are caused by hemispheric differences in the wave generation and propagation. The larger topography and land-sea contrasts in the NH excite more/larger planetary-scale Rossby waves that disturb the stratospheric vortex and push it farther from radiative equilibrium than in the SH. The hemispheric differences in the strength and, in particular, coldness of the polar vortices are extremely important for understanding ozone depletion, as explained below.

There are also significant hemispheric differences in the variability of the vortices, with the Antarctic vortex being less variable on both intraseasonal and interannual time scales. These differences can be seen in the evolution of minimum polar temperatures at 50 hPa, shown in Figure 2. Similar features are observed in other temperature diagnostics and in high-latitude zonal winds [e.g., *Randel and Newman*, 1998; *Yoden et al.*, 2002]. The climatological minimum temperatures (thick curves) in the Antarctic are lower and stay colder longer than in the Arctic. Also, there is much larger variability in the Arctic temperatures than in the Antarctic: In the Arctic, a large range of temperatures can be observed from fall to spring (November to April), whereas in the Antarctic there is a fairly narrow range of values except during late spring (October–November). The range and quartiles in Figure 2 show that the distribution of Arctic temperatures is non-Gaussian and highly skewed; see *Yoden et al.* [2002] for more discussion.

The large variability in the Arctic occurs on interannual, intraseasonal, and weekly time scales. Within a single winter, there can be periods with extremely low temperatures as well as periods with extremely high temperatures, and the transition between these events can occur rapidly. These extreme events, and in particular weak events (so called "stratospheric sudden warming"), are discussed further in section 4.

The differences in polar temperatures shown in Figure 2 explain hemispheric differences in polar ozone depletion. In the Antarctic, midwinter minimum temperatures are lower than threshold temperatures for formation of polar stratospheric clouds (PSCs) every year (horizontal lines in Figure 2), and formation of PSC, chemical processing, and widespread ozone depletion occur every year. In contrast, Arctic temperatures fall below the threshold for PSC formation less frequently, and, as a consequence, ozone depletion in the Arctic is much less frequent and widespread. See *Newman* [this volume] for more details.

The interannual variability of the vortices is due to external forcing of the atmospheric circulation, e.g., solar variations, volcanic eruptions, and anthropogenic changes in composition (e.g., ozone and greenhouse gases (GHGs)), as well as internal variations within the climate system, e.g., the quasi-biennial oscillation (QBO), El Niño–Southern Oscillation (ENSO), and internal variability due to nonlinearities. See *Gray* [this volume] and *Haigh* [this volume] for more discussion of the influence of the QBO and solar variation, respectively, on the variability of the vortices.

3. POTENTIAL VORTICITY DYNAMICS

While examination of zonal mean quantities yields information on the general structure and variability of the vortices, examination of the three-dimensional structure is required for greater insight into the synoptic variability and dynamics of the vortices. A quantity that is particularly useful for understanding the structure and dynamics of the polar vortices is potential vorticity (PV), i.e.,

$$PV = \rho^{-1}\zeta.\nabla\theta,$$

where ρ is the fluid density, ζ is the absolute vorticity, and $\nabla\theta$ is the gradient of the potential temperature. Several properties of PV make it useful for studying the polar vortices. First, PV is materially conserved for adiabatic,

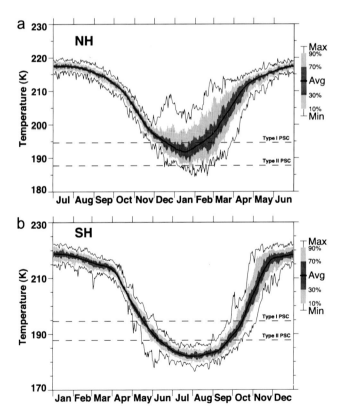

Figure 2. Time series of climatological daily minimum polar temperatures at 50 hPa for the (a) Arctic (50°–90°N) and (b) Antarctic (50°–90°S). The daily climatology is determined from the 1979–2008 period. The black line shows the average for each day of this 1979–2008 climatology. The grey shading shows the percentage range of those same values. Image courtesy of P. Newman.

frictionless flows. Second, other dynamical fields can be determined from the PV distribution (assuming an appropriate balance and given boundary conditions) using "PV inversion" [e.g., *Hoskins et al.*, 1985; *McIntyre*, 1992]. Finally, PV gradients provide the restoring mechanism for Rossby waves, so that the dynamics and propagation of Rossby waves are best understood by examining the distribution of PV [e.g., *Hoskins et al.*, 1985].

Maps of PV on isentropic surfaces provide useful information on the structure and evolution of polar vortices and the associated transport of trace gases. On such maps, a polar vortex appears as a roughly circular, coherent region of high PV, with steep PV gradients at the edge of the high-PV region. (Note that PV is negative in the SH, and when we refer to high PV we are referring to high absolute value of PV in the SH.) The steep PV gradients are colocated with strong westerly winds, and steeper gradients correspond to stronger winds. As PV is materially conserved over the time scale of several days to a week, maps of PV can be used to trace the evolution of the polar vortices and also the transport of trace gases.

Maps of climatological mean PV show, as discussed above, that the Antarctic vortex is larger and stronger (steeper gradients tend to correspond to more rapid westerlies) than the Arctic vortex; for example, see Figure 3, which shows maps of the climatological mean PV on the 850 K isentropic surface (~10 hPa) for the NH in January (Figure 3a) and the SH in July (Figure 3b). These maps also show significant differences in the zonal variations, with the climatological Antarctic vortex being more symmetrical and centered nearer the pole than the Arctic vortex [e.g., *Randel and Newman*, 1998; *Waugh and Randel*, 1999; *Karpetchko et al.*, 2005]. The climatological Arctic vortex is shifted toward the Eurasian continent, and there is, climatologically, a stationary

anticyclone over the Aleutian Islands [e.g., *Harvey et al.*, 2002]. The vortices are smaller and, especially in the NH, more distorted in the lower stratosphere than in the upper stratosphere [e.g., *Randel and Newman*, 1998; *Waugh and Randel*, 1999]. Even though the Arctic vortex is more disturbed than the Antarctic vortex, the scale of the disturbances is much larger than those in the troposphere (e.g., disturbances in the stratosphere are typically between zonal wave numbers 1 and 3, whereas tropospheric disturbances are waves 5 to 7). This is typically understood in terms of the Charney-Drazin theory that describes the filtering of waves propagating into the stratosphere [e.g., *Andrews et al.*, 1987].

As mentioned above, there can be large day-to-day variability in the structure of the Arctic vortex. Figure 4 illustrates this with PV maps for several days during January and February 1979. In mid-January, the Arctic vortex is centered near the pole, but by the end of the month the vortex has elongated and is displaced from the pole, and there is a "tongue" of PV extending from the vortex into the midlatitudes. The vortex has returned to a circular shape by early February, but it is again disturbed in middle to late February. During this latter period, the vortex again becomes elongated, but in contrast to the late January event the vortex remains centered near the pole and splits into two smaller regions of high PV. The February event corresponds to a "stratospheric sudden warming" (SSW) and has been examined in numerous studies and by *Andrews et al.* [1987]. The occurrence and general characteristics of SSWs are discussed further in section 4.

PV is useful not only for visualizing the structure and evolution of the vortices, but it is also provides valuable insight into the dynamics of the vortices. *McIntyre and Palmer* [1983, 1984] examined maps of PV for the same period shown in Figure 4 and interpreted features in these

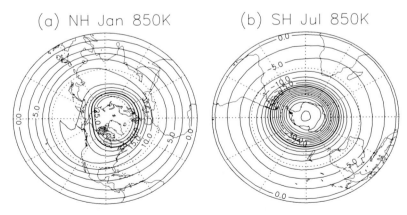

Figure 3. Maps of the climatological mean PV on the 850 K isentropic surface (~10 hPa) for (a) NH in January and (b) SH in July.

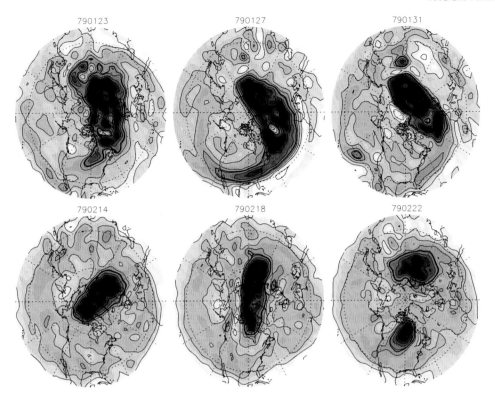

Figure 4. Maps of PV on the 750 K isentropic surfaces for several days during January and February 1979 from National Centers for Environmental Prediction/National Center for Atmospheric Research reanalyses.

maps in terms of Rossby waves. They associated reversible distortions of the vortex with propagating Rossby waves and contrasted these with irreversible deformations where air with high PV is pulled off the vortex and stirred into midlatitudes (e.g., late January 1979, Figure 4). McIntyre and Palmer referred to the latter process as Rossby "wave breaking" and to the region surrounding the vortex as the stratospheric "surf zone." Such Rossby wave breaking is very common at the vortex edge, occurs even when there are no SSW events, and has been documented in numerous observational studies [e.g., *Baldwin and Holton*, 1988; *Abatzoglou and Magnusdottir*, 2007].

Rossby wave breaking and accompanying formation of a surf zone has been extensively studied with numerical models. In a seminal study, *Juckes and McIntyre* [1987] performed high-resolution, single-layer simulations of an initially symmetric polar vortex disturbed by "wave 1" forcing (to mimic the impact of upward propagating Rossby waves). These simulations showed Rossby wave breaking at the edge of the vortex, resulting in filaments of vortex air being stripped from the vortex and stirred into middle latitudes (and filaments of tropical air also being entrained into midlatitudes), and formation of steep PV gradients. This is illustrated in Figure 5b, which shows the vortex evolution

for similar calculations in a spherical, shallow water model [see also *Juckes*, 1989; *Salby et al.*, 1990; *Norton*, 1994; *Polvani et al.*, 1995]. This Rossby wave breaking, stirring of filaments into middle latitudes, and vortex erosion are ubiquitous features of polar vortex simulations, and they have been reported in a large number of subsequent studies using a hierarchy of models, ranging from simple planar models where the vortex is represented by a single region of uniform PV (Figure 5a) [*Polvani and Plumb*, 1992] to three-dimensional models (Figure 5c) [see also *Haynes*, 1990; *Dritschel and Saravanan*, 1994; *Waugh and Dritschel*, 1999; *Polvani and Saravanan*, 2000]. Furthermore, classical two-dimensional vortex flows also show the formation of steep gradients at the edge of vortices (so called "vortex stripping" [e.g., *Legras et al.*, 2001]) and the stability of filaments in straining flows [*Dritschel*, 1989; *Dritschel et al.*, 1991]. The latter helps explain the robustness of filaments in stratospheric simulations.

Three-dimensional (3-D) simulations of a forced vortex show two classes of wave breaking: "remote" wave breaking where Rossby waves propagate up the vortex edge and break in upper levels (e.g., Figure 5c) and "local" breaking where the wave breaking occurs at lower levels of the vortex and inhibits further wave propagation into the upper levels

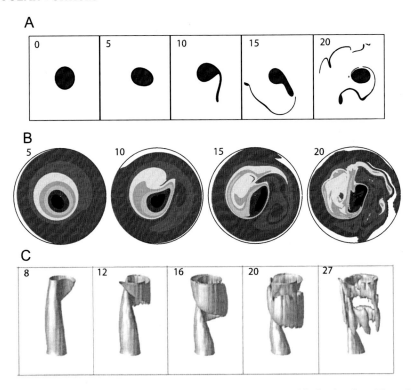

Figure 5. Simulations of polar vortices disturbed by stationary wave 1 topographic forcing for a hierarchy of models. (a) Planar, quasi-geostrophic model with vortex represented by a single discontinuity in PV [*Polvani and Plumb*, 1992]. (b) Spherical shallow water model with continuous PV distribution (as given by, e.g., *Polvani et al.* [1995]). (c) Three-dimensional, spherical primitive equation model [*Polvani and Saravanan*, 2000]. Numbers in Figures 5a–5c indicate the elapsed time, in days, since the start of the simulations.

[*Dritschel and Saravanan*, 1994; *Waugh and Dritschel*, 1999; *Polvani and Saravanan*, 2000]. The 3-D simulations also show the sensitivity of the wave propagation and wave breaking to the gradients at the edge of the vortex, with enhanced vertical propagation and breaking for steeper-edge gradients [*Polvani and Saravanan*, 2000; *Scott et al.*, 2004]. This has potential implications for how well the dynamics of polar vortices is modeled in low- and moderate-resolution climate models.

The focus of the above simulations, and analysis of observations, has primarily been on the impact of upward propagating Rossby waves. However, the recent study of *Esler and Scott* [2005] revisited the idea of resonant excitation of free modes considered earlier by *Tung and Lindzen* [1979] and *Plumb* [1981]. Esler and Scott showed that in an idealized 3-D quasi-geostrophic model, there are not only upward and downward propagating Rossby waves, but there is also a barotropic mode. This latter mode can be forced by transient forcing and can dominate over the upward propagating waves. In fact, they showed that a "barotropic" sudden warming occurs if this mode is resonantly excited.

Furthermore, some observed NH vortex-splitting major sudden warmings, e.g., in 1979 and 2009, exhibit a very similar structure to the modeled barotropic sudden warming.

Although the dynamical evolution of polar vortices is controlled by relatively large zonal wave numbers, numerical simulations show the rapid formation of fine-scale features, e. g., PV filaments and very steep PV gradients (e.g., Figure 5). As these fine-scale features cannot be resolved in low-resolution stratospheric satellite observations or meteorological analyses (e.g., Figure 4), their realism might, at first glance, be doubted However, high-resolution trajectory calculations performed using winds from meteorological analyses are able to produce fine-scale features, including filaments and steep gradients, that closely resemble those in the single and multilayer numerical simulations. For example, Figure 6 shows high-resolution simulations of the lower stratospheric vortex in January 1992. These maps show fine-scale features that are not seen in the analyzed PV fields, but they are consistent with features in the high-resolution dynamical simulations. Furthermore, the presence of these fine-scale features was confirmed by aircraft measurements of

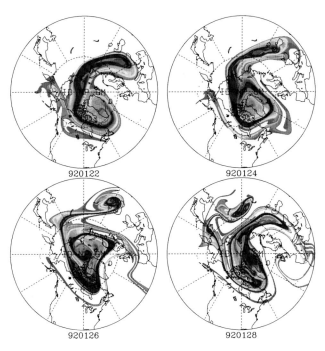

Figure 6. High-resolution trajectory simulation of a passive tracer on the 450 K surface in January 1992. The tracer was initialized with the PV field for 16 January 1992. See *Plumb et al.* [1994] for details.

chemical tracers made during this period [e.g., *Plumb et al.*, 1994; *Waugh et al.*, 1994].

The breaking of Rossby waves at the vortex edge is important not only for the dynamics of vortices, but it also plays an important role in the transport of trace gases; see *Plumb* [2002] and *Shepherd* [2007] for comprehensive reviews of stratospheric tracer transport. As discussed above, Rossby wave breaking strips air from the edges of the vortices, forms steep edge gradients, and stirs vortex air into middle latitudes. This transport plays a major role in determining the distribution of trace gases in the stratosphere. Latitude-height contour plots of long-lived trace gases such as N_2O, CH_4, and CFCs (which have tropospheric sources and upper stratospheric/mesospheric sinks) show that tracer isopleths are depressed within the polar vortices compared to middle latitudes [e.g., *Schoeberl and Douglass*, this volume, Figure 1]. While larger polar descent contributes to the difference in tracers inside and outside the vortex, the dominant cause is the above difference in horizontal stirring. There is rapid stirring in the surf zone and very little mixing across the vortex edge, which leads to homogenization of tracers within the surf zone and steep latitudinal tracer gradients at the edge of the polar vortices (there are also steep gradients at the subtropical edge of the surf zone).

Although the vortex edge is a barrier to mixing, it is not a perfect barrier. There is evidence, primarily from tracer-tracer

relationships, of mixing across the vortex edge [e.g., *Plumb*, 2007, and references therein]. The mechanisms for this mixing and when this mixing occurs is still a matter of debate. "Inward" Rossby wave-breaking events in which surf zone air is mixed into the vortex have been observed (e.g., Figure 6). However, these events are infrequent, and Rossby wave breaking that transports air from the vortex into middle latitudes is much more common.

4. EXTREME EVENTS

As discussed in section 2, there is large variability in the Arctic vortex during fall to spring, and there are periods when the vortex is anomalously strong and periods when it is weak (or even nonexistent), with rapid transitions between these states. The variability of the Arctic vortex and existence of extreme events is illustrated in Figure 7, which shows the time series of the Northern Annular Mode (NAM) index for 24 years [*Polvani and Waugh*, 2004]. The NAM is the dominant pattern of variability in the northern extratropical troposphere and stratosphere [*Thompson and Wallace*, 2000]. Since the polar vortex dominates the variability of the stratosphere, the NAM index at 10 hPa is a rough measure of the strength of the stratospheric vortex, with a positive NAM index corresponding to a strong vortex and a negative index corresponding to a weak vortex (e.g., the breakup of the Arctic vortex shown in Figure 4 corresponds to a period when the NAM is around −3).

There is virtually no variability in the NAM index during summer but large variability during winter and spring, with rapid transitions between strong and weak vortices. Extremely strong and weak events occur on average around once every other winter, although the occurrence of extreme events is not evenly spread: Two weak vortex events can occur in one winter (e.g., 1998/1999 winter), and there can be extended periods when there are few weak events but frequent strong events (e.g., early to mid-1990s [*Manney et al.*, 2005]). The probability distribution function of the NAM index in winter [*Baldwin and Dunkerton*, 2001; *Polvani and Waugh*, 2004] is close to Gaussian, and the frequency of strong/weak vortex events is consistent with expectations for a random process.

The occurrence of strong and weak vortex events has been linked to the upward wave activity entering the stratosphere. Case studies and composite analyses have shown that anomalously strong wave activity nearly always precedes weak vortex events (SSWs), and, conversely, anomalously weak wave activity precedes strong vortex events [e.g., *Christiansen*, 2001; *Polvani and Waugh*, 2004]. This is illustrated in Figure 7 where the time series of eddy heat flux at 100 hPa integrated over the prior 40 days (a measure of the time-integrated wave activity entering the stratosphere) is

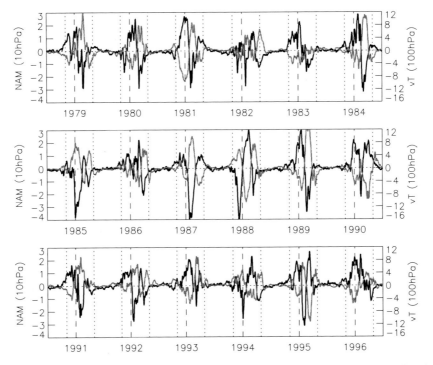

Figure 7. Daily values of NAM index at 10 hPa (black) and 40 day averaged heat flux anomalies at 100 hPa (grey) for 1 July 1978 to 31 December 1996. Adapted from *Polvani and Waugh* [2004].

shown to be anticorrelated with the NAM index at 10 hPa (correlation coefficient of -0.8). Theoretical support for this observed relationship is provided by *Newman et al.* [2001], who showed that stratospheric polar temperatures (and, via geostrophic balance, stratospheric winds) on a given day are related not to the instantaneous upward wave activity but to its weighted integral over several weeks prior to that day [see also *Esler and Scott*, 2005]. There is also a clear relationship on interannual time scales between the state of the stratosphere and the time-integrated wave activity upwelling from the troposphere, with strong upward wave activity during winters with a weaker, warmer vortex that breaks up earlier [e.g., *Waugh et al.*, 1999; *Newman et al.*, 2001; *Hu and Tung*, 2002].

Periods when the NAM index is less than -2.7 are generally associated with major SSW events, which are traditionally, and more simply, defined by the reversal of the zonal mean zonal winds at 60°N and 10 hPa. There has been considerable research into SSWs, and there is an extensive literature on the vortex evolution during individual SSWs. However, until recently, there were few studies of the climatological nature of SSWs. Using the traditional definition of SSWs based on zonal winds, *Charlton and Polvani* [2007] reported an average of around six SSWs every decade (29 SSWs in the 44 winters between 1957/1958 and 2001/2002). This study and those of *Limpasuvan et al.* [2004] and *Matthewman et al.*

[2009] also used composite-based analysis to examine the climatological nature of SSWs, including the evolution of temperature, zonal flow, and eddy fluxes during warming events as well as the three-dimensional structure of the vortices. (The climatological nature of strong vortex events has also been examined by *Limpasuvan et al.* [2005].)

Traditionally, SSWs were classified into wave 1 or wave 2 events depending on the amplitude of the longitudinal wave numbers. However, given the nonlinearity of the flow, such a classification can be misleading [see *Waugh*, 1997], and it is more appropriate to use a vortex-oriented classification. *Charlton and Polvani* [2007] and *Matthewman et al.* [2009] used such a classification and divided the SSWs into either "vortex displacement" or "vortex split" events. (The minor warming in January and major warming in February 1979, shown in Figure 4, are examples of vortex displacement and vortex split events, respectively.) They showed that the two types of events are dynamically different, with differences in the stratospheric and tropospheric flows before the events and in vortex evolution during the event. Although the life cycles of the displacement and split events are very different, there is much less variation between individual events of each class, both in terms of vertical structure and longitudinal orientation, and characteristics are well captured by the composite events [*Matthewman et al.*, 2009]. In particular, during splitting events the vortex deformations are highly barotropic, while for displacement events the vortex tilts westward with height.

The basic understanding of the dynamics of SSWs, based on the seminal *Matsuno* [1970, 1971] studies, involves the anomalous growth of upward propagating planetary-scale Rossby waves originating in the troposphere (see *Andrews et al.* [1987] for review). However, the cause of the rapid amplification of the Rossby wave and the role of the initial state of the stratospheric vortex are not fully understood. There is evidence that the vortex needs to be preconditioned for SSWs to occur [e.g., *McIntyre*, 1982; *Limpasuvan et al.*, 2004]. While most analyses of SSWs have focused on upward propagating Rossby waves, alternate theories include resonant excitation of free modes [*Tung and Lindzen*, 1979; *Plumb*, 1981; *Esler and Scott*, 2005; *Esler et al.*, 2006] and vortex interactions, in particular interactions between the polar vortex and Aleutian anticyclone [*O'Neill and Pope*, 1988; *Scott and Dritschel*, 2006].

Until recently, it was thought that sudden warmings were exclusively a NH phenomenon. However, a dramatic event occurred in the SH in September 2002, when the Antarctic vortex elongated and split into two. This is the only known SSW in the SH, and there has been considerable research into this event, much of which is summarized in the March 2005 special issue of the *Journal of Atmospheric Sciences* [see also *Baldwin et al.*, 2003]. Although there have been many studies into the dynamics of this event (see the above mentioned special issue), the exact cause remains unknown. Most of the focus has been on upward propagating Rossby waves, but *Esler et al.* [2006] provide evidence that the event may have been the result of a self-tuned resonance. *Kushner and Polvani* [2005] documented the spontaneous occurrence of a sudden warming, resembling the observed SH event, in a long numerical simulation of a simple troposphere-stratosphere general circulation model with no stationary forcing, suggesting that the 2002 event may have been just a rare, random ("natural") event.

There is large interannual variability not only in the midwinter vortices but also in the timing and characteristics of the final breakdown of the vortices (so called "final warming") [see, e.g., *Waugh and Rong*, 2002]. The stratospheric final warmings influence not only the stratospheric circulation (e. g., a transition from winter stratospheric westerlies to summer easterlies) but also strongly organize the tropospheric circulation, with a rapid weakening of high-latitude tropospheric westerlies occurring for both NH and SH final warmings [e.g., *Black and McDaniel*, 2007a, 2007b].

5. STRATOSPHERE-TROPOSPHERE COUPLING

Stratospheric vortices have, until recently, been considered an interesting middle-atmosphere phenomenon, with little attention paid to their possible impact on the troposphere. However, increasing observational and modeling evidence in the last decade suggests that polar stratospheric vortices can have a significant influence on the tropospheric flow for a range of time scales [e.g., *Baldwin and Dunkerton*, 2001; *Thompson and Solomon*, 2002; *Polvani and Kushner*, 2002; *Gillett and Thompson*, 2003; *Norton*, 2003; *Charlton et al.*, 2004].

Much of the evidence for a stratospheric impact on the troposphere focuses on the so-called annular modes: the Northern Annular Mode (NAM) and Southern Annular Mode (SAM) [e.g., *Thompson and Wallace*, 2000]. As discussed above, these modes are the dominant patterns of variability in the extratropical troposphere and stratosphere, and the NAM/ SAM index in the stratosphere is a measure of the vortex strength (see Figure 7). *Baldwin and Dunkerton* [1999] showed that anomalous values in the NAM index are found to appear in the stratosphere first and subsequently progress downward over periods of several weeks. Moreover, subsequent studies showed that extreme stratospheric events can be followed by anomalous weather regimes at the surface that persist for up to 2 months [*Baldwin and Dunkerton*, 2001; *Thompson et al.*, 2002]. The exact dynamical mechanism by which the stratosphere influences the troposphere is unknown, but there are several proposed theories, including direct PV inversion [*Hartley et al.*, 1998; *Ambaum and Hoskins*, 2002; *Black*, 2002], changes in refractive properties and Rossby wave propagation [*Hartmann et al.*, 2000] or wave reflection [*Perlwitz and Harnick*, 2004], and eddy-mediated feedbacks [*Kushner and Polvani*, 2004; *Song and Robinson*, 2004; *Chen and Held*, 2007]. See *Kushner* [this volume] for more detailed discussion of these mechanisms.

Although anomalous values of the annular mode index appear first in the upper stratosphere, it is important to note, as discussed in section 4, that these the extreme events are preceded by anomalous wave activity entering the stratosphere (see Figure 7). While the fact that stratospheric extreme events are preceded by anomalous wave activity might indicate that the stratosphere is slave to the troposphere, this is not necessarily the case. Numerous studies, using a hierarchy of models, have shown that internal variability can be generated within the stratosphere, with vacillation cycles of strong (westerly) and weak (easterly) polar winds [e.g., *Holton and Mass*, 1976; *Yoden*, 1987; *Scott and Haynes*, 2000; *Rong and Waugh*, 2003; *Scott and Polvani*, 2004]. Furthermore, the *Scott and Polvani* [2004] simulations show cycles in wave activity entering the stratosphere that resemble those in observations (e.g., Figure 7) even though all forcings in their simple model are completely time-independent. This therefore suggests that the stratosphere plays a role in determining the wave activity entering from the troposphere.

Couplings between the stratospheric vortices and tropospheric circulation have also been found in the SH. Observations show a strengthening of westerlies (and corresponding increase in the SAM) in both the stratosphere and troposphere over the past 2–3 decades. The largest stratospheric trends occur in spring months, whereas the largest tropospheric trends occur in the summer. This is consistent with Antarctic ozone depletion strengthening the stratospheric vortex (see section 6) and a time lag for stratospheric anomalies to descend to the surface [*Thompson and Solomon*, 2002; *Gillett and Thompson*, 2003]. By modifying the SAM, a strengthening (or weakening) Antarctic vortex also has the potential to impact other aspects of the tropospheric circulation, including subtropical jets, storm tracks, the Hadley cell width, and subtropical hydrology [*Perlwitz et al.*, 2008; *Son et al.*, 2008, 2009].

6. TRENDS

Given the key role the polar vortices play in ozone depletion and stratosphere-troposphere coupling, it is important to understand how polar vortices have changed over the past few decades and how they might change in the future.

Several studies have examined decadal variability and trends in polar vortices over the past 4 decades using meteorological reanalyses [e.g., *Waugh et al.*, 1999; *Zhou et al.*, 2000; *Langematz and Kunze*, 2006; *Karpetchko et al.*, 2005]. Because of the lack of stratospheric measurements in the SH in the presatellite era, reliable trends can only be determined from 1979 on. However, there are sufficient stratospheric observations in the 1960s and 1970s in the NH to perform trend analyses from 1960 to present day for the Arctic vortices. These studies have shown that there are significant trends in the springtime Antarctic vortex and that the vortex has become stronger, colder, and more persistent (i.e., breaks up later) since 1979; see Figure 8. The colder vortex and delay in breakup are attributed to the decreases in ozone within the vortex over this period (i.e., growth in the Antarctic ozone hole). As discussed above, these changes in the spring Antarctic vortex have caused changes in SH tropospheric circulation.

The large interannual variability of the Arctic vortex (e.g., Figure 2) makes detecting long-term trends very difficult. Although trends have sometimes been reported for shorter data records, there are no significant trends in the size or persistence in the Arctic vortex between 1958 and 2002 (Figure 8) [see also *Karpetchko et al.*, 2005]. However, trends in the area or volume of temperatures below the threshold for PSC formation have been noted [e.g., *Knudsen et al.*, 2004; *Karpetchko et al.*, 2005; *Rex et al.*, 2006]. The most intriguing result is the analysis of *Rex et al.* [2006] that indicates that

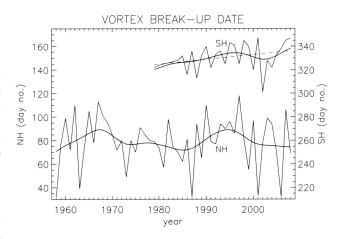

Figure 8. Variation in final breakdown date of Arctic (left axis) and Antarctic polar vortices (right axis). Dashed line shows linear trend for SH between 1979 and 2008. Updated from *Waugh et al.* [1999].

there is a significant increase in "PSC volume" if coldest winters within 5 year periods are considered. As there is a high correlation between PSC volume and ozone depletion, this cooling of cold winters implies an increase in Arctic ozone depletion. The cause of the increased PSC volume for coldest winters is not known, and similarly, it is not known whether this trend will continue into the future.

There is a great deal of interest in possible future trends in the polar vortices. As significant changes in concentrations of key radiative gases in the stratosphere are expected over the 21st century (ozone is expected to increase as the concentrations of ozone-depleting substances decrease back to 1960 levels, and GHGs are expected to continue to increase), some changes in the polar vortices may be expected. It is also possible that increases in GHGs could lead to changes in wave activity and propagation from the troposphere, which could then lead to changes in the vortices. There have been numerous modeling studies examining possible changes in the stratosphere over the 21st century, but the majority of these studies have focused on changes in stratospheric ozone [e.g., *Austin et al.*, 2003; *Eyring et al.*, 2007; *Shepherd*, 2008] or circulation [e.g., *Garcia and Randel*, 2008; *Oman et al.*, 2009; *Butchart et al.*, 2009], and there have been limited detailed studies of changes in the vortices. However, there has been analysis of monthly mean temperatures in these simulations that provide insight into possible changes in the vortices.

An early study by *Shindell et al.* [1998] indicated a significant cooling (and strengthening) of the Arctic vortex as GHGs increased, leading to significant ozone depletion and formation of an Arctic ozone hole. However, more recent simulations with more sophisticated chemistry-climate models (CCMs) that have a better representation of the dynamics and chemistry (and couplings between them) do not show a significant strengthening or formation of Arctic ozone holes during the 21st century [e.g., *Austin et al.*, 2003; *Eyring et al.*, 2007]. The long-term trends in Arctic temperatures in these chemistry-climate models are small, with no consistency as to whether the polar stratosphere will be warmer or colder [*Butchart et al.*, 2009]. The CCMs also predict a limited impact of increased GHGs on the Antarctic vortex during the 21st century. However, the same simulations all predict an increase in the tropical upwelling as GHGs increase, which has been attributed to changes in subtropical wave driving [e.g., *Garcia and Randel*, 2008; *Oman et al.*, 2009]. This change in tropical circulation appears not to be strongly connected with changes in polar regions [e.g., *McLandress and Shepherd*, 2009].

In terms of sudden warmings, it is important to note first that most state of the art chemistry-climate models severely underestimate their frequency [*Charlton et al.*, 2007]. As for what might be expected in the 21st century, only one study is available [*Charlton-Perez et al.*, 2008]: On the basis of several long integrations with a single model, the Charlton-Perez et al. suggest that the frequency of sudden warmings (currently six events per decade) might increase by one event per decade by the end of the 21st century. Needless to say, owing to the large interannual variability, such trends are very difficult to estimate, and the question remains largely open.

Finally, and perhaps most importantly, the recovery of Antarctic ozone is predicted to cause a positive trend in lower stratospheric temperatures and vortex strength in late spring to summer. As discussed in section 5, changes in the Antarctic lower stratospheric temperatures over the last 2 decades have been linked to changes in Southern Hemisphere climate. The ozone recovery over the next 4 to 5 decades is predicted to reverse these changes [e.g., *Son et al.*, 2008; *Perlwitz et al.*, 2008]. It is important to note that in the latter part of the 20th century, the impact of ozone depletion on the tropospheric circulation has been in the same sense as the impact of increasing GHGs. However, as ozone recovers, the stratospheric impact will oppose, and even reverse, some of the expected changes to increases in GHGs.

7. OUTSTANDING ISSUES

Over the past few decades, the combined use of theory, observations, and modeling has greatly advanced our understanding of, and ability to model, the dynamics of stratospheric polar vortices, their impact on the transport of chemical tracers, and stratosphere-troposphere couplings. Many of these advances have come from consideration of the vortex as a material entity. This has lead to an improved under-

standing of the three-dimensional propagation of Rossby waves, the impact of Rossby wave breaking on vortex dynamics and tracer transport, and the dynamical coupling between the stratosphere and troposphere. Two recent examples are the paper of *Waugh et al.* [2009], which showed the importance of zonal asymmetries in the Antarctic vortex and ozone hole in tropospheric climate change, and that of *Martius et al.* [2009], which showed a close connection between the type of atmospheric block and whether the stratospheric warming was a vortex displacement or splitting event. Nevertheless, some important questions remain unanswered.

The exact causes of the variability in the vortices, including the occurrence of SSWs, are not known. It has been shown that this variability is linked to wave activity entering the stratosphere, but the relative contributions from internal stratospheric, tropospheric, and coupled processes to the variability in wave activity are not known. Furthermore, the recent *Esler and Scott* [2005] and *Esler et al.* [2006] studies raise the possibility that the resonant excitation of free modes, as opposed to upward propagating Rossby waves, plays a larger role than previously thought. Most attention on extreme vortex events has been focused on the Northern Hemisphere, but the dramatic Southern Hemisphere SSW in 2002 has changed this. There is now a lot of focus on the cause and occurrence of SSWs in the SH.

Numerous recent observational and modeling studies have shown that changes in the stratospheric polar vortices can influence the tropospheric circulation, on both weather and climate time scales. However, there remains uncertainty in the precise dynamical processes involved. Several mechanisms have been proposed, including direct nonlocal dynamical effects, downward reflection of Rossby waves, and alteration of synoptic eddies in the upper troposphere, but more research is required to determine the relative importance of these processes.

A key issue for both the recovery of stratospheric ozone and the influence of the stratosphere on tropospheric climate is how the polar vortices will change, if at all, as greenhouse gases continue to increase. The stratosphere will cool because of the direct radiative effect of increased CO_2, but whether the polar vortices will be come stronger or weaker will likely depend on changes in wave activity entering the stratosphere. There is currently no agreement between climate models as to trends in either the wave activity entering the stratosphere or the strength of the polar vortex, although the trends are generally small in all models. It is unclear how much confidence can be put into the model projections of the vortices given that the models typically only have moderate resolution and that the climatological structure of the vortices in the models depends on the tuning of gravity wave parameterizations.

Given the above outstanding issues, there is need for continued research in the dynamics of the vortices and their representation in global models.

REFERENCES

Abatzoglou, J. T., and G. Magnusdottir (2007), Wave breaking along the stratospheric polar vortex as seen in ERA-40 data, *Geophys. Res. Lett.*, *34*, L08812, doi:10.1029/2007GL029509.

Ambaum, M. H. P., and B. J. Hoskins (2002), The NAO troposphere–stratosphere connection, *J. Clim.*, *15*, 1969–1978.

Andrews, D. G., J. R. Holton, and C. B. Leovy (1987), in *Middle Atmosphere Dynamics*, 489 pp., Academic, San Diego, Calif.

Austin, J., et al. (2003), Uncertainties and assessments of chemistry-climate models of the stratosphere, *Atmos. Chem. Phys.*, *3*, 1–27.

Baldwin, M. P., and T. J. Dunkerton (1999), Propagation of the Arctic Oscillation from the stratosphere to the troposphere, *J. Geophys. Res.*, *104*, 30,937–30,946.

Baldwin, M. P., and T. J. Dunkerton (2001), Stratospheric harbingers of anomalous weather regimes, *Science*, *294*, 581–584.

Baldwin, M. P., and J. R. Holton (1988), Climatology of the stratospheric polar vortex and planetary wave breaking, *J. Atmos. Sci.*, *45*, 1123–1142.

Baldwin, M. P., T. Hirooka, A. O'Neill, and S. Yoden (2003), Major stratospheric warming in the Southern Hemisphere in 2002: Dynamical aspects of the ozone hole split, *SPARC Newsl.*, *20*, 24–26.

Black, R. X. (2002), Stratospheric forcing of surface climate in the Arctic Oscillation, *J. Clim.*, *15*, 268–277.

Black, R. X., and B. A. McDaniel (2007a), The dynamics of Northern Hemisphere stratospheric final warming events, *J. Atmos. Sci.*, *64*, 2932–2946.

Black, R. X., and B. A. McDaniel (2007b), Interannual variability in the Southern Hemisphere circulation organized by stratospheric final warming events, *J. Atmos. Sci.*, *64*, 2968–2974.

Butchart, N., et al. (2009), Chemistry-climate model simulations of 21st century stratospheric climate and circulation changes, *J. Clim.*

Charlton, A. J., and L. M. Polvani (2007), A new look at stratospheric sudden warmings. Part I: Climatology and modeling benchmarks, *J. Clim.*, *20*, 449–469.

Charlton, A. J., A. O'Neill, W. A. Lahoz, and A. C. Massacand (2004), Sensitivity of tropospheric forecasts to stratospheric initial conditions, *Q. J. R. Meteorol. Soc.*, *130*, 1771–1792.

Charlton, A. J., L. M. Polvani, J. Perlwitz, F. Sassi, E. Manzini, S. Pawson, J. E. Nielsen, K. Shibata, and D. Rind (2007), A new look at stratospheric sudden warmings. Part II: Evaluation of numerical model simulations, *J. Clim.*, *20*, 471–488.

Charlton-Perez, A. J., L. M. Polvani, J. Austin, and F. Li (2008), The frequency and dynamics of stratospheric sudden warmings in the 21st century, *J. Geophys. Res.*, *113*, D16116, doi:10.1029/2007JD 009571.

Charney, J. G., and P. G. Drazin (1961), Propagation of planetary-scale disturbances from the lower into the upper atmosphere, *J. Geophys. Res.*, *66*, 83–109.

Chen, G., and I. M. Held (2007), Phase speed spectra and the recent poleward shift of Southern Hemisphere surface westerlies, *Geophys. Res. Lett.*, *34*, L21805, doi:10.1029/2007GL031200.

Christiansen, B. (2001), Downward propagation of zonal mean zonal wind anomalies from the stratosphere to the troposphere: Model and reanalysis, *J. Geophys. Res.*, *106*, 27,307–27,322.

Dritschel, D. G. (1989), On the stabilization of a two-dimensional vortex strip by adverse shear, *J. Fluid Mech.*, *206*, 193–221.

Dritschel, D. G., and R. Saravanan (1994), Three-dimensional quasi-geostrophic contour dynamics, with an application to stratospheric dynamics, *Q. J. R. Meteorol. Soc.*, *120*, 1267–1298.

Dritschel, D. G., P. H. Haynes, M. N. Juckes, and T. G. Shepherd (1991), The stability of a two-dimensional vorticity filament under uniform strain, *J. Fluid Mech.*, *230*, 647–665.

Esler, J. G., and R. K. Scott (2005), Excitation of transient Rossby waves on the stratospheric polar vortex and the barotropic sudden warming, *J. Atmos. Sci.*, *62*, 3661–3682.

Esler, J. G., L. M. Polvani, and R. K. Scott (2006), The Antarctic stratospheric sudden warming of 2002: A self-tuned resonance?, *Geophys. Res. Lett.*, *33*, L12804, doi:10.1029/2006GL026034.

Eyring, V., et al. (2007), Multimodel projections of stratospheric ozone in the 21st century, *J. Geophys. Res.*, *112*, D16303, doi:10.1029/2006JD008332.

Garcia, R. R., and B. Randel (2008), Acceleration of the Brewer-Dobson circulation due to increases in greenhouse gases, *J. Atmos. Sci.*, *65*, 2731–2739.

Gillett, N., and D. W. J. Thompson (2003), Simulation of recent Southern Hemisphere climate change, *Science*, *302*, 273–275.

Gray, L. J. (2010), Stratospheric equatorial dynamics, in *The Stratosphere: Dynamics, Transport, and Chemistry, Geophys. Monogr. Ser.*, doi:10.1029/2009GM000868, this volume.

Haigh, J. D. (2010), Solar variability and the stratosphere, in *The Stratosphere: Dynamics, Transport, and Chemistry, Geophys. Monogr. Ser.*, doi:10.1029/2010GM000937, this volume.

Hamilton, K. (1999), Dynamical coupling of the lower and middle atmosphere: Historical background to current research, *J. Atmos. Sol. Terr. Phys.*, *61*, 73–84.

Hartley, D. E., J. T. Villarin, R. X. Black, and C. A. Davis (1998), A new perspective on the dynamical link between the stratosphere and troposphere, *Nature*, *391*, 471–474.

Hartmann, D. L., J. M. Wallace, V. Limpasuvan, D. W. J. Thompson, and J. R. Holton (2000), Can ozone depletion and global warming interact to produce rapid climate change?, *Proc. Nat. Acad. Sci. U. S. A.*, *97*, 1412–1417.

Harvey, V. L., R. B. Pierce, T. D. Fairlie, and M. H. Hitchman (2002), A climatology of stratospheric polar vortices and anticyclones, *J. Geophys. Res.*, *107*(D20), 4442, doi:10.1029/2001JD001471.

Haynes, P. H. (1990), High-resolution three-dimensional modelling of stratospheric flows: Quasi-2D turbulence dominated by a single vortex, in *Topological FluidMechanics*, edited by H. K. Moffatt and A. Tsinober, pp. 345–354, Cambridge Univ. Press, Cambridge, U. K.

Holton, J. R., and C. Mass (1976), Stratospheric vacillation cycles, *J. Atmos. Sci.*, *33*, 2218–2225.

Hoskins, B. J., M. E. McIntyre, and A. W. Robertson (1985), On the use and significance of isentropic potential vorticity maps, *Q. J R. Meteorol. Soc.*, *111*, 877–946.

Hu, Y., and K. K. Tung (2002), Interannual and decadal variations of planetary wave activity, stratospheric cooling, and the Northern Hemisphere annular mode, *J. Clim.*, *15*, 1659–1673.

Juckes, M. N. (1989), A shallow water model of the winter stratosphere, *J. Atmos. Sci.*, *46*, 2934–2954.

Juckes, M. N., and M. E. McIntyre (1987), A high-resolution one-layer model of breaking planetary waves in the stratosphere, *Nature*, *328*, 590–596.

Karpetchko, A., E. Kyrö, and B. M. Knudsen (2005), Arctic and Antarctic polar vortices 1957–2002 as seen from the ERA-40 reanalyses, *J. Geophys. Res.*, *110*, D21109, doi:10.1029/2005JD006113.

Knudsen, B. M., N. R. P. Harris, S. B. Andersen, B. Christiansen, N. Larsen, M. Rex, and B. Naujokat (2004), Extrapolating future Arctic ozone losses, *Atmos. Chem. Phys.*, *4*, 1849–1856.

Kushner, P. J. (2010), Annular modes of the troposphere and stratosphere, in *The Stratosphere: Dynamics, Transport, and Chemistry, Geophys. Monogr. Ser.*, doi:10.1029/2009GM000924, this volume.

Kushner, P. J., and L. M. Polvani (2004), Stratosphere-troposphere coupling in a relatively simple AGCM: The role of eddies, *J. Clim.*, *17*, 629–639.

Kushner, P. J., and L. M. Polvani (2005), A very large, spontaneous stratospheric sudden warming in a simple AGCM: A prototype for the Southern Hemisphere warming of 2002?, *J. Atmos. Sci.*, *62*, 890–897.

Labiztke, K. G., and H. van Loon (1999), *The Stratosphere: Phenomena, History, and Relevance*, 179 pp., Springer, New York.

Langematz, U., and M. Kunze (2006), An update on dynamical changes in the Arctic and Antarctic stratospheric polar vortices, *J. Clim. Dyn.*, *27*, 647–660.

Legras, B., D. G. Dritschel, and P. Calliol (2001), The erosion of a distributed two-dimensional vortex in a background straining flow, *J. Fluid Mech.*, *441*, 369–398.

Limpasuvan, V., D. W. J. Thompson, and D. L. Hartmann (2004), The life cycle of Northern Hemisphere sudden stratospheric warmings, *J. Clim.*, *17*, 2584–2596.

Limpasuvan, V., D. L. Hartmann, D. W. J. Thompson, K. Jeev, and Y. L. Yung (2005), Stratosphere-troposphere evolution during polar vortex intensification, *J. Geophys. Res.*, *110*, D24101, doi:10.1029/2005JD006302.

Manney, G. L., K. Krüger, J. L. Sabutis, S. A. Sena, and S. Pawson (2005), The remarkable 2003–2004 winter and other recent warm winters in the Arctic stratosphere since the late 1990s, *J. Geophys. Res.*, *110*, D04107, doi:10.1029/2004JD005367.

Martius, O., L. M. Polvani, and H. C. Davies (2009), Blocking precursors to stratospheric sudden warming events, *Geophys. Res. Lett.*, *36*, L14806, doi:10.1029/2009GL038776.

Matsuno, T. (1970), Vertical propagation of stationary planetary waves in the winter Northern Hemisphere, *J. Atmos. Sci.*, *27*, 871–883.

Matsuno, T. (1971), A dynamical model of the stratospheric sudden warming, *J. Atmos. Sci.*, *28*, 1479–1494.

Matthewman, N. J., J. G. Esler, A. J. Charlton-Perez, and L. M. Polvani (2009), A new look at stratospheric sudden warmings. Part III: Polar vortex evolution and vertical structure, *J. Clim.*, *22*, 1566–1585.

McIntyre, M. E. (1982), How well do we understand the dynamics of stratospheric warmings?, *J. Meteorol. Soc. Jpn.*, *60*, 37–65.

McIntyre, M. E. (1992), Atmospheric dynamics: Some fundamentals, with observational implications, *Proc. Int. Sch. Phys. Enrico Fermi, CXV*, 313–386.

McIntyre, M. E., and T. N. Palmer (1983), Breaking planetary waves in the stratosphere, *Nature*, *305*, 593–600.

McIntyre, M. E., and T. N. Palmer (1984), The 'surf zone' in the stratosphere, *J. Atmos. Terr. Phys.*, *46*, 825–849.

McLandress, C., and T. G. Shepherd (2009), Simulated anthropogenic changes in the Brewer–Dobson circulation, including its extension to high latitudes, *J. Clim.*, *22*, 1516–1540.

Newman, P. A. (2010), Chemistry and dynamics of the Antarctic ozone hole, in *The Stratosphere: Dynamics, Transport, and Chemistry, Geophys. Monogr. Ser.*, doi:10.1029/2009GM000873, this volume.

Newman, P. A., and M. R. Schoeberl (2003), Middle atmosphere: Polar vortex, in *Encyclopedia of Atmospheric Sciences*, edited by J. R. Holton, J. Pyle, and J. A. Curry, pp. 1321–1328, Academic, San Diego, Calif.

Newman, P. A., E. R. Nash, and J. E. Rosenfield (2001), What controls the temperature of the Arctic stratosphere during the spring?, *J. Geophys. Res.*, *106*, 19,999–20,010.

Norton, W. A. (1994), Breaking Rossby waves in a model stratosphere diagnosed by a vortex-following coordinate system and a contour advection technique, *J. Atmos. Sci.*, *51*, 654–673.

Norton, W. A. (2003), Sensitivity of Northern Hemisphere surface climate to simulation of the stratospheric polar vortex, *Geophys. Res. Lett.*, *30*(12), 1627, doi:10.1029/2003GL016958.

Oman, L., D. W. Waugh, S. Pawson, R. S. Stolarski, and P. A. Newman (2009), On the influence of anthropogenic forcings on changes in the stratospheric mean age, *J. Geophys. Res.*, *114*, D03105, doi:10.1029/2008JD010378.

O'Neill, A., and V. D. Pope (1988), Simulations of linear and nonlinear disturbances in the stratosphere, *Q. J. R. Meteorol. Soc.*, *114*, 1063–1110.

Perlwitz, J., and N. Harnik (2004), Downward coupling between the stratosphere and troposphere: The relative roles of wave and zonal mean processes, *J. Clim.*, *17*, 4902–4909.

Perlwitz, J., S. Pawson, R. L. Fogt, J. E. Nielsen, and W. D. Neff (2008), Impact of stratospheric ozone hole recovery on Antarctic climate, *Geophys. Res. Lett.*, *35*, L08714, doi:10.1029/2008GL033317.

Plumb, R. A. (1981), Instability of the distorted polar night vortex: A theory of stratospheric warmings, *J. Atmos. Sci.*, *38*, 2514–2531.

Plumb, R. A. (2002), Stratospheric transport, *J. Meteorol. Soc. Jpn.*, *80*, 793–809.

Plumb, R. A. (2007), Tracer interrelationships in the stratosphere, *Rev. Geophys.*, *45*, RG4005, doi:10.1029/2005RG000179.

Plumb, R. A., D. W. Waugh, R. J. Atkinson, P. A. Newman, L. R. Lait, M. R. Schoeberl, E. V. Browell, A. J. Simmons, and M. Loewenstein (1994), Intrusions into the lower stratospheric Arctic vortex during the winter of 1991–1992, *J. Geophys. Res.*, *99*, 1089–1105.

Polvani, L. M., and P. J. Kushner (2002), Tropospheric response to stratospheric perturbations in a relatively simple general circulation model, *Geophys. Res. Lett.*, *29*(7), 1114, doi:10.1029/2001GL014284.

Polvani, L. M., and R. A. Plumb (1992), Rossby wave breaking, filamentation and secondary vortex formation: The dynamics of a perturbed vortex, *J. Atmos. Sci.*, *49*, 462–476.

Polvani, L. M., and R. Saravanan (2000), The three-dimensional structure of breaking Rossby waves in the polar wintertime stratosphere, *J. Atmos. Sci.*, *57*, 3663–3685.

Polvani, L. M., and D. W. Waugh (2004), Upward wave activity flux as precursor to extreme stratospheric events and subsequent anomalous surface weather regimes, *J. Clim.*, *17*, 3548–3554.

Polvani, L. M., D. W. Waugh, and R. A. Plumb (1995), On the subtropical edge of the stratospheric surf zone, *J. Atmos. Sci.*, *52*, 1288–1309.

Randel, W., P. A. Newman (1998), The stratosphere in the Southern Hemisphere, in *Meteorology of the Southern Hemisphere*, edited by D. J. Karoly and D. G. Vincent, *Meteor. Monogr.*, *27*, 243–282.

Rex, M., et al. (2006), Arctic winter 2005: Implications for stratospheric ozone loss and climate change, *Geophys. Res. Lett.*, *33*, L23808, doi:10.1029/2006GL026731.

Rong, R. R., and D. W. Waugh (2003), Vacillations in a shallow water model of the stratosphere, *J. Atmos. Sci.*, *61*, 1174–1185.

Salby, M. L., R. R. Garcia, D. O'Sullivan, and J. Tribbia (1990), Global transport calculations with an equivalent barotropic system, *J. Atmos. Sci.*, *47*, 188–214.

Schoeberl, M. R., and A. R. Douglass (2010), Trace gas transport in the stratosphere: Diagnostic tools and techniques, in *The Stratosphere: Dynamics, Transport, and Chemistry*, *Geophys. Monogr. Ser.*, doi:10.1029/2009GM000855, this volume.

Schoeberl, M. R., and D. L. Hartmann (1991), The dynamics of the stratospheric polar vortex and its relation to springtime ozone depletions, *Science*, *251*, 46–52.

Scott, R. K., and D. G. Dritschel (2006), Vortex-vortex interactions in the winter stratosphere, *J. Atmos. Sci.*, *63*, 726–740.

Scott, R. K., and P. H. Haynes (2000), Internal vacillations in stratospheric-only models, *J. Atmos. Sci.*, *57*, 3233–3250.

Scott, R. K., and L. M. Polvani (2004), Stratospheric control of upward wave flux near the tropopause, *Geophys. Res. Lett.*, *31*, L02115, doi:10.1029/2003GL017965.

Scott, R. K., D. G. Dritschel, L. M. Polvani, and D. W. Waugh (2004), Enhancement of Rossby wave breaking by steep potential vorticity gradients in the winter stratosphere, *J. Atmos. Sci.*, *61*, 904–918.

Shepherd, T. G. (2007), Transport in the middle atmosphere, *J. Meteorol. Soc. Jpn.*, *85B*, 165–191.

Shepherd, T. G. (2008), Dynamics, stratospheric ozone, and climate change, *Atmos. Ocean*, *46*, 117–138.

Shindell, D. T., D. Rind, and P. Lonergan (1998), Increased polar stratospheric ozone losses and delayed eventual recovery owing to increasing greenhouse-gas concentrations, *Nature*, *392*, 589–592, doi:10.1038/33385.

Solomon, S. (1999), Stratospheric ozone depletion: A review of concepts and history, *Rev. Geophys.*, *37*, 275–316.

Son, S.-W., L. M. Polvani, D. W. Waugh, H. Akiyoshi, R. Garcia, D. Kinnison, S. Pawson, E. Rozanov, T. G. Shepherd, and K. Shibata (2008), The impact of stratospheric ozone recovery on the Southern Hemisphere westerly jet, *Science*, *320*, 1486–1489.

Son, S.-W., N. F. Tandon, L. M. Polvani, and D. W. Waugh (2009), Ozone hole and Southern Hemisphere climate change, *Geophys. Res. Lett.*, *36*, L15705, doi:10.1029/2009GL038671.

Song, Y., and W. A. Robinson (2004), Dynamical mechanisms of stratospheric influences on the troposphere, *J. Atmos. Sci.*, *61*, 1711–1725.

Thompson, D. W. J., and S. Solomon (2002), Interpretation of recent Southern Hemisphere climate change, *Science*, *296*, 895–899.

Thompson, D. W. J., and J. M. Wallace (2000), Annular modes in the extratropical circulation. Part I: Month-to-month variability, *J. Clim.*, *13*, 1000–1016.

Thompson, D. W. J., M. P. Baldwin, and J. M. Wallace (2002), Stratospheric connection to Northern Hemisphere wintertime weather: Implications for predictions, *J. Clim.*, *15*, 1421–1428.

Tung, K. K., and R. S. Lindzen (1979), A theory of stationary long waves. Part I: A simple theory of blocking, *Mon. Weather Rev.*, *107*, 714–734.

Waugh, D. W. (1997), Elliptical diagnostics of stratospheric polar vortices, *Q. J. R. Meteorol. Soc.*, *123*, 1725–1748.

Waugh, D. W., and D. G. Dritschel (1999), The dependence of Rossby wave breaking on the vertical structure of the polar vortex, *J. Atmos. Sci.*, *56*, 2359–2375.

Waugh, D. W., and W. J. Randel (1999), Climatology of Arctic and Antarctic polar vortices using elliptical diagnostics, *J. Atmos. Sci.*, *56*, 1594–1613.

Waugh, D. W., and P. P. Rong (2002), Interannual variability in the decay of lower stratospheric Arctic vortices, *J. Meteorol. Soc. Jpn.*, *80*, 997–1012.

Waugh, D. W., et al. (1994), Transport out of the stratospheric Arctic vortex by Rossby wave breaking, *J. Geophys. Res.*, *99*, 1071–1088.

Waugh, D. W., W. J. Randel, S. Pawson, P. A. Newman, and E. R. Nash (1999), Persistence of the lower stratospheric polar vortices, *J. Geophys. Res.*, *104*, 27,191–27,201.

Waugh, D. W., L. Oman, P. A. Newman, R. S. Stolarski, S. Pawson, J. E. Nielsen, and J. Perlwitz (2009), Effect of zonal asymmetries in stratospheric ozone on simulated Southern Hemisphere climate trends, *Geophys. Res. Lett.*, *36*, L18701, doi:10.1029/2009GL040419.

Yoden, S. (1987), Dynamical aspects of stratospheric vacillations in a highly truncated model, *J. Atmos. Sci.*, *44*, 3683–3695.

Yoden, S., M. Taguchi, and Y. Naito (2002), Numerical studies on time variations of the troposphere-stratosphere coupled system, *J. Meteorol. Soc. Jpn.*, *80*, 811–830.

Zhou, S., M. E. Gelman, A. J. Miller, and J. P. McCormack (2000), An inter-hemisphere comparison of the persistent stratospheric polar vortex, *Geophys Res. Lett.*, *27*, 1123–1126.

L. M. Polvani, Department of Applied Physics and Applied Mathematics, Columbia University, New York, NY 10027, USA.

D. W. Waugh, Department of Earth and Planetary Sciences, Johns Hopkins University, Baltimore, MD 21218, USA. (waugh@jhu.edu)

Annular Modes of the Troposphere and Stratosphere

Paul J. Kushner

Department of Physics, University of Toronto, Toronto, Ontario, Canada

A pedagogical review of the phenomenology, dynamics, and simulation of the annular modes is presented. First, the manner in which the annular modes capture the principal features of variability in the zonal mean circulation of the extratropical stratosphere and troposphere is demonstrated. Next, elementary ideas about annular mode dynamics, in particular, the distinct eddy mean-flow interactions that govern the modes in the troposphere and stratosphere, are introduced. The review concludes with a discussion of the representation of the annular modes in simplified and comprehensive general circulation models in the context of stratosphere-troposphere interactions in intraseasonal variability and climate change.

1. INTRODUCTION

For the theory of the atmospheric general circulation, it is important to not only account for the circulation's principal climatological features, such as the jet streams of the troposphere and the polar night jet of the stratosphere, but also to develop a dynamical description of how the circulation varies about its climatological mean. In this chapter, we present a pedagogical review of the annular modes (hereafter AMs) [*Thompson and Wallace*, 2000; *Thompson et al.*, 2000; *Limpasuvan and Hartmann*, 1999, 2000], which are the principal modes of variation of the extratropical circulation of the troposphere and stratosphere on time scales greater than a few weeks. Because much of the observed extratropical variability is manifested in the AMs, explaining their dynamics represents an important theoretical task. Beyond this, the AMs are of applied importance because of their impacts on many climate processes and because of their relationship to climate change signals [e.g., *Shindell et al.*, 1999; *Thompson et al.*, 2000; *Thompson and Solomon*, 2002; *Kushner et al.*,

2001; *Russell and Wallace*, 2004; *Miller et al.*, 2006; *Butler et al.*, 2007].

In this chapter, we will discuss the observed AMs (section 2), which consist of the Southern Annular Mode (SAM) in the Southern Hemisphere and the Northern Annular Mode (NAM) in the Northern Hemisphere; present some elementary ideas on their dynamics and their representation in highly simplified models (section 3); and explore their central role in extratropical climate variability and change, with an emphasis on their behavior in more complex models (section 4). We will adopt the viewpoint that the AMs represent coherent variations of the zonal mean circulation that interact with synoptic-to-planetary scale atmospheric waves and the zonal stresses they induce [e.g., *Limpasuvan and Hartmann*, 1999, 2000]. We will find that, despite their importance, the AMs remain enigmatic because they are derived empirically and because there is no simple first-principles theory describing their structure and dynamics in the stratosphere and troposphere. Our current theoretical understanding is instead represented in simplified nonlinear models in which AM structures emerge as the leading modes of variability [e.g., *Vallis et al.*, 2004; *Holton and Mass*, 1976; *Scott and Polvani*, 2006]. and in detailed dynamical analyses of their temporal variability [e.g., *Robinson*, 2000; *Lorenz and Hartmann*, 2001, 2003]. In the absence of a predictive theory, and in light of the many ways in which AM responses can be stimulated by climate forcings, we need to rely on comprehensive climate

The Stratosphere: Dynamics, Transport, and Chemistry
Geophysical Monograph Series 190
Copyright 2010 by the American Geophysical Union
10.1029/2009GM000924

models to quantitatively simulate the AMs and predict their future behavior under climate change.

Before proceeding, we highlight some of the important papers from the late 1990s and early 2000s that defined the AMs and recognized their significance. These papers represent a starting point for further exploration for the interested student.

1. The original description of the AMs as the principal modes of variability of the extratropical circulation, for the stratosphere and the troposphere of both the Northern Hemisphere and Southern Hemisphere, their relationship to the North Atlantic Oscillation, and their impacts on weather variability is primarily due to Thompson and Wallace [see *Thompson and Wallace*, 1998, 2000; *Wallace*, 2000; *Thompson et al.*, 2000; *Thompson and Wallace*, 2001; among other studies]. The principal original work describing the eddy mean-flow interactions that maintain the AMs is that of *Limpasuvan and Hartmann* [1999, 2000] and *Lorenz and Hartmann* [2001, 2003]. This set of studies revitalized a line of work on variability of the zonal mean circulation that dates back several decades. (The historical literature is well referenced in the cited papers.) Regarding terminology, the term "annular modes" was first used by *Thompson and Wallace* [2000] (which was received by *Journal of Climate* in 1998) and *Limpasuvan and Hartmann* [1999] (which was received by *Geophysical Research Letters* in 1999). The NAM and SAM were originally named the Arctic Oscillation and the Antarctic Oscillation; these terms are still used in some papers. The relationship between the AMs and the classical zonal index, which is a dipolar mode of variability that corresponds to the tropospheric AM in zonal wind, is detailed by *Wallace* [2000].

2. The observational description of the time-dependent coupling between stratospheric and tropospheric variability through the daily evolution of the AMs and its implications for prediction of climate variability on seasonal time scales is primarily due to Baldwin, Dunkerton and collaborators [e.g., *Baldwin and Dunkerton*, 1999, 2001; *Baldwin et al.*, 2003]. These papers followed a long line of inquiry that identified influences that the state of the stratosphere might have on the tropospheric circulation, involving both zonal mean and regional circulation features [e.g., *Boville*, 1984; *Kodera et al.*, 1990; *Perlwitz and Graf*, 1995].

3. The connection between the annular modes and climate change in the extratropics and to Antarctic stratospheric ozone depletion is due to *Thompson and Wallace* [1998], *Shindel et al.* [1999], *Fyfe et al.* [1999], *Thompson et al.* [2000], *Kushner et al.* [2001], *Sexton* [2001], *Thompson and Solomon* [2002], *Polvani and Kushner* [2002], *Gillett and Thompson* [2003], among others.

The discovery of the AMs has spurred a large literature that extends well outside the atmospheric and physical climate sciences. We do not provide here a comprehensive review of this literature, but aim to highlight a select set of points intended to introduce the AMs to researchers new to the field.

2. OBSERVED AM STRUCTURE

The AMs are atmospheric "teleconnection patterns," that is, continental-to-planetary scale patterns involving statistical coherence between spatially remote points that are derived using a variety of methods (as recently reviewed by *Baldwin and Thompson* [2009] and *Baldwin et al.* [2009]). Typically, the AMs are derived using empirical orthogonal function (EOF) analysis of selected extratropical meteorological fields. EOF analysis represents a straightforward multivariate statistical approach, but one that risks artificially conflating spatially and dynamically distinct phenomena. In this section, we will show that the AMs indeed represent physically meaningful and statistically robust patterns that connect the zonal mean variability of the troposphere and stratosphere.

Using notation similar to that of *Baldwin et al.* [2009], EOF analysis decomposes an $n \times p$ data matrix $\mathbf{X}$ consisting of n temporal observations at p points into a series of $n \times p$ matrices according to

$$\mathbf{X} = \sum_{i=1}^{r} \mathbf{y}_i \mathbf{e}_i^T = \mathbf{y}_1 \mathbf{e}_1^T + \mathbf{y}_2 \mathbf{e}_2^T + \ldots + \mathbf{y}_r \mathbf{e}_r^T, \qquad (1)$$

where r is the rank of $\mathbf{X}$, which is never greater than the minimum of n and p, $\mathbf{y}_i$ is a column vector representing the ith n-point "principal component" (PC) time series, and $\mathbf{e}_i$ is a column vector representing the ith p-point spatial EOF. The EOFs are the eigenvectors of the $p \times p$ sample covariance matrix $\mathbf{S} = \mathbf{X}^T\mathbf{X}$, and are ranked in descending order of the size of the positive quantity $\mathbf{e}_i^T\mathbf{S}\mathbf{e}_i$; when divided by the total variance in the field, this quantity is known as the "variance explained" or "represented" by EOF i. The r PC time series $\mathbf{y}_i$ are mutually orthogonal, as are the r EOFs $\mathbf{e}_i$. A spatial pattern associated with another field $\tilde{\mathbf{X}}$ that is coherent with the jth EOF is found by regression according to

$$\tilde{\mathbf{X}}_j = \frac{\tilde{\mathbf{X}}^T \mathbf{y}_j}{\mathbf{y}_j^T \mathbf{y}_j}, \qquad (2)$$

where $\tilde{\mathbf{X}}$ is an $n \times \tilde{p}$ data matrix consisting of $\tilde{p}$ n-point time series. If we set $\tilde{\mathbf{X}} = \mathbf{X}$, which is the original data matrix, then the spatial pattern that results from equation (2) is the jth EOF: $\tilde{\mathbf{x}}_j = \mathbf{e}_j$. The effects of spatial weighting are straightforward to bring into this formalism [*Baldwin et al.*, 2009].

In this notation, the AM EOF corresponds to $\mathbf{e}_1$ for an appropriately chosen $\mathbf{X}$, and the AM index corresponds to the PC time series $\mathbf{y}_1$. To define the AMs, one must choose the

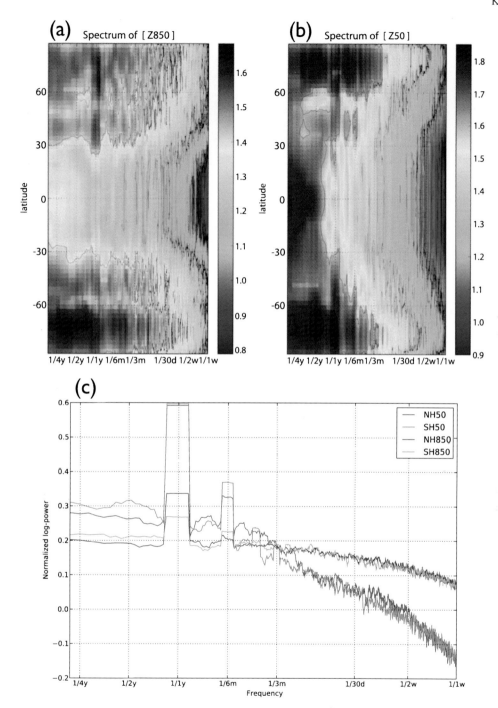

Plate 1. (a) Log of the power spectrum of the zonal mean geopotential anomaly $\tilde{Z}$ from equation (3) at $p = 850$ hPa. Units are arbitrary. The data used is National Centers for Environmental Prediction (NCEP)/National Center for Atmospheric Research reanalysis [*Kalnay et al.*, 1996] daily averaged geopotential from 1979 to 2008. A simple five-point running mean across frequencies has been applied to create smoother plots. (b) As in Plate 1a, for $p = 50$ hPa. (c) Meridional mean of the log-spectra in Plates 1a and 1b poleward of 22° latitude, scaled to have equal area under each curve.

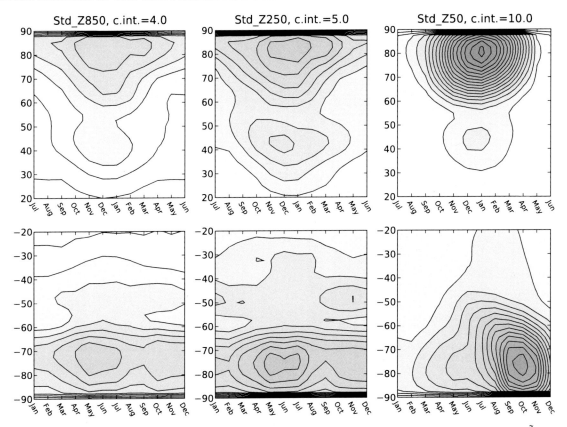

Plate 2. Seasonal cycle of the standard deviation of the seasonal time series of the zonal mean geopotential anomaly $\tilde{Z}_{\mathrm{mon}}$ at 850 hPa (left column) (contour interval 4 m), 250 hPa (middle column) (contour interval 5 m), and 50 hPa (right hand column) (contour interval 10 m), for the Northern and Southern Hemisphere extratropics. Calendar months are arranged so that winter solstice is at the center of each plot.

meteorological field **X** and the way to sample that field in space and time. A review of various approaches [e.g., *Gong and Wang*, 1999; *Thompson and Wallace*, 2000; *Wallace*, 2000; *Baldwin*, 2001; *Lorenz and Hartmann*, 2001; *Monahan and Fyfe*, 2006; *Kushner and Lee*, 2007], not all of which use EOF analysis as the basis for the AM index [e.g., *Gong and Wang*, 1999], shows that the different methods yield similar results in the sense that the resulting AM indexes are strongly temporally correlated. But this similarity can mask systematic underlying differences: for example, *Monahan and Fyfe* [2006] show that the meridional structure of the AM defined using geopotential height differs from that using geostrophic wind because the loading pattern of variance for a field differs from the loading pattern for the field's spatial gradient. Focusing on the vertical structure, *Baldwin and Thompson* [2009] show that the choice of input field affects the manner in which stratosphere-troposphere coupling (see below) is manifested on short time scales.

To help understand the nature of the variance represented by the AMs, we examine the spatial and temporal structure of

extratropical zonal mean variability to be captured by the EOF analysis. We start with the zonal mean geopotential $\bar{Z} = \bar{Z}(\theta, p, t)$, where the overbar indicates a zonal mean and (θ, p, t) are latitude, pressure, and time. We then remove its climatological mean; call this quantity Z^{a}, where the superscript "a" indicates an anomaly from the climatology. We then define

$$\tilde{Z}(\theta, p, t) = \bar{Z}^{a}(\theta, p, t)\sqrt{\cos\theta}, \qquad (3)$$

which, after some further processing, will provide the input to the EOF calculation. The geometric factor $\sqrt{\cos\theta}$ relates to the cosine-latitude weighting within the inner product for the EOF calculation [*Baldwin et al.*, 2009]; we will, for this analysis, retain it in the definition of $\tilde{Z}$ and now examine $\tilde{Z}$ as a measure of variability in the zonal mean circulation.

We first show the periodogram (i.e., the discrete Fourier temporal spectrum) of $\tilde{Z}$ as a function of latitude from South Pole to North Pole and of time scale from weeks to years for the lower troposphere in Plate 1a and for the lower

Regressions for Z850EOF1

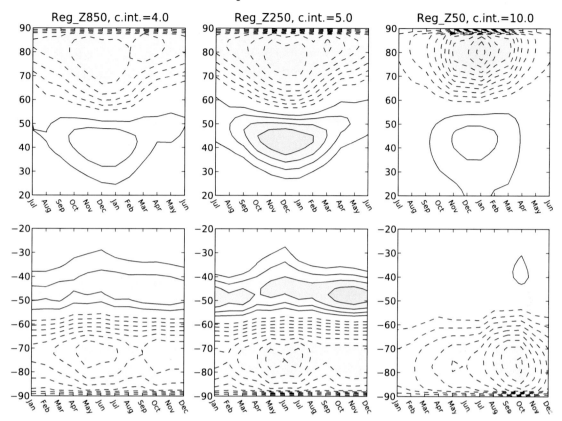

Plate 3. Regressions on the (top row) 850 hPa NAM and the (bottom row) 850 hPa SAM of $\tilde{\mathbf{X}} = \tilde{Z}_{\text{mon}}$ (left column) at 850 hPa, $\tilde{\mathbf{X}} = \tilde{Z}_{\text{mon}}$ (middle column) at 250 hPa, and $\tilde{\mathbf{X}} = \tilde{Z}_{\text{mon}}$ at (right column) 50 hPa. Contour intervals and format as in Plate 2.

stratosphere in Plate 1b. We also show the extratropical meridional mean of these spectra in Plate 1c. While the variance of $\tilde{Z}$ from equation (3), by construction, goes to zero at the poles, the variance of $\bar{Z}^a$ amplifies to a maximum at the poles (not shown). Because for time scales shorter than annual geopotential variance is generally stronger in the extratropics than the tropics (as expected from geostrophic balance), the definition of the AMs is relatively insensitive to the equatorial extent of the domain of analysis [see, e.g., *Baldwin and Thompson*, 2009]. In the troposphere, there is a broad band of variance poleward of about 30° latitude and a suggestion of a minimum in variance around 60° latitude in each hemisphere. Plate 1c shows that in the troposphere, the variance has, for the most part, saturated on time scales longer than the intraseasonal (multiple-week) time scale. In the stratosphere, the variance is relatively strong at higher latitudes and shifted to longer time scales; the variance continues to build up beyond intraseasonal time scales. The structure in latitude and frequency is similar in both hemispheres, suggesting that the dynamics of the AMs in each

hemisphere might be dynamically similar. This hemispheric symmetry is less evident for zonal wind than for $\tilde{Z}$ (not shown).

We now examine the seasonal dependence of the variability of the zonal mean circulation represented by $\tilde{Z}$. We calculate the monthly mean $\tilde{Z}$, remove that part of the result that is linearly coherent with the Monthly El Niño–Southern Oscillation (ENSO) Index (MEI defined by *Wolter and Timlin* [1993]) [see *Lorenz and Hartmann*, 2003], linearly detrend this result, and from this last result, create seasonal time series for three sequential months each year. (For example, we create a time series of the months January-February-March for each year from 1979 to 2008, representing a three-month "JFM" season, yielding a 90-record time series.) We denote by $\tilde{Z}_{\text{mon}}$ the result of this sequence of operations on $\tilde{Z}$. This captures seasonal and longer time scale variability and eliminates submonthly variability, for which, from Plate 1, the variability can be seen to be far from saturation.

We show the seasonal cycle of the standard deviation of monthly mean $\tilde{Z}_{\text{mon}}$ in the extratropics, as a function of

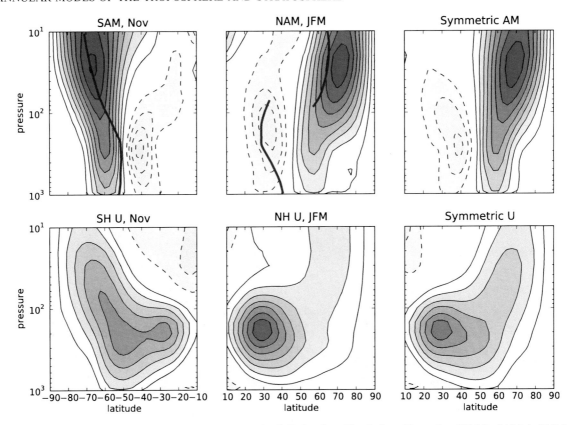

Plate 4. (top row) AM zonal wind regression patterns for (left) Southern Hemisphere November 850 hPa SAM, (middle) Northern Hemisphere January-February-March (JFM) 850 hPa NAM, and (right) the hemispherically symmetric component of these two patterns. (bottom row) Climatological mean zonal wind for (left) Southern Hemisphere November, (middle) Northern Hemisphere JFM, and (right) the hemispherically symmetric component of these two patterns. The red curves in the top left plots indicate the approximate location of the jet axis in each hemisphere. Contour interval in top row is 0.5 m s^{-1} and in bottom row is 5 m s^{-1}.

center month (February in the previous example) in Plate 2, for lower tropospheric, upper tropospheric, and lower stratospheric levels. The ENSO filtering removes variance at low latitudes (not shown). In the troposphere of both hemispheres, $\tilde{Z}_{\mathrm{mon}}$ has a strong wintertime maximum in the higher extratropics and a weaker maximum in the lower extratropics. These features are vertically aligned, suggesting that the underlying variability might be equivalent barotropic in the troposphere. The seasonal cycle is more pronounced in the Northern Hemisphere troposphere than in the Southern Hemisphere troposphere. The maximum in variance in the Southern Hemisphere troposphere arrives in late fall, earlier in the season than the corresponding maximum in the Northern Hemisphere troposphere. In the Northern Hemisphere stratosphere, the maximum variance is found in the winter, but in the Southern Hemisphere stratosphere, the maximum variance is delayed toward the spring. The period of maximum variance in the stratosphere represents the "active season" of the stratospheric polar vortex in each hemisphere, when the

polar stratospheric winds are eastward and display considerable variability; this season is delayed toward spring in the Southern Hemisphere [*Thompson and Wallace*, 2000; *Hartmann et al.*, 2000].

Plates 1 and 2 represent the variability that the AMs, as leading modes, should capture. To calculate the AMs, we set $\mathbf{X} = \tilde{Z}_{\mathrm{mon}}(|\theta| > 20^0, p = 850 \text{ hPa}, t)$ in equation (1), i.e., we use extratropical lower tropospheric $\tilde{Z}_{\mathrm{mon}}$ as input to the EOF analysis [*Thompson and Wallace*, 2000]. The NAM and SAM are the leading EOFs in the Southern and Northern hemispheres. We repeat this calculation for each calendar month and for each hemisphere. Plate 3 plots the regression of the AMs on $\tilde{Z}_{\mathrm{mon}}$, using equation (2), for the same vertical levels as in Plate 2. These are the patterns of variability coherent with the lower-troposphere-based AM in its positive phase at each level, for each calendar month and for each hemisphere.

Comparing Plates 2 and 3, which have equivalent contour intervals in each plot, we see that the leading EOF captures the principal features of the variance and its seasonal cycle throughout the troposphere and stratosphere, even though the AM definition involves only information from the lower troposphere. The tropospheric AM EOF consists of a dipolar structure in $\tilde{Z}_{\text{mon}}$ that reflects the high-extratropical and low-extratropical maxima in variance in the two hemispheres and the broad band of variance seen in Plate 1a. During the active season when the variance is large in the lower stratosphere, the AM anomaly pattern has a consistent sign between the stratosphere and the troposphere poleward of 60° latitude. This vertical coherence indicates that the stratosphere and troposphere are coupled in the active season.

Despite the evidence of coupling between the stratosphere and troposphere, Plate 3 also suggests that the main features of the tropospheric AM structure appear to be independent of the stratosphere. In particular, the dipolar tropospheric AM in Plate 3 persists throughout most of the year, especially in the Southern Hemisphere. The stratospheric AM, on the other hand, is strongly amplified during the active stratospheric season and is strongly attenuated in the inactive season when the stratospheric winds are easterly. This implies, as will be discussed in section 3, that, to some extent, the dynamics of the stratospheric and tropospheric AM are distinct and can be considered separately.

During the active season, when the variability of the stratosphere and troposphere is vertically coherent, the NAM and SAM are remarkably hemispherically symmetric [*Thompson and Wallace*, 2000]. Plate 4 shows the 850-hPa SAM and NAM signatures in the zonal wind (that is, $\tilde{\mathbf{X}}$ corresponds to $\bar{u}$ in equation (2), where $\bar{u}$ is the zonal mean zonal wind). In the troposphere, the AM in the wind in each hemisphere is strongly dipolar, while in the stratosphere, the AM is more monopolar. Although the node of the SAM is located a few degrees poleward of the node of the NAM, the patterns essentially represent a reflection of each other across the equator. To quantify the hemispheric symmetry, we also show the associated hemispherically symmetric component of the patterns, $\bar{u}_{\text{sym}}$, which is obtained by first reflecting SAM pattern $\bar{u}_{\text{SAM}}$ and then averaging it with the NAM pattern $\bar{u}_{\text{NAM}}$:

$$\bar{u}_{\text{sym}}(y) = \frac{1}{2}[\bar{u}_{\text{SAM}}(-y) + \bar{u}_{\text{NAM}}(y)],$$

where y is the Cartesian meridional coordinate that we will use in the rest of the chapter. The symmetric component explains 96% of the variance in the AMs (using a log-pressure weighting poleward of 10° latitude), and the symmetric pattern looks quite similar to both the SAM and the NAM. Applying such a construction to the zonal mean wind in each hemisphere (bottom row of Plate 4) does not give the same qualitative sense of hemispheric symmetry: the Northern Hemisphere jet is dominated by the subtropical jet, while the Southern Hemisphere jet structure is more split between its subtropical

and extratropical parts; the Northern Hemisphere surface jet is not as sharp, and the stratospheric jet is relatively weak. (The resulting symmetric component (bottom right plot of Plate 4) explains 88% of the variance of the total wind.)

The fact that the AMs are more hemispherically symmetric than the background circulation (as is also found for the tropospheric AMs outside the active season, not shown) suggests that the AMs are governed by dynamics that does not depend highly on the meridional structure of the background flow. It also means that the spatial relationship between the jets and the AMs is distinct in each hemisphere. In the Southern Hemisphere, the tropospheric jet maximum at each altitude (red solid line in the upper left plot of Plate 4) is located along the node of the SAM; the SAM in its positive phase, thus, describes a poleward shift of the jet. Given a shift δy of the jet, we expect a wind anomaly that is related to the vorticity of the jet according to

$$\bar{u}_{\text{SAM}} \sim \bar{u}(y - \delta y) - \bar{u}(y) \approx -\delta y\, \bar{u}_y = \delta y\, \bar{\zeta},$$

where $\bar{u}(y)$ represents the tropospheric jet profile as a function of meridional coordinate y, where coordinate subscripts indicate partial derivatives, and where $\bar{\zeta} = -\bar{u}_y$ is the zonal mean vorticity of the jet. This relationship suggests that spatial scale of the tropospheric SAM is linked to the spatial scale of the barotropic shear, i.e., of the vorticity of the jet, and the amplitude of the SAM to the degree of displacement of the jet. Taking extratropical tropospheric vorticity values of $\bar{\zeta} \sim 10^{-5}\text{s}^{-1}$ and $\bar{u}_{\text{SAM}} \sim 1\text{–}2\text{ m s}^{-1}$ for the SAM in the troposphere suggests that typical seasonal-to-interannual SAM fluctuations are related to seasonal-to-interannual variations in the position of the jet of about $\delta y \sim \bar{\zeta}/\bar{u}_{\text{SAM}} \sim$ 100–200 km; climate trends discussed in section 4 correspond to tropospheric jet shifts of this magnitude [*Kushner et al.*, 2001]. The relationship between the jet position and the SAM node holds at each longitude, when the regional structure of the SAM is analyzed [*Codron*, 2005]. In the stratosphere, the jet axis is aligned with the maximum of the SAM. Thus, in the Southern Hemisphere active season, the SAM in its positive phase is described as a poleward shift of the tropospheric jet and a strengthening of the stratospheric polar vortex.

This empirical description for the Southern Hemisphere does not generalize simply to the Northern Hemisphere. In the Northern Hemisphere, the relationship between the stratospheric vortex and the NAM is similar to that for the Southern Hemisphere (see red lines in Plate 4b). But in the troposphere, positive NAM conditions consist of a reduction in strength of the subtropical jet and an intensification on the poleward side of the jet, rather than a simple poleward shift of the jet. Thus, we do not find a simple connection between the barotropic shear of the zonal mean wind and the tropospheric NAM structure. The relationship between the Northern Hemisphere jet stream and the NAM varies regionally: in the North Atlantic sector, the NAM corresponds to a poleward shifted North Atlantic jet, whereas in the North Pacific sector, the NAM corresponds to a weaker subtropical jet [*Ambaum et al.*, 2001; *Eichelberger and Hartmann*, 2007]. We will return to this point in section 3.

While our observational analysis has so far focused on monthly and longer time scales, the dynamics of zonal mean extratropical variability, and of the AMs in particular, is controlled to a large degree on shorter time scales. A daily index of the AM is found by projecting the AM EOF e_1 (which is typically calculated, as we have, from seasonal mean or low pass filtered data) onto the daily circulation using $y_{1,daily} = Z_{daily}e_1/e_1{}^T e_1$ where Z_{daily} represents the daily time series of the geopotential anomaly at the p spatial points of the EOF. When AM variability is analyzed on daily-to-multiple week time scales, the vertically coherent static patterns in Plates 3 and 4 resolve into vertically propagating signals in which tropospheric AM anomalies lag stratospheric AM anomalies by periods of several weeks [*Baldwin and Dunkerton*, 1999, 2001; *Baldwin and Thompson*, 2009]. This relationship is diagnosed by calculating the AM and its principal component time series at each vertical level and then using lag-correlation or compositing techniques to determine the vertical variation of the time lags. As an example, we show in the top plot of Plate 5 a *Baldwin and Dunkerton* [2001] type composite over low NAM index events (weak or warm stratospheric vortex events). The figure shows a vertically coherent NAM signal that propagates down from the stratosphere into the troposphere and leads to like-signed NAM anomalies in the troposphere that persist for 40 days or more. These downward propagating signals are typically associated with extreme events in the stratosphere, such as stratospheric sudden warmings and vortex intensification events [*Limpasuvan et al.*, 2004, 2005]. For example, during stratospheric sudden warmings, a weak vortex anomaly (negative NAM event) is found to burrow down into the stratosphere; the NAM signal then makes its way into the troposphere, where it is manifested as an equatorward shift of the storm tracks and an equatorward intensification of the tropospheric jet [*Baldwin and Dunkerton*, 2001]. The bottom plot of Plate 5 shows a similar calculation to the top plot, but using the geopotential averaged over the 60°N–90°N polar cap. The similarity between the two plots suggests that we can use the polar cap geopotential as a simple proxy for the AM; we will do so in section 3.

The multiple week lag between stratospheric and tropospheric AM signals suggests that the stratosphere can provide a source of predictability that extends beyond the classical limit for tropospheric weather predictability [*Baldwin et al.*, 2003]. *Baldwin et al.* [2003] argue that the tropospheric AM is observed to be more persistent in winter and that this persistence results from stratosphere-troposphere coupling. A version of their analysis is presented in Plate 6, from *Gerber et al.* [2008a], which plots in the top row the seasonal cycle of the autocorrelation decay time scale of NAM and SAM variability from the National Centers for Environmental Prediction reanalysis. The characteristic time scales of the AM, consistently with Plate 1, is on the order of 1–2 weeks in the troposphere and several weeks in the stratosphere, with a strong peak during the active season for the SAM and NAM, suggesting a stratospheric connection for tropospheric persistence of these modes [*Baldwin et al.*, 2003]. The bottom row of Plate 6 shows the same diagnostic for the set of climate simulations in the Intergovernmental Panel on Climate Change's Fourth Assessment Report, which we will comment on in sections 3 and 4.

We conclude this section by discussing the regional, i.e., the longitudinal, structure of the AMs in the troposphere, which has been the subject of considerable debate [e.g., *Deser*, 2000; *Ambaum et al.*, 2001; *Wallace and Thompson*, 2002; *Vallis et al.*, 2004; *Kushner and Lee*, 2007]. If we regress the NAM index on a longitudinally varying field such as the two-dimensional (2-D) sea-level pressure, we typically obtain a pattern consisting of a low-pressure anomaly over the pole and high-pressure anomalies over the North Atlantic and North Pacific (see Figure 1a, from *Deser* [2000], which shows the regression of the sea-level pressure on the NAM index from *Thompson and Wallace* [1998]). The regressions have a predominantly zonally symmetric or "annular" appearance, although the amplitude of the patterns is greater over the oceans than over land. Similar predominantly zonally symmetric structures are found in both hemispheres whether or not the input to the EOF calculation (**X** in equation (1)) is zonally averaged [*Thompson and Wallace*, 2000; *Baldwin and Thompson*, 2009]. Thus, the zonal structure of the circulation that generates the AM is not critical to the zonal structure of the circulation that is coherent with the AM.

The point of debate about the zonal structure of the AMs centers on the fact that the zonal structure depends strongly on the analysis domain: for example, when the domain of the EOF calculation is restricted to the North Atlantic basin, the Pacific center of action is no longer present in the regression, and only the North Atlantic Oscillation type structure remains (Figure 1b); similarly, when the domain is restricted to the North Pacific basin, only a North Pacific center of action remains (Figure 1c). While all three patterns share a similar meridional structure, there is no coherence between the basins [*Deser*, 2000]. Consistently, on daily time scales, it is found that high AM index days are associated with regional (less than 90° scale) dipolar circulation anomalies [*Cash et al.*, 2002; *Vallis et al.*, 2004; *Kushner and Lee*, 2007]. This is in

Composite of Warm Vortex Events

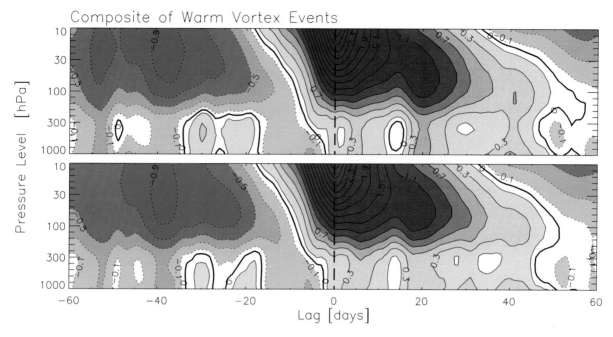

Plate 5. Composites of 50 warm vortex events based on NCEP reanalysis data from 1958 to 2007. For each winter season (November through March inclusive), the strongest geopotential height anomaly at 10 hPa was used. Contour intervals are spaced by 0.2. Index values between −0.1 and 0.1 are unshaded. The zero level is shown by the heavy black contour. The top plot shows a composite of nondimensional NAM index (based on geopotential height anomalies and using the zonal-mean EOF method described by *Baldwin and Thompson* [2009]). The bottom plot shows a composite for the time series of polar-cap averaged geopotential height (60°N–90°N), normalized by its standard deviation at each level. Figure courtesy of L. Mudryk.

contrast to what we found for the vertical structure: for example, the vertically coherent structures in Plate 3 involve only lower tropospheric information, and the vertically propagating stratosphere-troposphere signal in Plate 5 demonstrates a consistent statistical link between the zonal mean tropospheric and stratospheric circulation.

Kushner and Lee [2007] explore the connection between hemispheric AM variations and regional-scale variability. They find that day-to-day zonal-mean AM variability is consistent with the occurrence of zonally localized dipolar patterns that have a decoherence width of about 90° longitude and that propagate eastward with a phase speed of about

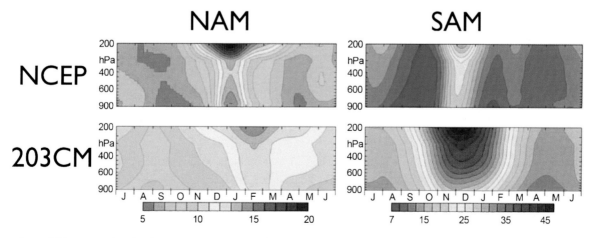

Plate 6. The seasonal cycle of NAM and SAM time scales, computed with the methods of *Baldwin et al.* [2003], for (top row) NCEP reanalysis and (bottom row) 20th century integrations of coupled climate models in the Fourth Assessment Report of the Intergovernmental Panel on Climate Change. Adapted from *Gerber et al.* [2008a].

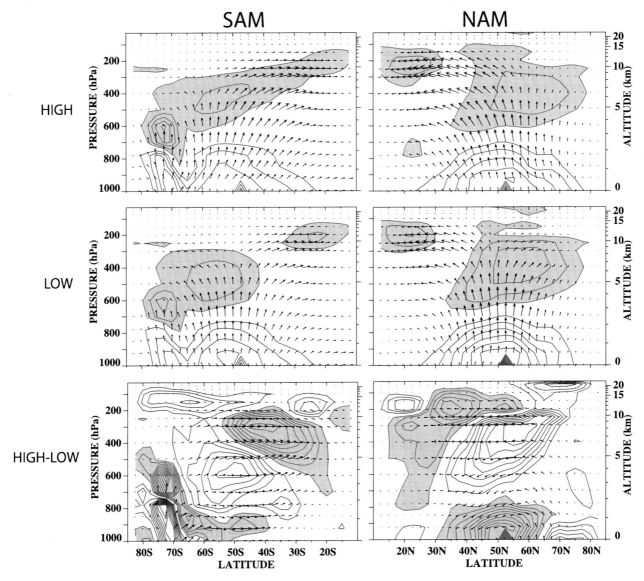

Plate 7. Composite Eliassen-Palm (EP) flux cross-sections for high and low AM conditions in each hemisphere and their difference. Contouring indicates regions of EP flux divergence/convergence, with shaded regions indicating EP flux convergence. Contour interval for the top two rows is 1.5 m s^{-1} d^{-1}, and for the bottom row is 0.25 m s^{-1} d^{-1}. Negative values (associated with westward acceleration) are shaded and the zero contour is omitted. EP flux vectors divided by the background density are shown; the longest vector is about 1.4 × 10^8 m^3 s^{-2}. Figure courtesy of V. Limpasuvan, adapted from *Limpasuvan and Hartmann* [2000]. Copyright American Meteorological Society.

8° longitude per day. In Figure 2, the regression pattern in surface circulation and the longitude-time propagation structure for these dipolar patterns in the Southern Hemisphere are plotted. Kushner and Lee argue that these patterns, which are similar in the Northern Hemisphere and Southern Hemisphere, are propagating versions of NAO type variability; outside the North Atlantic sector, they are not dominant but secondary patterns of variability. These regression

patterns have a westward phase tilt with height, suggesting that they are propagating and moderately baroclinic long waves. Kushner and Lee propose that these waves arise spontaneously and randomly, and their projection onto the zonal mean produces a zonal mean anomaly with the meridional structure of the zonal-mean AM. These ideas are not fully explored and have yet to be reconciled with the better established viewpoint that AM dynamics involves an interaction between synoptic eddies and the zonal mean flow. This viewpoint will be discussed in the next section.

3. ELEMENTARY IDEAS ON ANNULAR MODE DYNAMICS

If the AMs represent the dominant fluctuations of the extratropical zonal mean circulation, then "eddy mean flow interactions," in which the zonal flow controls the eddies and the eddies exert a return control on the zonal mean flow, are required to explain AM dynamics. To see why, consider the role of eddies in classical quasigeostrophic (QG) dynamics [e.g., *Andrews et al.*, 1987]. For the QG scaling that the extratropical circulation is observed to satisfy, the zonal mean circulation evolves according to (see the Appendix)

$$\bar{q}_t + (\overline{v'q'})_y = \bar{S}, \qquad (4)$$

where coordinate subscripts represent partial derivatives, overbars represent the zonal mean, primes represent eddy (zonally asymmetric) quantities, v is the meridional geostrophic velocity, q is the QG potential vorticity, and S is a forcing and dissipation operator that includes the effects of momentum dissipation and diabatic heating. In this notation, $\overline{v'q'}$ represents the meridional flux of eddy potential vorticity. Consider how temporal anomalies in the zonal mean potential vorticity, which we denote, $\bar{q}^a$, are related to temporal anomalies in the zonal mean potential vorticity flux, $(\overline{v'q'})^a$. Using the principal of inversion of potential vorticity [*Hoskins et al.*, 1985] from the zonal mean potential vorticity anomaly $\bar{q}^a$ and related surface information, we can diagnose the zonal mean geopotential anomaly and, hence, the zonal mean wind anomaly from geostrophic balance and the zonal mean temperature anomaly from hydrostatic balance. Thus, equation (4) represents the evolution of the entire zonal mean state.

If we remove the climatological mean from equation (4), multiply by $\bar{q}^a$, and take a density-weighted volume average (denoted by angle brackets), we obtain a variance equation for the zonal mean potential vorticity anomaly:

$$\frac{d}{dt}\left\langle \frac{1}{2}(\bar{q}^a)^2 \right\rangle = -\left\langle (\overline{v'q'})_y^a \cdot \bar{q}^a \right\rangle + \left\langle \bar{S}^a \cdot \bar{q}^a \right\rangle, \qquad (5)$$

where $(\overline{v'q'})_y^a$ and $\bar{S}^a$ represent the eddy potential vorticity flux divergence anomaly and the dissipation anomaly. In the Appendix, we argue that $\langle \bar{S}^a \cdot \bar{q}^a \rangle \leq 0$ to within a boundary term of indeterminate sign. If this boundary term can be neglected, this implies that $\bar{S}^a$ acts to attenuate the amplitude of zonal mean potential vorticity anomalies. This, in turn, would mean that extratropical dynamics cannot sustain fluctuations of the zonal mean unless, averaged over the spatial domain, the eddy potential vorticity flux convergence anomaly $-(\overline{v'q'})_y^a$ is positively correlated with the potential vorticity anomaly $\bar{q}^a$. This suggests that AM dynamics requires the eddy potential vorticity flux anomalies to amplify rather than attenuate local potential vorticity anomalies, over some part of the domain, in order to maintain AM variability. (A mathematically similar argument for the variance of the eddy potential vorticity leads to the classical result that the eddy potential vorticity flux must be down the meridional potential vorticity gradient, i.e., $\langle \overline{v'q'} \cdot \bar{q}_y \rangle < 0$, in order to maintain eddy variance. The sum of the eddy and zonal mean variance equations is related to conservation of total potential enstrophy.) We will argue that the means by which AM variability is maintained by eddy forcing is distinct in the stratosphere and troposphere.

In Plate 5, we showed that the AM is represented well by the geopotential averaged over the polar cap; *Baldwin and Thompson* [2009] discuss the strengths and weaknesses of using the polar cap as a proxy for the AM. For dynamical purposes, the polar cap is a useful control volume to work with. In the context of QG dynamics, following a similar

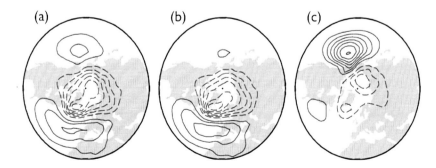

Figure 1. Regression of SLP on principal component time series from empirical orthogonal function (EOF) analysis of SLP in three domains: (a) the domain poleward of 20°N for the Northern Hemisphere, (b) as in Figure 1a, for the North Atlantic sector, and (c) as in Figure 1a, for the North Pacific sector. Adapted from *Deser* [2000].

approach to *Hu and Tung* [2002], we consider the integrated potential vorticity for a polar cap bounded at the top and bottom by $z = z_T$ and $z = z_B$ and bounded meridionally by the North Pole $y = y_N$ (90°N) and by $y = y_S$. This will lead to relationships between circulation, polar cap isentropic thickness, eddy fluxes, and dissipation. If we denote a density-weighted average over this polar cap by $<>_p$, we find that the integrated potential vorticity anomaly tendency satisfies (see Appendix)

$$< \overline{q_t^a} >_p = \int_{z_B}^{z_T} \mathrm{d}z(\rho_0 \overline{u_t^a})\Big|_{y=y_S} + \int_{y_S}^{y_N} \mathrm{d}y(f_0\rho_0 \overline{\theta_t^a}/\theta_{0z})\Big|_{z=z_B}^{z=z_T}. \quad (6)$$

Here, the $|_{y=y_S}$ notation represents the integral evaluated at $y = y_S$, and the $|_{z=z_B}^{z=z_T}$ notation represents the integral evaluated at $z = z_T$ minus the integral evaluated at $z = z_B$. Locally, the potential vorticity, which is approximated by the QG potential vorticity, represents the ratio of the absolute vorticity to the thickness between isentropic surfaces or the product of the absolute vorticity and static stability. A positive potential vorticity tendency requires either (or both) a positive tendency in vorticity or a negative tendency in thickness (which is a positive tendency in static stability). Analogously, a positive polar cap-averaged potential vorticity tendency involves either (or both) a positive tendency in circulation around the polar cap (acceleration of the zonal wind $\bar{u}$ around the polar cap), as represented by the first term on the right hand side of equation (6) or a negative tendency in average thickness, as represented by an increase in the vertical difference of the term proportional to θ^a on the right hand side of equation (6). These effects are coupled via the residual circulation that is extensively discussed by *Andrews et al.* [1987] and *Haynes et al.* [1991].

Polar cap potential vorticity anomalies can be induced by anomalies in eddy fluxes and in dissipation. We can express the QG potential vorticity flux in terms of the QG Eliassen-Palm (EP) flux:

$$\overline{v'q'} = \rho_0^{-1}[-(\rho_0 \overline{v'u'})_y + (\rho_0 f_0 \overline{v'\theta'}/\theta_{0z})_z] \\ = \rho_0^{-1}\nabla \cdot \mathbf{F}, \quad (7)$$

where

$$\mathbf{F} = (F_{(y)}, F_{(z)}) = (-\rho_0 \overline{v'u'}, \rho_0 f_0 \overline{v'\theta'}/\theta_{0z})$$

is the EP flux. We find that

$$< \overline{q_t^a} >_p = \int_{z_B}^{z_T} \mathrm{d}z\rho_0(\overline{v'q'})^a\Big|_{y=y_S} + < \overline{S^a} >_p \\ = \int_{z_B}^{z_T} \mathrm{d}z(F_{(y)}^a)_y\Big|_{y=y_S} + F_{(z)}^a\Big|_{z=z_B, y=y_S}^{z=z_T, y=y_S} + < \overline{S^a} >_p \quad (8)$$

The polar cap potential vorticity can thus be increased by an eddy flux of potential vorticity into the cap. This eddy potential vorticity flux can involve either (or both) eddy momentum flux convergence at the cap boundary, which is equivalent to a vorticity flux into the cap, since $F_{(y)} = (-\overline{u'v'})_y = \overline{v'\zeta'}$ or a transport of thickness (vertically varying meridional heat transport) into the meridional boundary of the cap, via the terms involving the vertical difference in $F_{(z)} \propto \overline{v'\theta'}$, which are evaluated only at the polar cap boundary. We note that the vertical EP flux at a given latitude drives potential vorticity anomalies poleward of that latitude. Through the EP flux divergence, both eddy thickness and momentum fluxes can thus induce a PV tendency, which corresponds to a tendency in circulation or in stratification.

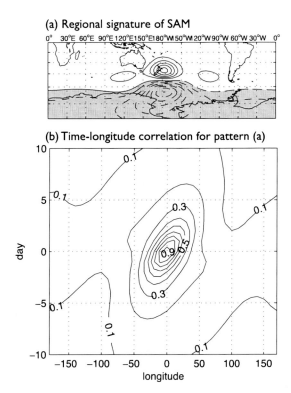

(a) Regional signature of SAM

(b) Time-longitude correlation for pattern (a)

Figure 2. (a) Composite regression pattern for local dipole index on daily surface pressure. The local dipole index represents the principal component time series for the second EOF of surface pressure at a single longitude. The regression with two-dimensional surface pressure is calculated for each longitude, shifted to a common longitude for plotting purposes, and averaged. Thus, the geographic boundaries are provided for a sense of scale only. Contour interval: 1 hPa. (b) Longitude-time lag correlation plot of Southern Hemisphere dipole indexes associated with the pattern in Figure 2a. The dipoles propagate eastward as coherent packets in both hemispheres with a characteristic speed of about 8° longitude d^{-1}. Contour interval 0.1. From *Kushner and Lee* [2007]. Copyright American Meteorological Society.

The connection of the vertical EP flux (meridional heat flux) to stresses on the mean flow is related to the concept of "form stress," which involves the correlation between fluctuations in pressure and in the height of isentropic surfaces for a system in geostrophic and hydrostatic balance [e.g., *Vallis*, 2006].

We consider two special cases of the relationships in equation (8). If we take $z_B \to 0$, where $z = 0$ is the Earth's surface (expressed approximately in log-pressure coordinates) and $z_T \to \infty$, we obtain the well-known vertical mean momentum balance between acceleration, mechanical forcing, and eddy momentum fluxes (see Appendix):

$$\int_0^\infty dz\rho_0[\overline{u_t^a} + (\overline{u'v'})_y^a - \overline{X}^a] = 0 \text{ at } y = y_S, \qquad (9)$$

where X represents x-momentum dissipation. This expression provides a mechanism for eddy momentum fluxes to produce the dipolar wind anomalies that characterize the tropospheric AM. Eddy momentum flux divergence (or vorticity flux) out of one latitude band into another will be balanced by dipolar wind structures via either the $\bar{u}_t$ term or the X term. This description suggests that simple layer models that are dominated by vorticity fluxes will be useful to understand the vertically coherent AM dynamics. The vertical integral averages out the vertical coupling by the mean meridional circulation: when we look in more detail, we see that the eddy momentum transfer occurs mainly in the upper troposphere and is communicated vertically to the surface via a thermally indirect eddy-driven mean meridional circulation, which is balanced by frictional dissipation [*Limpasuvan and Hartmann*, 2000]. The mean meridional circulation that is coherent with eddy momentum fluxes is clearly seen in AM regressions [*Thompson and Wallace*, 2000].

As another special case, it is observed that for the lower stratosphere, the contribution from the meridional component of the EP flux to the total eddy driving is small compared to the vertical contribution [*Newman et al.*, 2001; *Hu and Tung*, 2002; *Polvani and Waugh*, 2004]. Thus, we expect stratospheric AM variability, which involves fluctuations in the strength of the polar vortex, to be dominated by variability in the vertical EP flux component and thus by the thickness flux. For example, on intraseasonal to interannual time scales, the vertical component of the EP flux correlates very well with polar cap stratospheric temperatures [*Newman et al.*, 2001; *Polvani and Waugh*, 2004]. This suggests that a useful simple model of the stratosphere will focus on the forcing of stratospheric winds by vertical EP fluxes.

Before discussing simple theoretical models, we return to observations to examine the eddy fluxes that are coherent with the AM. *Limpasuvan and Hartmann* [1999, 2000] calculate the contributions to $\mathbf{F}$ and $\nabla \cdot \mathbf{F}$ (in primitive equation form, which is approximated by equation (7)) that are coherent with the SAM and NAM index via a composite technique, using all-year data. The main focus of the paper is on the tropospheric AMs in observations and general circulation models. Plate 7 shows composite EP flux cross sections for high AM index conditions (top row), for low AM index conditions (middle row), and their difference (bottom row), adapted from the work of *Limpasuvan and Hartmann* [2000]. There is some degree of hemispheric symmetry in these patterns. In both hemispheres, high AM index conditions correspond to a deflection of EP flux away from the poles and toward the tropics. This figure also shows similar behavior in the stratosphere in both hemispheres: high AM index conditions correspond to positive EP flux divergence anomalies in the high-latitude stratosphere of each hemisphere, indicating reduced wave driving in the high-latitude stratosphere. The pattern of wave driving is consistent with the AM structure in the manner we just described: equatorward deflection of wave activity, since $F_{(y)} \propto -\overline{u'v'}$ corresponds to an increased poleward eddy flux of eastward momentum, which exerts a westward stress on the equatorward side and an eastward stress on the poleward side. This is consistent with the dipolar structure of the tropospheric wind anomaly (e.g., Plate 4). Reduced wave driving in the stratosphere allows polar stratospheric temperatures to relax closer to radiative equilibrium and hence allows the polar vortex to strengthen. A key difference between the two hemispheres is the relative importance of stationary versus transient eddies in the EP fluxes: stationary eddies are dominant in the eddy momentum flux driving in the Northern Hemisphere, while transient eddies dominate the eddy momentum flux driving in the Southern Hemisphere *Limpasuvan and Hartmann* [2000].

The balance between eddy fluxes, circulation anomalies, and dissipation does not address the fundamental questions of how the AMs arise, why they are the dominant mode of variability, what determines their evolution and dynamics, and what couples their stratospheric and tropospheric parts. Much insight can be gained by careful observational analysis. For example, *Lorenz and Hartmann* [2001, 2003] investigate the role of the interactions between eddy momentum fluxes and the AM zonal mean wind anomalies in detail and quantify the positive feedback between the two. In this feedback, the zonal mean flow organizes the eddies in such a way as to enhance the momentum flux driving of the AM, and hence its persistence. But to answer our fundamental questions, it is also important to consider simplified theoretical models within a model hierarchy [*Held*, 2005]. We first consider the tropospheric AM, which, as we learned from our observational analysis, is not critically dependent on coupling to the stratosphere (see the discussion of Plate 3) and for which

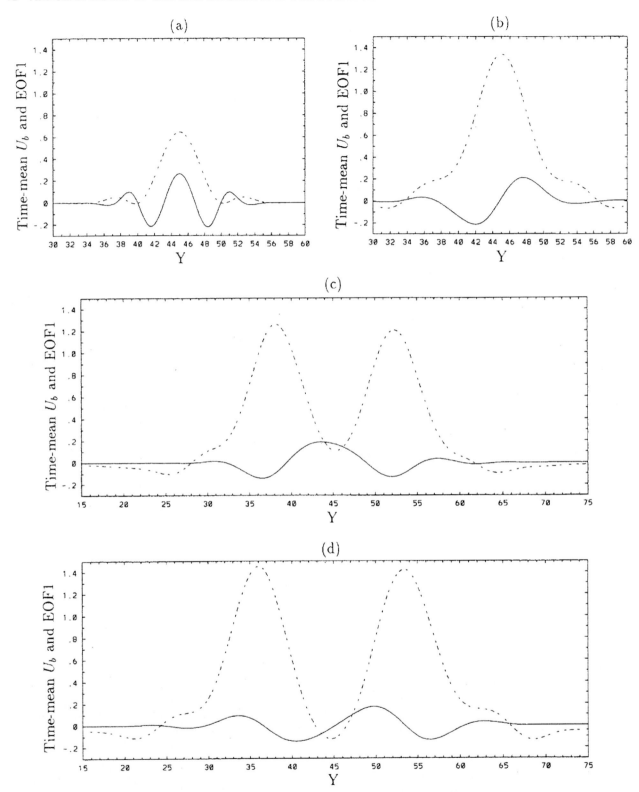

Figure 3. Two-layer quasigeostrophic model simulation of dependence of leading EOF of zonal wind variability (solid) on width of baroclinic zone as represented by the zonal mean zonal wind (dotted). (a)–(d) The baroclinic zone is progressively widened. Between Figures 3b and 3c, the zonal mean wind transitions from a single-jet to a two-jet structure. From *Lee and Feldstein* [1996]. Copyright American Meteorological Society.

vorticity fluxes associated with horizontal propagation of EP flux will play a prominent role. For the stratospheric AM, we will turn to models in which the dominant eddy driving comes from form stresses associated with the meridional eddy flux of heat; for these models, vertical propagation of EP flux dominates.

Fundamentally, variability in the extratropical tropospheric zonal mean circulation reflects eddy mean-flow interactions involving baroclinic and stationary eddies; these give rise to considerable variability even with time-independent boundary conditions [e.g., *Robinson*, 1991]. Coupling to the ocean, to the land surface, to the stratosphere, etc., might modulate this dynamics but does not exert a leading order control. For example, one of the main conclusions of *Limpasuvan and Hartmann* [2000] is that an atmospheric general circulation model (AGCM) with a prescribed sea surface temperatures (SST) field that varies seasonally, but not from year to year, produces a realistic NAM and SAM structure with eddy fluxes that closely resemble the features seen in Plates 4 and 7. This also holds true for coupled ocean atmosphere general circulation models (GCMs) in which SSTs are predicted, confirming that coupling to the ocean is not a critical issue for AM variability [*Miller et al.*, 2006]. An important caveat, which we will return to in section 4, is that these GCMs produce AMs that are too persistent. This is evident from the bias toward the long time scales for the climate model simulations seen in the bottom row of Plate 6 [adapted from *Gerber et al.*, 2008a]. AM variability appears in AGCMs with simplified boundary conditions (such as an aquaplanet) [*Cash et al.*, 2002] and in "simplified GCMs," which will be discussed in section 4.1, with linear representations of diabatic heating and mechanical dissipation [*Yu and Hartmann*, 1993; *Held and Suarez*, 1994; *Polvani and Kushner*, 2002; *Song and Robinson*, 2004; *Gerber and Polvani*, 2009; *Ring and Plumb*, 2008]. But an AGCM, in any configuration, still represents a highly complex system, and it is useful to have in hand even simpler models with which to test ideas.

The simplest model in which realistic tropospheric AM variability is spontaneously generated is a two-layer QG model with the density relaxed to a prescribed baroclinically unstable reference distribution. The width of the region over which this reference distribution is baroclinically unstable determines the width of the baroclinic zone. In one limit, the baroclinic zone is infinitely wide, and zonal mean variability consists of multiple jets that exhibit a slow and unpredictable meridional wander [e.g., *Panetta*, 1993]; the jet and eddy scales are internally determined by differential rotation (β) and the eddy energy, which is related to the imposed baroclinicity. *Lee and Feldstein* [1996] show that the emergence of more realistic AM type variability, which they

describe in terms of zonal index variability, is linked to the width of the baroclinic zone. Some of the possible regimes are seen in Figure 3: when the jet is narrow (Figure 3a), the leading mode is a monopolar "pulsing" mode and represents an alternate sharpening and flattening of the jet profile. As the jet widens (Figure 3b), a dipolar AM EOF emerges that corresponds to a meridionally shifting jet. Once the baroclinic zone is sufficiently wide, a double jet structure emerges, and the leading EOF represents the degree to which the two jets are merged together or separated (Figure 3c). Finally, for an even wider zone, the leading mode involves AM dipoles for both jets (Figure 3d). In all these regimes, the zonal index or AM EOF is present, but it is not always the dominant mode [see also *Feldstein*, 2000].

An arguably even simpler model of tropospheric AM variability is the stirred barotropic model of *Vallis et al.* [2004], in which the forcing from baroclinic eddies is not internally generated but is externally imposed as a stochastic forcing. The nonlinear barotropic potential vorticity dynamics organizes this forcing into a jet whose variability can be tuned by adjusting the structure of the forcing. Similarly to the work of *Lee and Feldstein* [1996], the transition from a narrow-jet "pulsing" regime to a wide-jet "wobbling" regime is governed by the width of the stirring region (Figure 4). This model can also be used to investigate the effect of storm-track localization on the formation of the North Atlantic Oscillation, thus linking in a simple framework the transition from hemispheric-scale to regional-scale variability. *Gerber and Vallis* [2005] strip the dynamics down even further by studying AMs using stochastic models in which angular momentum conservation is built in as a global meridional constraint and coupling in longitude can be introduced in a controlled way.

We return to the observation in section 2 that the NAM and SAM are more similar to each other in the troposphere than the Northern and Southern Hemisphere tropospheric jets. Classically, the observed zonal mean zonal wind reflects two types of jets: the "subtropical jet" for which axisymmetric angular momentum transport from the tropics are important, and the higher latitude "eddy-driven jet," which is dominated by eddy mean-flow interactions we have just described [e.g., *Lee and Kim*, 2003]. (This classical viewpoint is evolving with the realization that eddy potential vorticity fluxes are also important for the subtropical jet [e.g., *Kim and Lee*, 2001; *Walker and Schneider*, 2009].) The Southern Hemisphere November troposphere zonal wind displays aspects of both jets: there is an upper tropospheric (baroclinic) subtropical jet core near 30°S in the upper troposphere and a vertically deep (equivalent barotropic) eddy-driven jet near 50°S (bottom row, left plot of Plate 4). The SAM corresponds to meridional shifts of the latter feature. In the Northern Hemisphere,

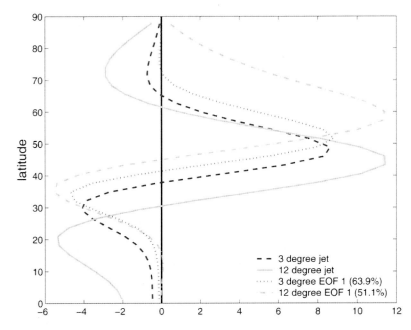

Figure 4. Latitude dependence of the zonal mean zonal wind ("jet") and the leading mode of variability of the zonal mean zonal wind ("EOF 1") in a stochastically stirred barotropic model. Results from simulations in which the stirring region is 3° latitude wide and 12° latitude wide are shown. The narrow 3° jet has a "pulsing" EOF; the wider 12° jet has "wobbling" EOF that corresponds to the annular mode (AM). From *Vallis et al.* [2004]. Copyright American Meteorological Society.

by contrast, the zonal mean jet in JFM (bottom row, middle plot of Plate 4) is dominated by the subtropical jet near 30°N. But for the tropospheric zonal mean wind sector averaged over the North Atlantic region (not shown), a higher latitude eddy-driven jet is seen, and the NAM corresponds to poleward and equatorward shifts of this feature [*Ambaum et al.*, 2001; *Eichelberger and Hartmann*, 2007]. Thus, a similar connection between the eddy-driven jet and the AM can be found in both hemispheres, although the connection is more regional in the Northern Hemisphere. *Eichelberger and Hartmann* [2007] use a simplified GCM (section 4) to explain dynamically the local relationships between Northern Hemisphere jet structure and NAM variability.

Unlike the tropospheric AM, which is manifested as a jet meander, the stratospheric AM is manifested as a strengthening and weakening of the jet. Eddy forcing is important in stratospheric AM variability, but stratospheric waves are primarily generated from tropospheric sources that are external to the stratosphere. The model of *Holton and Mass* [1976] that has been further explored and developed by many researchers [e.g., *Yoden*, 1987; *Plumb and Semeniuk*, 2003; *Scott and Polvani*, 2006] provides the simplest formulation for capturing stratospheric AM variability. In this model, the zonal mean stratospheric circulation is relaxed toward a prescribed radiative equilibrium profile. Planetary Rossby waves are generated by prescribing a zonally asymmetric

lower boundary, conventionally chosen to represent the stationary wavefield in the upper troposphere. The upwelling wave activity flux can be calculated in various ways; in the simplest formulation, it is parameterized using WKB theory. In Figure 5, from *Plumb and Semeniuk* [2003], the wind and geopotential variability resembles the observed downward propagating AM signals in the stratosphere. The model in the work of *Plumb and Semeniuk* [2003] is effectively 1-D in the vertical and represents in a simple way the interaction between stationary Rossby waves with upward group velocity and the zonal mean flow. The resulting downward propagating signal in geopotential and winds is a consequence of enhanced wave activity absorption (EP flux convergence) where the winds are weaker, and the reestablishment of the eastward winds at levels shielded from upward propagating wave activity.

The Holton-Mass model exhibits distinct circulation regimes that depend mainly on the amplitude of the zonally asymmetric lower boundary forcing. For example, in the weakly forced subcritical regime that was explored by *Plumb and Semeniuk* [2003] and is illustrated in Figure 5, the period and evolution of the stratospheric flow is determined entirely by the zonally asymmetric forcing of the lower boundary. *Plumb and Semeniuk* [2003] show that downward propagation of $\overline{Z}^a$ does not imply downward propagation of stratospheric information, since the source of the wave driving

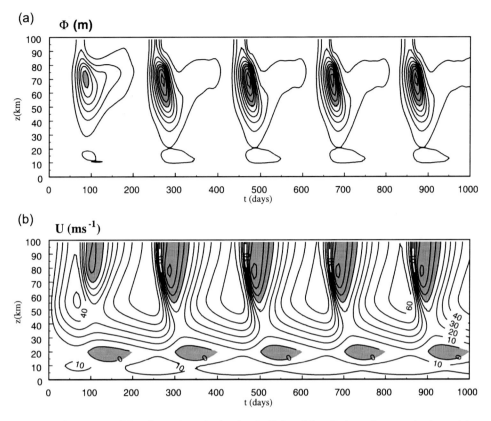

Figure 5. Response in geopotential and zonal wind of a simple Holton-Mass β-channel stratospheric model to periodic forcing from the troposphere. The model represents extratropical eddy mean flow interactions from upward propagating Rossby waves generated by a prescribed zonally asymmetric lower boundary. The prescribed forcing has a period of 200 days. From *Plumb and Semeniuk* [2003].

and the wave activity absorption are determined from the prescribed boundary evolution and state information below a given level. This example shows how some aspects of stratospheric variability might be largely determined by tropospheric dynamics. In the strongly forced supercritical regime [e.g., *Yoden*, 1987; *Scott and Polvani*, 2006], the zonal mean Holton-Mass model exhibits spontaneous vacillations of the zonal wind even for time-independent lower boundary forcing. This regime provides an example of a more active stratosphere that might not depend on the detailed time evolving state of the troposphere, provided that the overall wave forcing from the troposphere is sufficiently strong. Thus, within this simple model, we see a wide range of possibilities for the way in which the stratosphere might depend on the troposphere.

To summarize, we have learned that tropospheric and stratospheric AM variability can be isolated in truncated models whose dynamics are complementary. Tropospheric AM models require relatively fine horizontal resolution to capture the synoptic scale eddies that are involved in eddy

momentum flux transport, but require only coarse vertical resolution involving even only one or two vertical layers. Stratospheric AM models require relatively fine vertical resolution to capture the downward propagating wave activity absorption region, but require only coarse horizontal resolution to capture planetary Rossby waves. But neither type of simple model can realistically simulate the observed coupling, via the AMs, between the troposphere and stratosphere that is depicted in Plate 5. One might imagine constructing a model that couples the two types of truncated systems, but this direction has not been explored, to our knowledge. To simulate realistic AM variability, we turn in the next section to more realistic GCMs with sufficient vertical and horizontal resolution to capture the dynamics as a whole.

4. ANNULAR MODES IN GENERAL CIRCULATION MODELS

Given the limitations of the truncated models discussed in section 3, we turn to models of greater complexity in the model hierarchy and consider the simulation of AM variability and AM responses to climate forcings in simplified GCMs and comprehensive climate models. We will also discuss recently developed ideas related to the fluctuation dissipation theorem, whose application to climate simulation helps elucidate the simulated AM response to climate forcings. Along the way, we will consider the general question of "stratospheric influence" on the troposphere, in which the AMs are strongly implicated.

4.1. Simulation of Intraseasonal Variability

As discussed in section 3, modern climate models generally simulate the AM spatial structure credibly, especially in the troposphere [*Limpasuvan and Hartmann*, 2000; *Miller et al.*, 2006]. Indeed, realistic tropospheric AM structures appear to be a basic feature of any model that realistically represents the zonal mean circulation and tropospheric large-scale waves (baroclinic and stationary waves). It was also established soon after the original work of *Baldwin and Dunkertson* [1999, 2001] that comprehensive GCMs with sufficient stratospheric resolution are capable of capturing intraseasonal stratosphere-troposphere coupling in free-running climate [e.g., *Christiansen*, 2001] and seasonal forecast [e.g., *Charlton et al.*, 2004] settings. The simulations of a free running climate model (with no information about the observed atmospheric state) of *Christiansen* [2001] are able to spontaneously produce realistic downward propagating NAM anomalies that progress from the stratosphere into the troposphere. In current models with good stratospheric representation [e.g., *Baldwin et al.*, 2010], downward propagating stratosphere-troposphere signals appear to be a generic feature that is easy to capture in some form.

Charlton et al. [2004] demonstrate a practical application of this capability by using a numerical weather prediction model to investigate whether stratospheric information can provide useful predictive information in the troposphere beyond the limit of deterministic weather prediction. They initialize ensembles of simulations with different polar vortex strengths and obtain a downward propagating NAM response, which penetrates into the troposphere over the course of several weeks (see Figure 6). Practically, this study shows that accurate stratospheric information and representation can potentially improve seasonal forecasts. The question of stratospheric representation on capturing tropospheric AM variability arises in a variety of contexts, including the response to stratospheric sudden warmings, ENSO-related variability, and to variability in seasonal snow cover [e.g., *Gong et al.*, 2002; *Scaife and Knight*, 2008; *Fletcher et al.*, 2009; *Ineson and Scaife*, 2009; *Bell et al.*, 2009; *Cagnazzo and Manzini*, 2009]. But these studies, and similar ones which

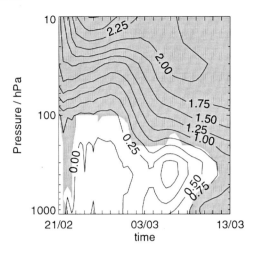

Figure 6. Ensemble mean Northern Annular Mode (NAM) response to switch on strengthening of stratospheric polar vortex, introduced by means of changing stratospheric initial state. The initial tropospheric response in the first 2 days arises from an adjustment to a new balanced state in the stratosphere. The positive NAM response, which occurs after a delay of about 2 weeks, is the result of changes to the stratospheric state. From *Charlton et al.* [2004]. Copyright 2004 Royal Meteorological Society, reprinted with permission.

examine the downward influence of altered stratospheric initial states [e.g., *Hardiman and Haynes*, 2008], do not unambiguously address the question of whether the stratosphere exerts a downward influence on the troposphere. This is because the prescribed stratospheric state might depend on the prior state of the troposphere, in particular, the history of wave activity fluxes from the troposphere to the stratosphere [*Plumb and Semeniuk*, 2003; *Polvani and Waugh*, 2004; *Scott and Polvani*, 2006]. We will show a more direct example of stratospheric influence in connection with climate responses to stratospheric cooling and ozone depletion below.

Having established that comprehensive climate models capture the downward propagating AM signals, the dynamical question of what the minimum ingredients are to capture these signals remains open. To answer this, we turn to "simplified GCMs," by which we mean models that integrate the primitive fluid equations of motion on the sphere at relatively fine horizontal and vertical resolution but that include only the simplest representations of radiative transfer and subgrid scale momentum dissipation and that do not include moisture [e.g., *Scinocca and Haynes*, 1997; *Polvani and Kushner*, 2002; *Taguchi and Yoden*, 2002; *Gerber and Polvani*, 2009]. Simplified GCMs typically represent diabatic heating by a term $-(T - T_{eq})/\tau_r$ in the thermodynamic equation, which corresponds to Newtonian relaxation to a prescribed equilibrium temperature profile T_{eq} on a radiative time scale τ_r and

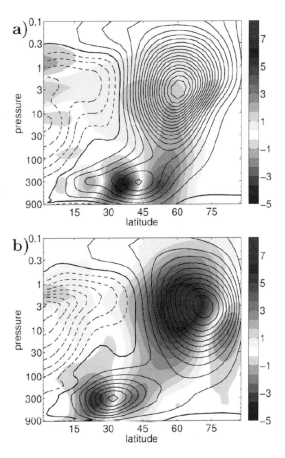

Plate 8. Climatological zonal wind (contours) and regression of AM in zonal wind (shading) for two versions of a simplified GCM with the same reference temperature profile in the troposphere and stratosphere but with (a) no topography and (b) wave-2 topography of amplitude 3000 m. The topography affects the position and strength of the jets and the AM structure; the AM exhibits stratosphere-troposphere coupling representative of the active season only in the second case. From *Gerber and Polvani* [2009]. Copyright American Meteorological Society.

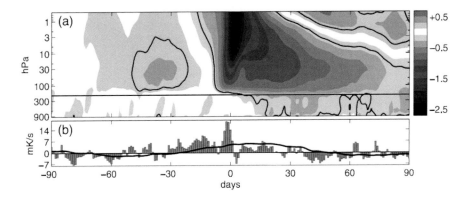

Plate 9. (a) AM index composited on strong negative NAM events in the stratosphere, which correspond to stratospheric warmings, for the simplified GCM configured as in Plate 8b. (b) Composite of a measure of the vertical component of the wave activity (bars) and its mean for the previous 40 days (solid line) for the same NAM index. In the configuration where the simplified GCM has a realistic AM structure, it is able to represent downward propagating NAM events in a realistic manner, although the persistence of the NAM in the troposphere and stratosphere is generally longer than observed. From *Gerber and Polvani* [2009]. Copyright American Meteorological Society.

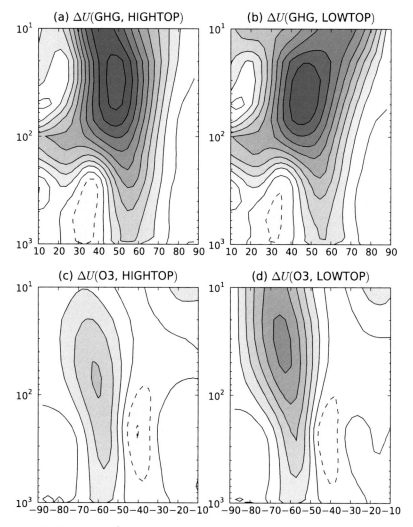

Plate 10. (a) Northern Hemisphere December-January-February zonal mean zonal wind response to CO_2 doubling in the high-top version of the Canadian Middle Atmosphere Model (CMAM) with a lid at 10^{-4} hPa. Contour interval is $1\,\mathrm{m\,s^{-1}}$ and dashed contours are negative. (b) As in Plate 10a, but for the low-top version of CMAM with a lid at 10 hPa but otherwise configured to be as consistent as possible with the high-top version. (c) Southern Hemisphere December-January zonal mean zonal wind response to Antarctic ozone depletion in the high-top CMAM. (d) As in Plate 10c, for the low-top CMAM. It is apparent that the model lid has little impact on the tropospheric AM response in these models. Adapted from *Sigmond et al.* [2008] and *Shaw et al.* [2009].

represent momentum dissipation by a simple mechanical drag $-\mathbf{u}/\tau_\mathrm{d}$, where $\mathbf{u}$ is the horizontal velocity and τ_d is a time scale for momentum dissipation. The reference temperature profile T_eq is baroclinically unstable in the troposphere and leads to an eddy-driven jet stream qualitatively similar to observed. In the stratosphere, T_eq is constructed to produce a polar night jet. We emphasize that these models are only "relatively" simple compared to comprehensive GCMs; in absolute terms, they represent a highly complex dynamical system of partial differential equations. Still, compared to comprehensive GCMs,

they have the advantage of being relatively easy to reproduce in a manner that allows them to provide the basis for benchmark calculations across different research groups [*Held and Suarez*, 1994; *Galewsky et al.*, 2004].

In simplified GCMs, stratospheric variability depends in a complicated way on tropospheric boundary conditions and dynamics. For example, models with time-independent and zonally symmetric boundary conditions can generate some realistic aspects of stratospheric variability [*Scinosa and Haynes*, 1997], and the stratospheric mean state and variability are sensitive to changes in topographic amplitude and structure [*Taguchi and Yoden*, 2002; *Gerber and Polvani*, 2009]. By simply varying the topographic forcing, these models can be tuned to represent the stratospheric variability of either hemisphere [*Taguchi and Yoden*, 2002].

Our interest in this subsection is to determine the minimum requirements for generating the downward propagating AM signals. *Gerber and Polvani* [2009] addressed this question by varying topographic scale and amplitude. For a given time-independent T_{eq}, they are able to tune their model from a stronger polar vortex regime without topography in which the AMs are tropospherically trapped, to a weaker polar vortex regime with topography in which the AMs have large amplitude in both the troposphere and stratosphere (Plate 8). In this regime, the model is capable of producing realistic coupled stratosphere-troposphere events (Plate 9a), including the buildup of wave activity flux into the stratosphere that precedes a negative stratospheric NAM event (Plate 9b). This regime is very sensitive: if the topographic amplitude is increased, the vortex becomes relatively weak and only produces small amplitude NAM events; if the topographic amplitude is reduced, the vortex becomes too strong and robust to be significantly weakened by wave activity pulses; if the wave number of the topography is changed, the character of the variability of the polar vortex also changes qualitatively. As for the other examples we will describe in this section, none of these sensitivities were predicted by preceding theoretical work, which highlights how simplified GCMs remain a primary theoretical tool to explore coupled stratosphere-troposphere variability.

To conclude this subsection on intraseasonal AM variability, we consider the tropospheric origin of stratospheric NAM anomalies in connection with surface forcing, including forcing from ENSO-related SSTs in the tropical Pacific [*Ineson and Scaife*, 2009; *Bell et al.*, 2009; *Cagnazzo and Manzini*, 2009; *Garfinkel and Hartmann*, 2008] and from snow cover anomalies over Eurasia [e.g., *Cohen and Entekhabi*, 1999; *Gong et al.*, 2002; *Cohen et al.*, 2007; *Fletcher et al.*, 2009]. Although the stratosphere is capable of generating internal vacillations independently of the state of the troposphere, as discussed in section 3, it is likely that some portion of stratospheric variability should potentially originate from tropospheric variability. Recent work by *Garfinkel et al.* [2010] shows that this should occur, for ENSO and for Eurasian snow forcing, when the wave component of circulation anomalies related to the surface forcing align with the climatological stationary wave and thereby reinforce or attenuate the wave activity flux into the stratosphere. The lagged nature of the response implies that surface forcing can further extend the window of seasonal predictability via stratosphere-troposphere interactions. ENSO forcing provides a Rossby wave source from the tropics to the extratropics, and the work by *Ineson and Scaife* [2009] and *Cagnazzo and Manzini* [2009] shows that stratospheric representation makes a difference to how this wave activity drives zonal mean NAM anomalies. Surface forcing by Eurasian snow cover anomalies in fall can similarly enhance the tropospheric wave activity flux into the stratosphere, thereby inducing a dynamical warming and a downward propagating NAM signal in Northern Hemisphere winter. This variability is simulated in comprehensive GCMs in which the snow cover forcing is prescribed [*Gong et al.*, 2002]. The effect is sensitive to stratospheric details [*Fletcher et al.*, 2009] and is not captured as an intrinsic mode of variability in the Coupled Model Intercomparison Project 3 prepared for the Intergovernmental Panel on Climate Change Report 4 (CMIP3/IPCC AR4) models [*Hardiman et al.*, 2008]. This work suggests that obtaining improved predictability of AM events of this kind will require considerable improvements in the simulation of stationary waves and land-surface processes.

4.2. AM Climate Trends and Climate Responses

Much of the original interest in the AMs stemmed from the fact that circulation trends through to the end of the 20th century projected strongly onto the AMs, in a manner corresponding to a positive NAM trend in the Northern Hemisphere and a positive SAM trend in the Southern Hemisphere [*Thompson and Wallace*, 1998; *Thompson et al.*, 2000; *Hartmann et al.*, 2000; *Thompson and Solomon*, 2002]. For example, *Thompson et al.* [2000] show that the pattern of land-surface temperature trends in the Northern Hemisphere in the late 20th century are largely consistent with a positive NAM trend. In the Southern Hemisphere, *Thompson and Solomon* [2002] reported that the seasonality and pattern of SAM trends suggested a direct influence by ozone depletion on the tropospheric circulation of the Southern Hemisphere (see Figures 7a–7b, from *Karpechko et al.* [2008]). Since the NAM's peak in the late 1990s, the NAM trend has weakened and reversed [*Overland and Wang*, 2005; *Cohen and Barlow*, 2005], but the long-term SAM trend remains robust [*Fogt et al.*, 2009]. Although current trends in the AMs are relatively ambiguous, understanding the cause and expected future of these trends remains a compelling research question. In this subsection, we will investigate AM responses to climate change in simplified and comprehensive GCMs, and in subsection 4.3 highlight some theoretical issues related to these responses.

To see how circulation trends might relate to AM variability, suppose that a circulation-related field-like sea-level pressure exhibits a long-term trend. In the notation of section 2, we can express a data matrix $\tilde{\mathbf{F}}$ of n observations at

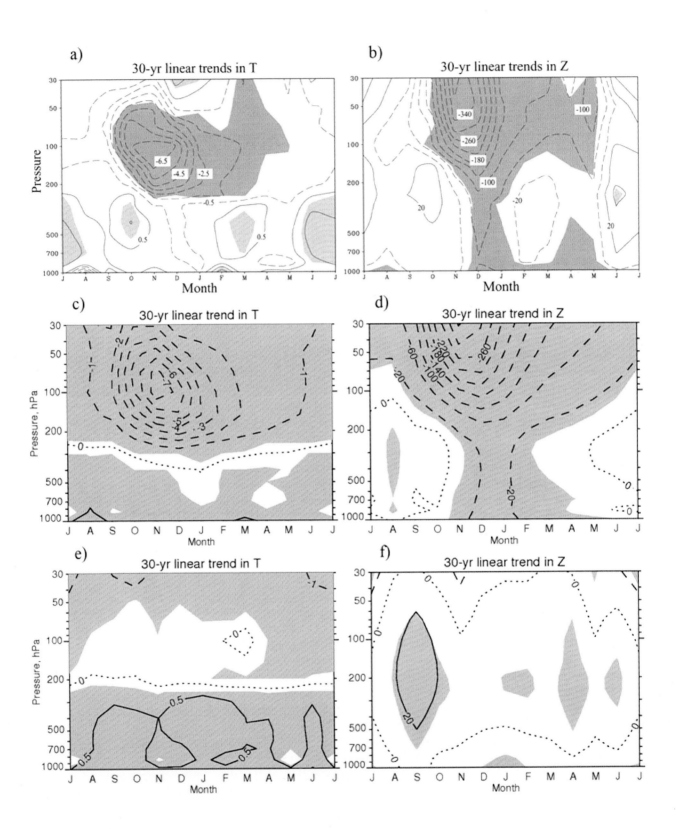

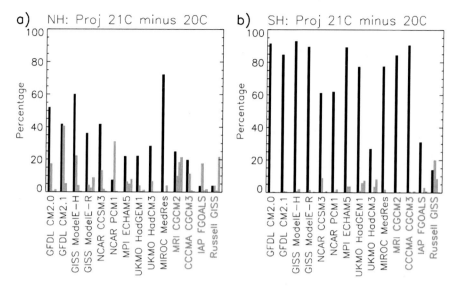

Figure 8. (a) Percentage of variance represented by projection of SLP trends in the 20th century onto four leading modes of extratropical variability in the CMIP3/IPCCAR4 models, for the Northern Hemisphere. (b) As in Figure 8a, for the Southern Hemisphere. From *Miller et al.* [2006].

p spatial points, which has zero time mean at each point, as the sum of a linear trend term and a residual term:

$$\tilde{\mathbf{F}} = \mathbf{t}\mathbf{f}_\Delta^T + \mathbf{R}, \tag{10}$$

where $\mathbf{t}$ is an n-point column of times of observation and $\mathbf{R}$ is a residual term, which has no linear trend. The spatial vector $\mathbf{f}_\Delta$ is a p-point column of trend coefficients, obtained by linear regression. (Alternatively, $\mathbf{f}_\Delta$ can represent simply the difference in the field between two time periods, e.g., 21st and 20th centuries, in which case no linear regression is carried out.) We are interested in what determines the spatial structure of this quantity and whether it is related to natural patterns of variability. For example, its projection on the AM is $\mathbf{f}_{\Delta,1} = \mathbf{f}_\Delta^T \mathbf{f}_1 / \mathbf{f}_1^T \mathbf{f}_1$, where $\mathbf{f}_1 = \tilde{\mathbf{F}}^T \mathbf{y}_1 / \mathbf{y}_1^T \mathbf{y}_1$ is the projection of the data matrix onto the PC time series of the AM. The projection coefficient $\mathbf{f}_{\Delta,1}$ provides the sign and degree of coherence between the trend pattern and the AM pattern; the quantity $(\mathbf{f}_{\Delta,1}^T \mathbf{f}_{\Delta,1})/(\mathbf{f}_\Delta^T \mathbf{f}_\Delta)$ represents the fraction of variance represented by the AM-related component of the trend.

Figure 8, from *Miller et al.* [2006], shows the fraction of variance represented by the projection of the sea level pressure (SLP) response to climate change, diagnosed as the difference between the late 21st century and the early 20th century, onto the AM and other leading modes of extratropical variability, in the simulations included in the CMIP3/IPCC AR4. "Climate change" here includes the effects of anthropogenic greenhouse gas loading, ozone-depleting substances like chlorofluoro-carbons implicated in photochemical ozone loss, and aerosol pollution. In the Southern Hemisphere, the figure shows that the SAM is the dominant pattern of SLP response, and *Miller et al.* [2006] show that the models produce a positive SAM response that agrees at least in sign with current trends. In the Northern Hemisphere, the situation is much more complicated: the projection onto the NAM is weak, and *Miller et al.* [2006] show that there is disagreement on the sign of the projection onto the NAM as well. The situations in the two hemispheres are sufficiently different that they should be considered separately.

The most robust and well-characterized AM response to climate forcing occurs in the Southern Hemisphere. In

Figure 7. (opposite) Antarctic trends for 1968–1999 in (left) temperature (K per 30 years) and (right) geopotential height (m per 30 years) as a function of (a) and (b) calendar month and pressure for observations, (c) and (d) ensemble mean of Coupled Model Intercomparison Project 3 prepared for the Intergovernmental Panel on Climate Change Report 4 (CMIP3/IPCC AR4) models that include the effects of ozone depletion, and (e) and (f) ensemble mean of the models that do not include the effects of ozone depletion. Averages use points corresponding to the stations from *Thompson and Solomon* [2002]. This and other studies [e.g., *Cai and Cowan*, 2007] confirm that ozone depletion is responsible for at least half of the 1968–1999 trend in the Southern Annular Mode (SAM) in the troposphere. From *Karpechko et al.* [2008].

particular, there is now extensive evidence from observations and models that photochemical Antarctic ozone loss has induced a strong positive tropospheric SAM response. This is borne out in several studies involving comprehensive GCMs [e.g., *Sexton*, 2001; *Gillett and Thompson*, 2003; *Miller et al.*, 2006; *Cai and Cowan*, 2007; *Son et al.*, 2008; *Shaw et al.*, 2009]. The middle and bottom rows of Figure 7, from *Karpechko et al.* [2008], illustrate this effect: they show the Antarctic temperature and geopotential response for the CMIP3/IPCC AR4 models that include ozone depletion and those that do not. Models with ozone depletion consistently obtain a response in temperature and geopotential qualitatively similar to the observed trends, while those models without ozone depletion do not capture this response. Although this is not a clean comparison because the sets of models represented in each row of Figure 7 are different, the main conclusion that ozone depletion is required to explain tropospheric circulation trends in the Southern Hemisphere in the late 20th century has been confirmed in several studies. The seasonal cycle of the response to ozone forcing is a downward propagating signal reminiscent of transient intraseasonal variability and suggests a strong connection between the two; this might be expected from the perspective of fluctuation dissipation theory discussed in the next subsection, which connects internal modes of variability to climate responses. The effects on the troposphere of ozone depletion are expected to reverse, as ozone-depleting substance emissions are reduced, and the ozone layer recovers into the mid-21st century [*Son et al.*, 2008; *Perlwitz et al.*, 2008]. This negative SAM response is expected to be accompanied by an equatorward shift of the tropospheric jet in Southern Hemisphere summer.

In simplified GCMs, the ozone-depletion effect on the tropospheric annular mode can be modeled by examining the stratosphere-troposphere response to cooling imposed in the stratosphere [*Polvani and Kushner*, 2002; *Kushner and Polvani*, 2004; *Gerber and Polvani*, 2009; *Chan and Plumb*, 2009]. *Polvani and Kushner* [2002] and *Kushner and Polvani* [2004] carry out such a simulation using a simplified GCM with zonally symmetric boundary conditions. In the simplified GCM, for sufficiently strong stratospheric cooling, a dramatic poleward shift in the tropospheric jet occurs that is manifested as a positive tropospheric AM response that is consistent in sign with the effects of ozone depletion. These papers provide an unambiguous example of stratospheric influence on the tropospheric circulation in a simplified dynamical setting. As was the case for using simplified GCMs to understand intraseasonal variability, the models also provide dynamical insight. These and subsequent studies using simplified GCMs with various thermal and mechanical forcings [*Song and Robinson*, 2004; *Ring and Plumb*, 2007,

2008; *Chan and Plumb*, 2009] suggest that the same eddy mean-flow interactions that generate AM variability are also responsible for amplifying the response to imposed stratospheric forcing.

The simplified GCMs are sufficiently computationally inexpensive that wide parameter regimes can be explored. Such explorations show that many potential sensitivities exist that can generate non-robust responses to stratospheric cooling. The situation is nicely captured in Figure 9 [from *Chan and Plumb*, 2009], which shows that the strength of the poleward shift of the tropospheric jet (the positive SAM change) in response to stratospheric cooling (a strengthening polar vortex) is contingent on the existence of multiple regime behavior for the tropospheric jet. When the tropospheric equator to pole temperature gradient is weak (in the first three columns of Figure 9), the tropospheric jet can be found in a subtropical position or can be in a state where it jumps between subtropical and extratropical positions on a very long time scale. Under these conditions, as the stratospheric polar vortex is strengthened (proceeding down the rows in Figure 9), the tropospheric jet position can be shifted poleward. This behavior characterizes the regime found in the works of *Polvani and Kushner* [2002] and *Kushner and Polvani* [2004]. As the tropospheric equator-to-pole temperature gradient is strengthened, the subtropical jet regime vanishes, and the sensitivity to stratospheric cooling is greatly reduced. Along the same lines, the tropospheric AM response to stratospheric cooling is greatly attenuated by the presence of zonal asymmetries [*Gerber and Polvani*, 2009]. Thus, we see a connection between AM responses to climate perturbations and the ease with which the tropospheric jet can fluctuate between different regimes. The implications of this behavior for present day, future, and past climates, has yet to be explored.

4.3. Perspective from Fluctuation Dissipation Theory

Why should we expect climate trends to be related to the AMs at all? The idea comes from a series of papers that have appealed to "fluctuation dissipation theory" in statistical mechanics [*Leith*, 1975; *Palmer*, 1999; *Gritsun and Branstator*, 2007; *Ring and Plumb*, 2008]. The theory postulates that, under general conditions, the principal pattern of response to climate forcings will project strongly onto persistent modes of internal variability of the system whose spatial structure is similar to that of the forcing. This means that many aspects of the response to climate change can be predicted from knowledge of the system's unforced climate variability. Formal techniques for diagnosing fluctuation-dissipation are described in the works of *Ring and Plumb* [2008] and *Gritsun and Branstator* [2007]. Even without

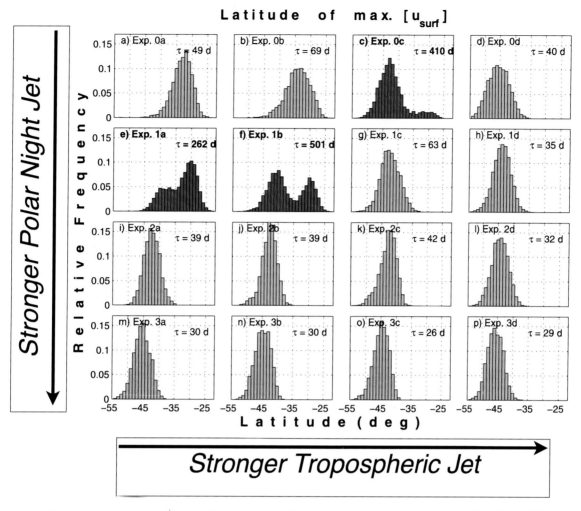

Figure 9. Frequency distribution of meridional location of the tropospheric jet maximum in a simplified GCM for increasingly cold polar stratosphere (increasingly strong polar vortex), going left to right in each row, and for increasingly strong tropospheric meridional temperature gradient, going from top to bottom in each column. Going from weak to strong tropospheric temperature gradient and stratospheric polar vortex, the jet variability passes from a subtropical regime to an extratropical regime, through a mixed bimodal regime. In the mixed regime, the jet can be found in either an extratropical or a subtropical location and can persist in one of these locations for a long period of time; the darker bars indicate high-persistence regimes with the marked time scales τ. In addition, in the mixed regime, the jet is relatively sensitive to stratospheric cooling. In the extratropically trapped regime, the sensitivity to the stratosphere is attenuated or vanishes. Adapted from *Chan and Plumb* [2009]. Copyright American Meteorological Society.

the formal approach, we can obtain useful information by empirically decomposing a response pattern into parts coherent with and unrelated to the internal variability [*Thompson et al.*, 2000; *Kushner et al.*, 2001; *Deser et al.*, 2004]. For example, if we consider the Southern Hemisphere response to greenhouse forcing [*Kushner et al.*, 2001], we can decompose the zonal wind response and the eddy momentum flux convergence response (Figure 10, top row) into a part that is coherent with the SAM (middle row) and a residual (bottom row). The residual is interpreted as the "direct" response to

climate change, while the eddy-driven AM response is interpreted as the "indirect" response. The analysis of *Deser et al.* [2004] of the Northern Hemisphere circulation response to high-latitude surface forcing used this approach to show that the direct response to sea ice and SST forcing in a GCM is localized to the forcing region and that the indirect response is an eddy-driven hemispheric teleconnection pattern. In a more controlled setting, *Ring and Plumb* [2007] show explicitly how the coherence of an applied forcing (in this case, a zonally symmetric momentum forcing applied in the vicinity of the

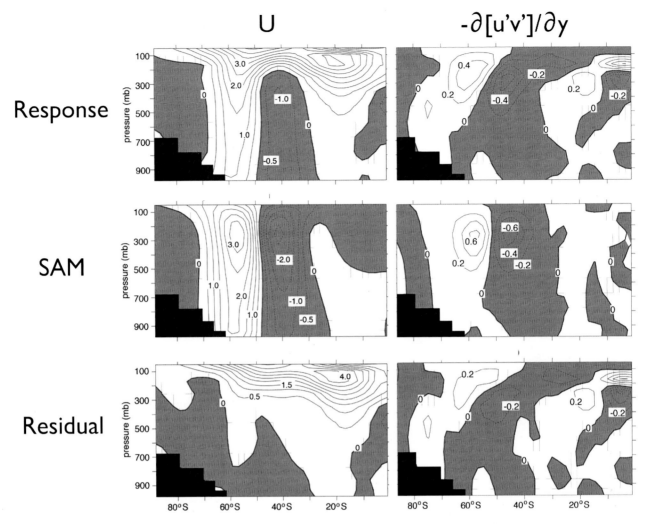

Figure 10. Analysis of zonal wind response (left) and eddy momentum flux convergence response (right) to greenhouse warming in the GFDL R30 model. The top row shows the response, the middle row shows the part of the response that is coherent with SAM variability based on the model's lower tropospheric winds, and the bottom row shows the residual (top row minus middle row). The residual is confined to the upper troposphere and lower stratosphere, which *Kushner et al.* [2001] suggest is the location of the strongest forcing of the meridional temperature gradients in the response to climate change; they thus interpret this as the "direct" response to climate change. The SAM-related response, on the other hand, involves the eddy-driven SAM variability that is interpreted to be stimulated by the direct forcing; they thus call this the "indirect" response to climate change. From *Kushner et al.* [2001]. Copyright American Meteorological Society.

jet in a simplified GCM) with the AM determines the sign and strength of the AM response.

Decomposing circulation responses into direct and indirect parts is useful for a broad brush characterization but is qualitative and has little predictive value. More quantitative analyses based on fluctuation-dissipation theory have been carried out in simplified GCMs [e.g., *Ring and Plumb*, 2008; *Gerber et al.*, 2008b] or in very long runs of comprehensive GCMs [e.g., *Gritsun and Branstator*, 2007, which deals with the global response to tropical heating]. A key prediction of the theory is that the expected pattern of response $\mathbf{f}_\Delta \sim \Sigma_i R_i f_i \tau_i$,

where R_i is the projection of the forcing onto mode f_i, and τ_i is the decorrelation time scale of the unforced mode. Thus, we expect that more persistent modes will dominate the response, other factors being equal. Pursuing this idea, *Gerber et al.* [2008b] emphasize how AM time scales determine the amplitude of the response to external forcings: models with long AM time scales have an AM response that is proportionately greater, in agreement with the predictions of the theory. *Ring and Plumb* [2008] use fluctuation dissipation theory in even more detail in a simplified GCM forced mechanically and thermally. The responses in these simulations scale linearly with the projection of the forcing onto the internal modes in general agreement with the theory, but the results do not agree quantitatively with its predictions.

There is thus fairly strong evidence to support the conclusion from fluctuation dissipation theory that the AM time scale is a key parameter controlling the strength of the indirect or free response to climate forcings. But the AM time scale, from the current modeling evidence, turns out to be highly non-robust and difficult to simulate accurately. For example, *Gerber et al.* [2008b] show that AM time scales in simplified GCMs are highly sensitive to horizontal resolution, vertical resolution, numerical diffusion, and other factors. In comprehensive GCMs, the IPCC AR4 models produce AM time scales that are too long in both the Northern Hemisphere and Southern Hemisphere, as seen in Plate 6. In simplified GCMs, we have already seen how some aspects of the sensitivity of the AM time scales are related to the structure of the jets; in particular, the existence of multiple regimes for the jet, in which the jet remains in a subtropical or in an extratropical location for long periods of time, greatly lengthens the time scale [*Chan and Plumb*, 2009]. In the IPCC AR4 models, *Kidston and Gerber* [2010] find a connection, similar to Chan and Plumb in the simplified GCMs, between the climatological bias, the AM time scale, and climate responses: those models with an equatorward bias in the Southern Hemisphere jet have a more persistent SAM and shift poleward more under climate change. From the analysis of simplified and comprehensive GCMs, in light of fluctuation dissipation theory, we conclude that AM responses might be too strong in present day GCMs. In addition, we conclude that the responses will be sensitive to many factors, even if the spatial structure and amplitude of the AMs are realistically represented in the models.

4.4. Stratospheric Representation and AM Responses

With this background, we are in a better position to address the question of stratospheric influence on the AM response to tropospheric climate change. The seminal paper in this area is the study of *Shindell et al.* [1999], who argue that accurate stratospheric representation is necessary to reproduce the NAM trend of the late 20th century. The argument involves a comparison between two versions of comprehensive GCMs forced by anthropogenic greenhouse gas loading: a GCM with very few levels in the stratosphere (which we will call "low top") and a model with a relatively more stratospheric resolution ("high top"). Of the two models, only the high-top model produces a positive NAM response similar to the observed trend. *Shindell et al.* [1999, 2001] argue that only in the high-top model is the effect of planetary wave refraction properly accounted for: the positive NAM response in the stratosphere involves refraction of wave activity away from the polar lower stratosphere into low latitudes. This work emphasizes that if the climate response truly projects onto the coupled stratosphere-troposphere AM, then the wave-driving response is intrinsic to the AM response and must be properly represented. While subsequent work has debated the findings and interpretation of the Shindell et al. work, it remains highly significant for the subsequent investigation it has stimulated.

The idea that stratospheric representation is important to tropospheric climate dates back to the work of *Boville* [1984] and *Boville and Cheng* [1988]. This work shows that the configuration of the model stratosphere (including factors influencing the polar night jet-like stratospheric diffusion, mechanical drag, and vertical resolution) has a significant influence on the tropospheric circulation. The patterns of sensitivity to stratospheric representation, in both the winds and the wave driving, resemble the AM structures. But the extension of this idea, that stratospheric representation is important to tropospheric climate change, is not currently straightforward to resolve. In apparent contradiction to *Shindell et al.* [1999], it is clear that low-top models are capable of obtaining AM responses similar to the observed trends [*Fyfe et al.*, 1999; *Kushner et al.*, 2001; *Miller et al.*, 2006]. But several factors confound simple generalizations: as we have said, the model responses are not robust (Figure 8); such responses might be exaggerated because of the long AM-time scale bias, and some observed AM trends are not robust. Thus, the issue of stratospheric representation on the response to climate change needs to be addressed separately for each climate response and for each model.

A necessary step to address the issue of stratospheric representation is to systematically vary stratospheric resolution, while minimizing other changes. When model lid height is lowered and other changes are minimized, at least two models show relatively little sensitivity in their AM responses to climate forcing [*Gillett et al.*, 2002; *Sigmond et al.*, 2008; *Shaw et al.*, 2009]. For example, *Sigmond et al.* [2008] and *Shaw et al.* [2009] present a specialized low-top version of a high-top GCM (the Canadian Middle Atmosphere Model) constructed so as to minimize inconsistencies between the two. (Plate 10 summarizes some of the results from these studies.) The lid of the low-top GCM is located at 10 hPa, which is lower than the lid of most present-day climate models. Yet both the low- and high-top models exhibit a strong positive NAM response to greenhouse warming [*Sigmond et al.*, 2008] and a strong positive SAM response to ozone depletion [*Shaw et al.*, 2009]; see Plate 10. Thus, stratospheric resolution, by itself, is not a key factor in controlling the AM response to climate forcings, at least in this model.

While in clean tests, stratospheric resolution does not necessarily influence tropospheric AM responses, the responses can be surprisingly sensitive to other factors. In particular, the models in Plate 10 are sensitive to the details of the parameterization of momentum dissipation by subgrid-scale gravity waves. In the work of *Shaw et al.* [2009], for example, if the flux of momentum from parameterized "nonorographic" gravity waves is not deposited in the top model level in a manner that satisfies momentum conservation in the vertical, but instead allowed to figuratively "escape to space," the tropospheric SAM response to ozone depletion is suppressed. In the work of *Sigmond et al.* [2008], the positive tropospheric NAM response to greenhouse warming can be suppressed by increasing the strength of the orographic gravity wave flux from below. These sensitivities are relevant to current modeling practice, in which an escape-to-space boundary condition on the gravity wave fluxes near the model lid is often used and in which a range of tunings of the orographic gravity wave flux are used. It would be interesting to see how much of the spread in AM responses in current GCMs could be attributed to inconsistencies in the implementation of gravity wave parameterizations, since these schemes are becoming better constrained observationally and better characterized numerically [*Alexander et al.*, 2009; *McLandress and Scinocca*, 2005].

5. CONCLUSION

The identification and characterization of the AMs, due in large part to the University of Washington "school" with its hallmark of dynamically motivated observational analysis, has represented a significant advance in atmospheric science that has strengthened the link between our field and the environmental sciences as a whole. This is because the AMs provide a unifying framework for large-scale variability in the extratropics that links stratospheric and tropospheric dynamics, that highlights common strands between Northern and Southern Hemisphere variability, and that connects intraseasonal variability to climate–time scale fluctuations. Even though it is incomplete, our current statistical and dynamical understanding of the AMs can be applied in a variety of environmental contexts, including investigation of temperature extremes [*Thompson and Wallace*, 2001], oceanic circulation, the global carbon cycle in the Northern and Southern hemispheres [e.g., *Russell and Wallace*, 2004; *Butler et al.*, 2007], Antarctic surface climate [e.g., *Thompson and Solomon*, 2002], and many other diverse topics that are outside the scope of this chapter. Because the AMs represent a characteristic pattern of climate response as well as climate variability, further attention will continue to be warranted as we move forward with attempts to predict the future of evolution of the Earth system.

A key issue that remains unresolved in this discussion is the mechanisms underlying stratosphere-troposphere AM cou-

pling and the coupling of AMs to the land and ocean surface. We have attempted to clarify those aspects of tropospheric and stratospheric AM variability that can be understood in isolation and those for which surface and stratosphere coupling are more critical. The stratospheric and tropospheric AMs can be regarded as dynamically distinct, but they are linked by eddy-driven processes that we are only beginning to understand and quantify. Likewise, we are only beginning to understand the role of snow, ocean surface temperatures in the tropics and extratropics, and other surface forcings in generating AM variability. In the effort to understand the various drivers of AM variability, simplified GCMs with sufficient horizontal and vertical resolution currently provide our main tool to develop theoretical understanding. At each stage, both comprehensive and simplified GCMs have furnished surprising results, especially in relation to AM time scales and AM climate responses. These developments are currently leading to fundamental insights that, we expect, will solidify our understanding of the observed AMs and their climate responses.

Regarding the climate responses themselves, an intensive research effort on the Southern Hemisphere extratropical circulation response to climate change is warranted at this point. A positive Southern Annular Mode (SAM) response to greenhouse gas increases and ozone depletion appears to be a robust signal common to many climate models and seems to be in agreement with observations in the second half of the 20th century. But there are reasons to expect, based on the long AM time scale bias identified in section 4, that the models will tend to be correspondingly biased toward overly large amplitude AM responses. Accurately representing the SAM response will become more challenging and critical as the ozone layer recovers, since the climate and ozone forcing signals are expected to produce opposite signed SAM responses [e.g., *Son et al.*, 2010]. This issue will require a systematic attack using the full model hierarchy before it can be fully resolved.

APPENDIX A

We recall some of the main results of quasigeostrophic (QG) theory [*Andrews et al.*, 1987, chap. 3]. The QG equation in log pressure and tangent plane linear Cartesian coordinates is

$$D q = -X_y + Y_x + f_0 \rho_0^{-1} (\rho_0 Q/\theta_{0z})_z, \qquad (A1)$$

where the QG potential vorticity (also known as the pseudo potential vorticity) q is defined by

$$q \equiv f_0 + \beta y + \psi_{xx} + \psi_{yy} + \rho_0^{-1} (\rho_0 f_0^2 N^{-2} \psi_z)_z,$$

the QG approximation to the advective derivative acting on the QG potential vorticity q is

$$D = \frac{\partial}{\partial t} + u\frac{\partial}{\partial x} + v\frac{\partial}{\partial y},$$

where the static stability $N^2(z)$ is associated with the reference potential temperature $\theta_0(z)$. Some of the remaining notation is explained as follows:

ψ	geostrophic streamfunction;
$(u,v) = (-\psi_y, \psi_x)$	geostrophic velocity;
$\theta_e = f_0\theta_{0z}\psi_z/N^2$	potential temperature perturbation from reference θ_{0z};
(X, Y)	momentum dissipation;
Q	diabatic heating on potential temperature.

If we take the zonal mean of equation (A1) and remove the climatological mean of the result, we obtain equation (4) with

$$\overline{S} = -(\overline{X^a})_y + f_0\rho_0^{-1}(\rho_0\overline{Q^a}/\theta_{0z})_z. \qquad (A2)$$

To show why we expect the density-weighted volume average of $\overline{S^a}\overline{q^a}$ to be negative, we assume that the momentum forcing is linear drag with time scale τ_m and the diabatic heating is relaxational on time scale τ_r. If the damping time scales are equal, with $\tau_m = \tau_r \equiv \tau$, then $\overline{S^a} = -\tau\overline{q^a}$ and $\overline{S^a}\cdot\overline{q^a} = -\tau(\overline{q^a})^2 \le 0$ at each point. But if the momentum and thermal damping time scales are different, it is more difficult to find a general result. We find

$$\overline{S^a} = \tau_m^{-1}(\overline{u^a})_y - f_0\tau_r^{-1}\rho_0^{-1}(\rho_0\overline{\theta_e^a}/\theta_{0z})_z$$
$$= -\tau_m^{-1}(\overline{\psi^a})_{yy} - \tau_r^{-1}\rho_0^{-1}\left[\rho_0 f_0^2 N^{-2}(\overline{\psi^a})_z\right]_z$$

and

$$\overline{q^a}\cdot\overline{S^a} = -\left\{\tau_m^{-1}(\overline{\psi^a})_{yy} + \tau_r^{-1}\rho_0^{-1}\left[\rho_0 f_0^2 N^{-2}(\overline{\psi^a})_z\right]_z\right\}$$
$$\cdot\left\{(\overline{\psi^a})_{yy} + \rho_0^{-1}\left[\rho_0 f_0^2 N^{-2}(\overline{\psi^a})_z\right]_z\right\}$$
$$= -\left\{\tau_m^{-1}\left[(\overline{\psi^a})_{yy}\right]^2\right.$$
$$+(\tau_m^{-1} + \tau_r^{-1})\rho_0^{-1}(\overline{\psi^a})_{yy}\left[\rho_0 f_0^2 N^{-2}(\overline{\psi^a})_z\right]_z$$
$$\left.+\tau_T^{-1}\left[\rho_0^{-1}(\rho_0 f_0^2 N^{-2}(\overline{\psi^a})_z)_z\right]^2\right\}$$

Now the cross-term in this expression is proportional to

$$(\overline{\psi^a})_{yy}(\rho_0 f_0^2 N^{-2}(\overline{\psi^a})_z)_z = \nabla\cdot\mathbf{G} + \rho_0 f_0^2 N^{-2}[(\overline{\psi^a})_{yz}]^2,$$

where

$$\mathbf{G} = [-\rho_0 f_0^2 N^{-2}(\overline{\psi^a})_y(\overline{\psi^a})_{zz}, \rho_0 f_0^2 N^{-2}(\overline{\psi^a})_{yy}(\overline{\psi^a})_z]$$

Density-weighted averaging corresponds to $\langle f\rangle = \int\rho_0 dV$, and so

$$\langle \overline{q^a}\cdot\overline{S^a}\rangle = -\langle\tau_m^{-1}[(\overline{\psi^a})_{yy}]^2 + (\tau_m^{-1} + \tau_r^{-1})f_0^2 N^{-2}[(\overline{\psi^a})_{yz}]^2$$
$$+\tau_r^{-1}[\rho_0^{-1}(\rho_0 f_0^2 N^{-2}(\overline{\psi^a})_z)_z]^2\rangle + (\tau_m^{-1} + \tau_r^{-1})\int\nabla\cdot\mathbf{G}dV.$$

Thus, $\langle\overline{q^a}\cdot\overline{S^a}\rangle$ evaluates to a negative quadratic term plus a flux divergence that integrates out to the boundaries. We have not found a simple justification for zeroing out this flux integral, particularly its vertical contribution, and thus, we must allow that boundary terms might provide a source of variance of the zonal mean PV independent of eddy feedbacks in equation (5). But if we choose boundary conditions that zero out this flux, for example, considering the set of solutions for which $(\overline{\psi^a}) = 0$ at the boundary, then $\langle\overline{q^a}\cdot\overline{S^a}\rangle$ will be negative.

The polar cap density-weighted average potential vorticity tendency in equation (6) is found from equation (4)

$$<\overline{q_t^a}>_p = \int_{z_B}^{z_T}\int_{y_S}^{y_N} dydz[-\rho_0(\overline{u^a})_{yt} + f_0\rho_0(\overline{\theta_e^a})_t/\theta_{0z}]$$

using $\overline{u^a} = 0$ at $y = y_N$, which corresponds to the North Pole. Equation (8) for the eddy fluxes and dissipation is derived similarly, provided we set the meridional component of the EP flux $F_{(y)} = 0$ at the North Pole. To obtain the vertically integrated momentum balance (equation (9)), we take $\rho_0\theta_e/\theta_{0z} \to 0$ as $z_T \to \infty$ and use the zonal mean of the standard QG lower-boundary condition

$$D\theta_e = Q \text{ at } z_B = 0;$$

topographic effects are neglected in this expression.

Acknowledgments. The support of the Natural Sciences and Engineering Research Council of Canada and the Canadian Foundation for Climate and Atmospheric Sciences is gratefully acknowledged. Thanks go to R.A. Plumb and E.P. Gerber for careful reviews; to K. Smith and D.W.J. Thompson for additional helpful comments; and to L. Mudryk for providing Plate 5, V. Limapsuvan for providing Plate 7, and T. Shaw and M. Sigmond for providing the data for Plate 10.

REFERENCES

Alexander, M. J., S. D. Eckermann, D. Broutman, and J. Ma (2009), Momentum flux estimates for South Georgia Island mountain waves in the stratosphere observed via satellite, *Geophys. Res. Lett.*, *36*, L12816, doi:10.1029/2009GL038587.

Ambaum, M. H. P., B. J. Hoskins, and D. B. Stephenson (2001), Arctic Oscillation or North Atlantic Oscillation, *J. Atmos. Sci.*, *14*, 3495–3507.

Andrews, D. G., J. R. Holton, and C. B. Leovy (1987), *Middle Atmosphere Dynamics*, 489 pp., Academic Press, New York.

Baldwin, M. (2001), Annular modes in global daily surface pressure, *Geophys. Res. Lett.*, *28*, 4114–4118.

Baldwin, M., et al. (2010), Effects of the stratosphere on the troposphere, in *SPARC CCMVal Report on the Evaluation of Chemistry-Climate Models*, edited by M. Baldwin, T. G. Shepherd, and D. W. Waugh, *SPARC Rep. 5, WCRP-132, WMO/TD-No. 1526,* World Clim. Res. Program, World Meteorol. Organ., Geneva, Switzerland.

Baldwin, M. P., and T. J. Dunkerton (1999), Propagation of the Arctic Oscillation from the stratosphere to the troposphere, *J. Geophys. Res.*, *104*, 30,937–30,946.

Baldwin, M. P., and T. J. Dunkerton (2001), Stratospheric harbingers of anomalous weather regimes, *Science*, *294*, 581–584.

Baldwin, M. P., and D. W. J. Thompson (2009), A critical comparison of stratosphere-troposphere coupling indices, *Q. J. R. Meteorol. Soc.*, *135*(644), 1661–1672.

Baldwin, M. P., D. B. Stephenson, D. W. J. Thompson, T. J. Dunkerton, A. J. Charlton, and A. O'Neill (2003), Stratospheric memory and skill of extended-range weather forecasts, *Science*, *301*(5633), 636–640.

Baldwin, M. P., D. B. Stephenson, and I. T. Jolliffe (2009), Spatial weighting and iterative projection methods for EOFs, *J. Clim.*, *22*, 234–243.

Bell, C. J., L. J. Gray, A. J. Charlton-Perez, M. M. Joshi, and A. A. Scaife (2009), Stratospheric communication of El Niño teleconnections to European winter, *J. Clim.*, *22*, 4083–4096.

Boville, B. A. (1984), The influence of the polar night jet on the tropospheric circulation in a GCM, *J. Atmos. Sci.*, *41*, 1132–1142.

Boville, B. A., and X. H. Cheng (1988), Upper boundary effects in a general-circulation model, *J. Atmos. Sci.*, *45*, 2591–2606.

Butler, A. H., D. W. J. Thompson, and K. R. Gurney (2007), Observed relationships between the Southern Annular Mode and atmospheric carbon dioxide, *Global Biogeochem. Cycles*, *21*, GB4014, doi:10.1029/2006GB002796.

Cagnazzo, C., and E. Manzini (2009), Impact of the stratosphere on the winter tropospheric teleconnections between ENSO and the North Atlantic and European region, *J. Clim.*, *22*, 1223–1238.

Cai, W. J., and T. Cowan (2007), Trends in Southern Hemisphere circulation in IPCC AR4 models over 1950–99: Ozone depletion versus greenhouse forcing, *J. Clim.*, *20*, 681–693.

Cash, B. A., P. J. Kushner, and G. K. Vallis (2002), The structure and composition of the annular modes in an aquaplanet general circulation model, *J. Atmos. Sci.*, *59*, 3399–3414.

Chan, C. J., and R. A. Plumb (2009), The response to stratospheric forcing and its dependence on the state of the troposphere, *J. Atmos. Sci.*, *66*, 2107–2115.

Charlton, A. J., A. O'Neill, W. A. Lahoz, and A. C. Massacand (2004), Sensitivity of tropospheric forecasts to stratospheric initial conditions, *Q. J. R. Meteorol. Soc.*, *130*(600), 1771–1792.

Christiansen, B. (2001), Downward propagation of zonal mean zonal wind anomalies from the stratosphere to the troposphere: Model and reanalysis, *J. Geophys. Res.*, *106*, 27,307–27,322.

Codron, F. (2005), Relation between annular modes and the mean state: Southern Hemisphere Summer, *J. Clim.*, *18*(2), 320–330.

Cohen, J., and M. Barlow (2005), The NAO, the AO, and global warming: How closely related?, *J. Clim.*, *18*, 4498–4513.

Cohen, J., and D. Entekhabi (1999), Eurasian snow cover variability and Northern Hemisphere climate predictability, *Geophys. Res. Lett.*, *26*, 345–348.

Cohen, J., M. Barlow, P. J. Kushner, and K. Saito (2007), Stratosphere-troposphere coupling and links with Eurasian land surface variability, *J. Clim.*, *20*, 5335–5343.

Deser, C. (2000), On the teleconnectivity of the Arctic Oscillation, *Geophys. Res. Lett.*, *27*, 779–782.

Deser, C., G. Magnusdottir, R. Saravanan, and A. Phillips (2004), The effects of North Atlantic SST and sea ice anomalies on the winter circulation in CCM3. Part II: Direct and indirect components of the response, *J. Clim.*, *17*, 877–889.

Eichelberger, S. J., and D. L. Hartmann (2007), Zonal jet structure and the leading mode of variability, *J. Clim.*, *20*(20), 5149–5163.

Feldstein, S. B. (2000), Is interannual zonal mean flow variability simply climate noise?, *J. Clim.*, *13*, 2356–2362.

Fletcher, C. G., S. C. Hardiman, P. J. Kushner, and J. Cohen (2009), The dynamical response to snow cover perturbations in a large ensemble of atmospheric GCM integrations, *J. Clim.*, *22*, 1208–1222.

Fogt, R. L., J. Perlwitz, A. J. Monaghan, D. H. Bromwich, J. M. Jones, and G. J. Marshall (2009), Historical SAM variability. Part II: Twentieth-century variability and trends from reconstructions, observations, and the IPCC AR4 models, *J. Clim.*, *22*, 5346–5365.

Fyfe, J. C., G. J. Boer, and G. M. Flato (1999), The Arctic and Antarctic oscillations and their projected changes under global warming, *Geophys. Res. Lett.*, *26*, 1601–1604.

Galewsky, J., R. K. Scott, and L. M. Polvani (2004), An initial-value problem for testing numerical models of the global shallow-water equations, *Tellus, Ser. A*, *56*, 429–440.

Garfinkel, C. I., and D. L. Hartmann (2008), Different ENSO teleconnections and their effects on the stratospheric polar vortex, *J. Geophys. Res.*, *113*, D18114, doi:10.1029/2008JD009920.

Garfinkel, C. I., D. L. Hartmann, and F. Sassi (2010), Tropospheric precursors of anomalous Northern Hemisphere stratospheric polar vortices, *J. Clim.*, *23*, 3282–3298.

Gerber, E. P., and L. M. Polvani (2009), Stratosphere-troposphere coupling in a relatively simple AGCM: The importance of stratospheric variability, *J. Clim.*, *22*, 1920–1933.

Gerber, E. P., and G. K. Vallis (2005), A stochastic model of the spatial structure of the annual patterns of variability and the NAO, *J. Clim.*, *18*, 2102–2118.

Gerber, E. P., L. M. Polvani, and D. Ancukiewicz (2008a), Annular mode time scales in the Intergovernmental Panel on Climate Change Fourth Assessment Report models, *Geophys. Res. Lett.*, *35*, L22707, doi:10.1029/2008GL035712.

Gerber, E. P., S. Voronin, and L. M. Polvani (2008b), Testing the annular mode autocorrelation time scale in simple atmospheric general circulation models, *Mon. Weather Rev.*, *136*, 1523–1536.

Gillett, N. P., and D. Thompson (2003), Simulation of recent Southern Hemisphere climate change, *Science*, *302*, 273–275.

Gillett, N. P., M. R. Allen, and K. D. Williams (2002), The role of stratospheric resolution in simulating the Arctic Oscillation response to greenhouse gases, *Geophys. Res. Lett.*, *29*(10), 1500, doi:10.1029/2001GL014444.

Gong, D., and S. Wang (1999), Definition of Antarctic Oscillation index, *Geophys. Res. Lett.*, *26*, 459–462.

Gong, G., D. Entekhabi, and J. Cohen (2002), A large-ensemble model study of the wintertime AO-NAO and the role of interannual snow perturbations, *J. Clim.*, *15*, 3488–3499.

Gritsun, A., and G. Branstator (2007), Climate response using a three-dimensional operator based on the fluctuation-dissipation theorem, *J. Atmos. Sci.*, *64*, 2558–2575.

Hardiman, S. C., and P. H. Haynes (2008), Dynamical sensitivity of the stratospheric circulation and downward influence of upper level perturbations, *J. Geophys. Res.*, *113*, D23103, doi:10.1029/2008JD010168.

Hardiman, S. C., P. J. Kushner, and J. Cohen (2008), Investigating the ability of general circulation models to capture the effects of Eurasian snow cover on winter climate, *J. Geophys. Res.*, *113*, D21123, doi:10.1029/2008JD010623.

Hartmann, D., J. Wallace, V. Limpasuvan, D. Thompson, and J. Holton (2000), Can ozone depletion and greenhouse warming interact to produce rapid climate change?, *Proc. Natl. Acad. Sci. U. S. A.*, *97*, 1412–1417.

Haynes, P. H., M. E. McIntyre, T. G. Shepherd, C. J. Marks, and K. P. Shine (1991), On the "downward control" of extratropical diabatic circulations by eddy-induced mean zonal forces, *J. Atmos. Sci.*, *48*, 651–680.

Held, I. M. (2005), The gap between simulation and understanding in climate modeling, *Bull. Am. Meteorol. Soc.*, *86*, 1609–1614.

Held, I. M., and M. J. Suarez (1994), A proposal for the intercomparison of the dynamical cores of atmospheric general circulation models, *Bull. Am. Meteorol. Soc.*, *75*, 1825–1830.

Holton, J. R., and C. Mass (1976), Stratospheric vacillation cycles, *J. Atmos. Sci.*, *33*, 2218–2225.

Hoskins, B. J., M. E. McIntyre, and A. W. Robertson (1985), On the use and significance of isentropic potential vorticity maps, *Q. J. R. Meteorol. Soc.*, *111*(470), 877–946.

Hu, Y., and K. K. Tung (2002), Interannual and decadal variations of planetary-wave activity, stratospheric cooling, and the Northern-Hemisphere annular mode, *J. Clim.*, *15*, 1659–1673.

Ineson, S., and A. A. Scaife (2009), The role of the stratosphere in the European climate response to El Niño, *Nat. Geosci.*, *2*, 32–36.

Kalnay, E., et al. (1996), The NCEP/NCAR 40-year reanalysis project, *Bull. Am. Meteorol. Soc.*, *77*(3), 437–471.

Karpechko, A. Y., N. P. Gillett, G. J. Marshall, and A. A. Scaife (2008), Stratospheric influence on circulation changes in the Southern Hemisphere troposphere in coupled climate models, *Geophys. Res. Lett.*, *35*, L20806, doi:10.1029/2008GL035354.

Kidston, J., and E. Gerber (2010), Intermodel variability of the poleward shift of the austral jet stream in the CMIP3 integrations linked to biases in 20th century climatology, *Geophys. Res. Lett.*, *37*, L09708, doi:10.1029/2010GL042873.

Kim, H., and S. Lee (2001), Hadley cell dynamics in a primitive equation model. Part II: Nonaxisymmetric flow, *J. Atmos. Sci.*, *58*, 2859–2871.

Kodera, K., K. Yamazaki, M. Chiba, and K. Shibata (1990), Downward propagation of upper stratospheric mean zonal wind perturbation to the troposphere, *Geophys. Res. Lett.*, *17*, 1263–1266.

Kushner, P. J., and G. Lee (2007), Resolving the regional signature of the annular modes, *J. Clim.*, *20*, 2840–2852.

Kushner, P. J., and L. Polvani (2004), Stratosphere-troposphere coupling in a relatively simple AGCM: The role of eddies, *J. Clim.*, *17*, 629–639.

Kushner, P. K., I. M. Held, and T. L. Delworth (2001), Southern-Hemisphere atmospheric circulation response to global warming, *J. Clim.*, *14*, 2238–2249.

Lee, S., and S. B. Feldstein (1996), Mechanism of zonal index evolution in a two-layer model, *J. Atmos. Sci.*, *53*, 2232–2246.

Lee, S., and H. Kim (2003), The dynamical relationship between subtropical and eddy-driven jets, *J. Atmos. Sci.*, *60*, 1490–1503.

Leith, C. E. (1975), Climate response and fluctuation dissipation, *J. Atmos. Sci.*, *32*, 2022–2026.

Limpasuvan, V., and D. L. Hartmann (1999), Eddies and the annular modes of climate variability, *Geophys. Res. Lett.*, *26*(20), 3133–3136.

Limpasuvan, V., and D. L. Hartmann (2000), Wave-maintained annular modes of climate variability, *J. Clim.*, *13*, 4414–4429.

Limpasuvan, V., D. W. J. Thompson, and D. L. Hartmann (2004), The life cycle of the Northern Hemisphere sudden stratospheric warmings, *J. Clim.*, *17*, 2584–2596.

Limpasuvan, V., D. L. Hartmann, D. W. J. Thompson, K. Jeev, and Y. L. Yung (2005), Stratosphere-troposphere evolution during polar vortex intensification, *J. Geophys. Res.*, *110*, D24101, doi:10.1029/2005JD006302.

Lorenz, D. J., and D. L. Hartmann (2001), Eddy-zonal flow feedback in the Southern Hemisphere, *J. Atmos. Sci.*, *58*, 3312–3327.

Lorenz, D. J., and D. L. Hartmann (2003), Eddy-zonal flow feedback in the Northern Hemisphere winter, *J. Clim.*, *16*(8), 1212–1227.

McLandress, C., and J. F. Scinocca (2005), The GCM response to current parameterizations of nonorographic gravity wave drag, *J. Atmos. Sci.*, *62*, 2394–2413.

Miller, R. L., G. A. Schmidt, and D. T. Shindell (2006), Forced annular variations in the 20th century Intergovernmental Panel on Climate Change Fourth Assessment Report models, *J. Geophys. Res.*, *111*, D18101, doi:10.1029/2005JD006323.

Monahan, A. H., and J. C. Fyfe (2006), On the nature of zonal jet EOFs, *J. Clim.*, *19*, 6409–6424.

Newman, P. A., E. R. Nash, and J. Rosenfield (2001), What controls the temperature of the Arctic stratosphere during the spring?, *J. Geophys. Res.*, *106*, 19,999–20,010.

Overland, J. E., and M. Wang (2005), The Arctic climate paradox: The recent decrease of the Arctic Oscillation, *Geophys. Res. Lett.*, *32*, L06701, doi:10.1029/2004GL021752.

Palmer, T. N. (1999), A nonlinear dynamical perspective on climate prediction, *J. Clim.*, *12*, 575–591.

Panetta, R. (1993), Zonal jets in wide baroclinically unstable regions: Persistence and scale selection, *J. Atmos. Sci.*, *50*, 2073–2106.

Perlwitz, J., and H.-F. Graf (1995), The statistical connection between tropospheric and stratospheric circulation of the Northern Hemisphere in winter, *J. Clim.*, *8*, 2281–2295.

Perlwitz, J., S. Pawson, R. L. Fogt, J. E. Nielsen, and W. D. Neff (2008), Impact of stratospheric ozone hole recovery on Antarctic climate, *Geophys. Res. Lett.*, *35*, L08714, doi:10.1029/2008GL033317.

Plumb, R., and K. Semeniuk (2003), Downward migration of extratropical zonal wind anomalies, *J. Geophys. Res.*, *108*(D7), 4223, doi:10.1029/2002JD002773.

Polvani, L. M., and P. Kushner (2002), Tropospheric response to stratospheric perturbations in a relatively simple general circulation model, *Geophys. Res. Lett.*, *29*(7), 1114, doi:10.1029/2001GL014284.

Polvani, L. M., and D. W. Waugh (2004), Upward wave activity flux as a precursor to extreme stratospheric events and subsequent anomalous surface weather regimes, *J. Clim.*, *17*, 3548–3554.

Ring, M. J., and R. A. Plumb (2007), Forced annular mode patterns in a simple atmospheric general circulation model, *J. Atmos. Sci.*, *64*, 3611–3626.

Ring, M. J., and R. A. Plumb (2008), The response of a simplified GCM to axisymmetric forcings: Applicability of the fluctuation-dissipation theorem, *J. Atmos. Sci.*, *65*, 3880–3898.

Robinson, W. A. (1991), The dynamics of low-frequency variability in a simple-model of the global atmosphere, *J. Atmos. Sci.*, *48*, 429–441.

Robinson, W. A. (2000), A baroclinic mechanism for the eddy feedback on the zonal index, *J. Atmos. Sci.*, *57*, 415–422.

Russell, J. L., and J. M. Wallace (2004), Annual carbon dioxide drawdown and the Northern Annular Mode, *Global Biogeochem. Cycles*, *18*, GB1012, doi:10.1029/2003GB002044.

Scaife, A. A., and J. R. Knight (2008), Ensemble simulations of the cold European winter of 2005–2006, *Q. J. R. Meteorol. Soc.*, *134*, 1647–1659.

Scinocca, J. F., and P. Haynes (1997), Dynamical forcing of stratospheric planetary waves by tropospheric baroclinic eddies, *J. Atmos. Sci.*, *55*, 2361–2392.

Scott, R. K., and L. M. Polvani (2006), Internal variability of the winter stratosphere. Part I: Time-independent forcing, *J. Atmos. Sci.*, *63*, 2758–2776.

Sexton, D. M. H. (2001), The effect of stratospheric ozone depletion on the phase of the Antarctic Oscillation, *Geophys. Res. Lett.*, *28*, 3697–3700.

Shaw, T. A., M. Sigmond, T. G. Shepherd, and J. F. Scinocca (2009), Sensitivity of simulated climate to conservation of momentum in gravity wave drag parameterization, *J. Clim.*, *22*, 2726–2742.

Shindell, D. T., R. L. Miller, G. Schmidt, and L. Pandolfo (1999), Simulation of recent Northern winter climate trends by greenhouse-gas forcing, *Nature*, *399*, 452–455.

Shindell, D. T., G. A. Schmidt, R. L. Miller, and D. Rind (2001), Northern Hemisphere winter climate response to greenhouse gas, ozone, solar, and volcanic forcing, *J. Geophys. Res.*, *106*, 7193–7210.

Sigmond, M., J. F. Scinocca, and P. J. Kushner (2008), Impact of the stratosphere on tropospheric climate change, *Geophys. Res. Lett.*, *35*, L12706, doi:10.1029/2008GL033573.

Son, S.-W., et al. (2008), The impact of stratospheric ozone recovery on the Southern Hemisphere westerly jet, *Science*, *320*, 1486–1489.

Son, S.-W., et al. (2010), Impact of stratospheric ozone on Southern Hemisphere circulation change: A multimodel assessment, *J. Geophys. Res.*, *115*, D00M07, doi:10.1029/2010JD014271.

Song, Y., and W. A. Robinson (2004), Dynamical mechanisms for stratospheric influences on the troposphere, *J. Atmos. Sci.*, *61*, 1711–1725.

Taguchi, M., and S. Yoden (2002), Internal interannual variations of the troposphere-stratosphere coupled system in a simple global circulation model. Part I: Parameter sweep experiment, *J. Atmos. Sci.*, *59*, 3021–3036.

Thompson, D. W. J., and S. Solomon (2002), Interpretation of recent Southern Hemisphere climate change, *Science*, *296*, 895–899.

Thompson, D. W. J., and J. M. Wallace (1998), The Arctic Oscillation signature in the wintertime geopotential height and temperature fields, *Geophys. Res. Lett.*, *25*, 1297–1300.

Thompson, D. W. J., and J. M. Wallace (2000), Annular modes in the extratropical circulation. Part I: Month-to-month variability, *J. Clim.*, *13*, 1000–1016.

Thompson, D. W. J., and J. M. Wallace (2001), Regional climate impacts of the Northern Hemisphere annular mode, *Science*, *293*, 85–89.

Thompson, D. W. J., J. M. Wallace, and G. C. Hegerl (2000), Annular modes in the extratropical circulation. Part II: Trends, *J. Clim.*, *13*, 1018–1036.

Vallis, G. K. (2006), *Atmospheric and Oceanic Fluid Dynamics*, 745 pp., Cambridge Univ. Press, Cambridge, U. K.

Vallis, G. K., E. P. Gerber, P. J. Kushner, and B. A. Cash (2004), A mechanism and simple dynamical model of the North Atlantic Oscillation and annular modes, *J. Atmos. Sci.*, *61*, 264–280.

Walker, C. C., and T. Schneider (2009), Eddy influences on Hadley circulations: Simulations with an idealized GCM, *J. Atmos. Sci.*, *63*, 3333–3350.

Wallace, J. M. (2000), North Atlantic Oscillation/annular mode: Two paradigms—One phenomenon, *Q. J. R. Meteorol. Soc.*, *126*(564), 791–805.

Wallace, J. M., and D. W. J. Thompson (2002), The Pacific center of action of the Northern Hemisphere annular mode: Real or artifact?, *J. Clim.*, *15*(14), 1987–1991.

Wolter, K., and M. Timlin (1993), Monitoring ENSO in COADS with a seasonally adjusted principal component index, *Proc. of the 17th Climate Diagnostics Workshop*, pp. 52–57, NOAA/NMC/CAC, NSSL, Oklahoma Clim. Survey, CIMMS and the School of Meteor., Univ. of Oklahoma, Norman, Okla.

Yoden, S. (1987), Dynamical aspects of stratospheric vacillations in a highly truncated model, *J. Atmos. Sci.*, *44*, 3683–3695.

Yu, J.-Y., and D. L. Hartmann (1993), Zonal flow vacillation and eddy forcing in a simple GCM of the atmosphere, *J. Atmos. Sci.*, *50*, 3244–3259.

P. J. Kushner, Department of Physics, University of Toronto, 60 St. George Street, Toronto, ON M5S 1A7, Canada. (paul.kushner@utoronto.ca)

Stratospheric Equatorial Dynamics

Lesley J. Gray

Department of Meteorology, University of Reading, Reading, UK

Observations of the main features of the stratospheric equatorial region are described, and our current theoretical understanding is outlined. The quasi-biennial oscillation (QBO) and semiannual oscillation in zonal winds and temperatures are described, including time series and meridional structure. The impacts of the QBO on both the extratropics and the troposphere are covered and also the impact on tracer transport and distributions of ozone and other chemical species. Finally, observations and theory of the interaction of the QBO with the solar cycle are reviewed.

1. INTRODUCTION

The quasi-biennial oscillation (QBO) and semiannual oscillation (SAO) dominate the variability of the equatorial stratosphere. The oscillations consist of alternating easterly and westerly winds at equatorial latitudes that descend through the atmosphere with time. This is well illustrated in Figure 1, which shows observations of equatorial zonal winds over nearly 50 years. In the vertical range ~5–100 hPa, the evolution is dominated by the QBO with a period of approximately 28 months. Above this, the SAO dominates, with an approximate period of 6 months.

2. ZONAL WINDS

In the region 5–100 hPa the alternating pattern of QBO westerly and easterly winds descends with an average rate of about 1 km month^{-1} [*Ebdon*, 1960; *Reed et al.*, 1961; *Baldwin et al.*, 2001]. The easterly phase has greater amplitude and is generally of longer duration, especially above ~70 hPa. The westerly phase shear zone exhibits a faster and more regular descent than the easterly phase shear zone. While the period of the QBO is determined by the amplitudes of the waves that drive it (as discussed further below), and not by the annual

cycle itself, there does appear to be some evidence for synchronization with the annual cycle. Figure 2 shows a histogram of the month of onset of the QBO phases at 44 hPa. Although there are slight variations between different studies, depending on the choice of QBO level and method of calculating the transition, they show a tendency for the transitions to occur during Northern Hemisphere (NH) late spring and summer, as evident in Figure 2. This is especially true of the onset of the easterly phase shear zone (i.e., west-to-east transition).

In the region above 5 hPa, it is evident from Figure 1 that the zonal wind variations are much less regular. The SAO is the dominant oscillation at these levels, but there are other influences, for example, from the QBO and the 11-year solar cycle. Observational coverage of this region is much poorer than at lower levels, with rather sparse rocketsonde observations in early years and only relatively recent satellite observations providing global coverage. To better illustrate the nature of the SAO, Figure 3 shows the annual cycle of equatorial winds for the period 1979–2008. While the QBO at lower altitudes has been averaged out, a persistent SAO pattern is seen at upper levels, with maximum SAO westerlies near the stratopause (~1 hPa) in April/May and October. There is also a marked asymmetry in the strength of this westerly phase, with the April/May phase almost double the amplitude of the October phase. Similarly, the easterly SAO phase in January/February is approximately double the amplitude of that in July/August.

The latitudinal structure of the zonal wind QBO is symmetric about the equator with a meridional half-width

The Stratosphere: Dynamics, Transport, and Chemistry
Geophysical Monograph Series 190
Copyright 2010 by the American Geophysical Union.
10.1029/2009GM000868

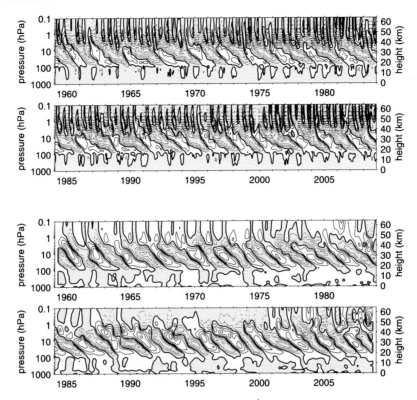

Figure 1. (top) Time-height section of monthly mean zonal wind (m s^{-1}) with the seasonal cycle removed for 1958–2008. Values are from the ERA-40 reanalysis up to 2001 and European Centre for Medium-Range Weather Forecasts operational data set thereafter. Contour intervals are 5 m s^{-1}, and dotted contours denote easterlies. (bottom) Band-pass-filtered data. Only periods between 9 and 48 months have been retained. After *Gray et al.* [2001].

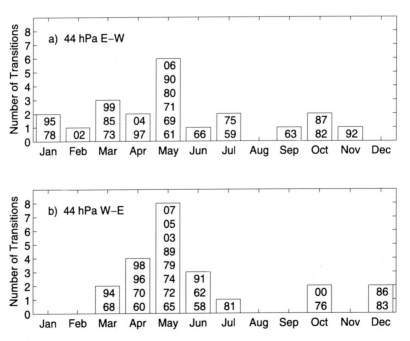

Figure 2. Histogram showing the distribution of quasi-biennial oscillation (QBO) phase transitions at 44 hPa from (a) east to west and (b) from west to east. The numbers denote the year. Updated from *Pascoe et al.* [2005].

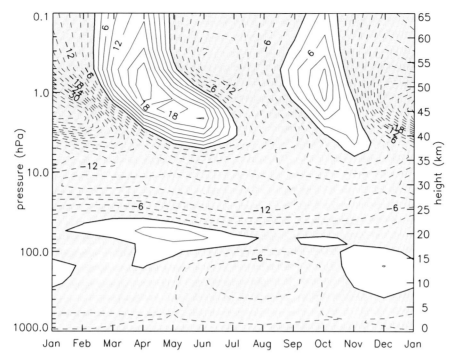

Figure 3. Annual cycle of equatorial zonal mean zonal wind (1979–2008). Contour interval is 3 m s^{-1}.

of approximately 12°, as shown in Figure 4a. The latitudinal structure of the SAO similarly has its maximum amplitude at the equator but is broader in structure (Figure 4b). Taking a latitudinal slice at the level of the QBO maximum (15 hPa, Figure 4c), the QBO and SAO amplitudes may be compared with that of the annual cycle, which is almost zero over the equator but increases with increasing latitude. The height structures of the QBO and SAO oscillations at the equator are illustrated in Figure 4d and confirms the dominance of the QBO variability at heights below 10 hPa and the SAO above 5 hPa. The amplitude estimates in Figure 4 have been derived using "assimilated analyses" in which global circulation model fields are nudged toward observations. In this way, a global representation is achieved in which the model fills in the regions with sparse data coverage. The main observations of zonal winds and temperatures come from satellite observations at heights above 10 hPa and a mixture of satellite and radiosonde observations below this. The spatial resolution of the satellite observations is relatively coarse, which means that amplitudes derived from these analyses are probably an underestimate of those in reality.

3. QBO AND SAO MECHANISMS

It is now well established that the QBO and SAO are caused by momentum transport associated primarily with vertically propagating equatorial waves, as first suggested by *Lindzen*

and Holton [1968]. There is a broad spectrum of waves that influence the tropics. The waves are generated by convection in the tropical troposphere and then propagate vertically into the stratosphere. The vertical propagation depends both on the properties of the wave and on their interaction with the local background flow. The oscillatory nature of the QBO and SAO is due to an internal two-way feedback mechanism between the waves and the background flow. The propagation, and hence momentum transport, of the waves depends on the background flow, but the background flow is also influenced by the wave momentum fluxes. Simple models have shown that the period of the oscillations is controlled by the wave momentum fluxes and not by the annual cycle, although the latter may have a small influence through its influence on the background flow.

The basic mechanism of the QBO oscillation can be understood by considering the behavior of two vertically propagating internal gravity waves of equal amplitude, but opposite zonal phase speeds in a steady, hydrostatic, quasi-linear flow unaffected by rotation and subjected to linear damping. *Plumb* [1977] noted that although a zero mean flow state is a possible equilibrium state, it is a very unstable state unless vertical diffusion is extremely strong. An oscillatory flow is much more likely. The vertical propagation of the waves depends on the vertical group velocity and the background flow. As each wave propagates vertically, its amplitude is diminished by damping, generating a force on the

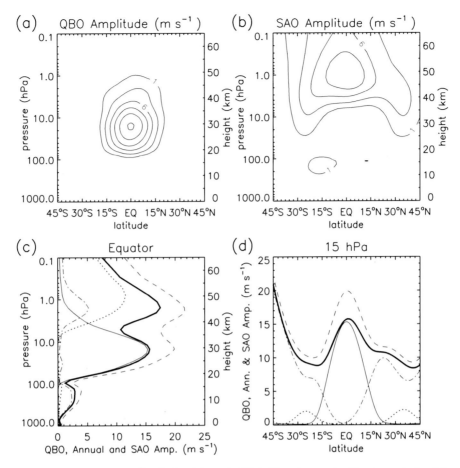

Figure 4. Amplitude of the QBO (m s^{-1}) calculated from the ERA-40 analyses using Fourier analysis of the zonal mean zonal wind, following the method of *Pascoe et al.* [2005] and updated to include 1958–2008. (a) QBO and (b) semiannual oscillation (SAO) amplitudes. Contours are drawn at 1 m s^{-1} and at multiples of 3 m s^{-1} thereafter. (c) Vertical profiles of the equatorial QBO (thin solid line), SAO (dotted line), and annual (dash-dotted line) zonal wind amplitudes. The sum of the QBO, SAO, and annual amplitudes (thick solid line) can be compared to the standard deviation of the zonal wind (dashed line). (d) The 15-hPa (~30 km) latitudinal profiles of the QBO, SAO, and annual amplitudes, annotation as in Figure 4c.

mean flow due to convergence of the vertical flux of zonal momentum, which will locally accelerate the mean flow in the direction of the wave's zonal phase propagation. The momentum flux convergence depends on the vertical structure of the background wind and is most strongly damped when it encounters a shear zone where $|u–c|$ is small (where u is the zonal wind and c is the wave phase velocity), as shown schematically in Figure 5a. Plumb showed that the maximum acceleration occurs below the maximum u, and hence, the jet maximum will descend with time. As the shear zone descends, the layer of westerly winds becomes sufficiently shallow that it can be destroyed by viscous diffusion, allowing the penetration of the westerly travelling wave to higher levels through the easterly mean flow (Figure 5b). Dissipation at the higher levels and the resulting transfer of momentum to the background flow means that a new westerly regime builds

up, which then propagates downward in a similar fashion (Figures 5c and 5d). This process then repeats, but with the shallow region of easterly winds in Figure 5d destroyed by viscous diffusion, allowing the easterly travelling wave to penetrate to high levels through the westerly background mean flow, thus setting up the next westerly phase regime at the upper levels.

While Plumb's simple schematic model of the oscillation is illustrative, in reality, there are important complicating factors, most notably that the rotating atmosphere introduces asymmetry between waves with westerly and easterly phase speeds, which (among other things) affects the meridional structure of the various waves. *Holton and Lindzen* [1972] refined the ideas of *Lindzen and Holton* [1968] and achieved a one-dimensional (1-D) model simulation of the QBO with the westerly phase and easterly phase driven by Kelvin and

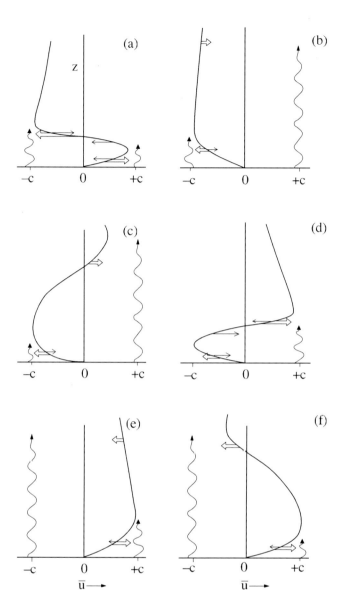

Figure 5. Schematic representation of the evolution of the mean flow in *Plumb*'s [1984] analog of the QBO. Four stages of a half cycle are shown. Double arrows show wave-driven acceleration, and single arrows show viscously driven acceleration. Wavy lines indicate penetration of eastward (i.e., westerly) and westward (i.e., easterly) waves. After *Plumb* [1984].

Rossby-gravity waves, respectively. Their conceptual model was also supported by the laboratory experiments of *Plumb and McEwan* [1978], who used a salt-stratified solution in a large rotating annulus and simulated vertically propagating gravity waves to produce a wave-induced mean flow regime with downward progressing reversals of the mean flow, similar to the observed QBO. The same mechanism, in which the penetration to higher levels of waves with phase propagation

in one direction is "shielded" by a background wind flow of the opposite direction, as described above, is also a feature of the stratospheric SAO, particularly the westerly phase.

In the tropics, the background vertical motion associated with the equatorial arm of the Brewer-Dobson circulation is in the opposite direction to the descending QBO phases, with generalized upwelling of approximately 1 km month^{-1}. This means that for the QBO wind regimes to propagate downward at the observed rate requires substantially more momentum transport than was originally assumed. Although the observed amplitudes of Kelvin and mixed Rossby-gravity waves were sufficient to drive a QBO in the simple 1-D model, *Gray and Pyle* [1989] found they were insufficient in a more realistic 2-D model simulation that included the influence of the background upwelling associated with the Brewer-Dobson circulation and also took into account the meridional structure of the oscillations and the feedback of the associated temperature changes on to the mean flow. They found that a QBO could only be achieved by increasing the amplitude of the modeled Kelvin and mixed Rossby-gravity wave forcing by a factor of three. *Dunkerton* [1997] suggested that in reality, this additional momentum flux is probably supplied by a broad spectrum of gravity waves similar to those originally suggested by *Lindzen and Holton* [1968].

It is now generally accepted that the momentum fluxes for the generation of the QBO are provided by a combination of vertically propagating waves that can be organized into three categories: (1) Kelvin and Rossby-gravity waves, which are westerly and easterly propagating, respectively; equatorially trapped; zonal wave numbers ~1–4; zonal wavelengths ≥10,000 km; periods ≥3 days. (2) inertia-gravity waves, which can travel in either direction; not necessarily equatorially trapped; zonal wave numbers ~4–40 km; zonal wavelengths ~1000–10,000 km; period ~1–3 days. (3) gravity waves, which can travel in either direction; ubiquitous at all latitudes; zonal wave numbers >40; zonal wavelengths ~10–10,000 km, period ≤1 day. A more detailed description of the characteristics of these waves is provided by *Baldwin et al.* [2001] and G.-Y. Yang et al., The behaviour of equatorial waves in opposite QBO phases, submitted to *Journal of the Atmospheric Sciences* [2010] [see also *Alexander*, this volume].

The vertically propagating waves that force the QBO are either those that propagate slowly and undergo absorption due to radiative or mechanical damping at such a rate that their momentum is deposited at the QBO altitudes or those that propagate with a faster vertical group propagation but are absorbed at critical levels that lie within the range of the QBO wind speeds. The SAO, conversely, is forced by waves with fast vertical group velocity or with phase speeds lying outside the range of the QBO wind speeds, so that they can propagate through the underlying QBO winds. In the easterly SAO phase, there are added contributions from planetary (Rossby) waves propagating equatorward from the winter hemisphere and also contributions due to the advection of summer easterlies by the Brewer-Dobson circulation (see *Andrews et al.* [1987] for a more detailed discussion of this, and also of the mesospheric SAO). The influences from planetary waves and advection are also likely to be a major factor in determining the period of the SAO, since there is a strong semiannual variation in strength of the planetary wave fluxes and Brewer-Dobson circulation associated with wave propagation from each winter hemisphere. Thus, unlike the QBO, the SAO period appears to be strongly influenced by the annual cycle. The planetary wave momentum fluxes and Brewer-Dobson circulation strength also have a strong annual cycle because NH winter fluxes are much larger than SH fluxes, which could explain the stronger easterly SAO phase in January/February compared with June/July in Figure 3.

4. MERIDIONAL STRUCTURE

While the simple mechanisms described above explain the oscillatory nature of the QBO and SAO, they do not explain why the phenomena are confined to low latitudes, as illustrated in Figure 4. An early assumption was that the oscillations are confined to equatorial latitudes because they are driven solely by equatorially trapped waves, but we now know that this is probably not the case. The *Lindzen and Holton* [1968] model experiments showed that the Coriolis torque reduces the amplitude of the wind oscillation away from the equator. More recently, *Haynes* [1998] and *Scott and Haynes* [1998] proposed that the latitudinal width of the QBO is determined by the latitudinal dependence of the response to a wave-induced force. They noted that the response to an applied force is not a straightforward acceleration but depends also on the Coriolis force. At high latitudes, part of the applied force is balanced by the Coriolis torque, and most of the response to the applied force appears as a mean meridional circulation. This circulation will induce temperature anomalies that are then thermally damped, which effectively limits the response amplitude. In the tropics, on the other hand, where rotational effects are weak, the response is almost entirely in the acceleration term with relatively little temperature response, and radiative damping has minimal effect. In this way, the tropical response lasts much longer, and thus, a long-period oscillation such as the QBO is possible at tropical latitudes but cannot be sustained at higher latitudes.

Because of its long period and equatorial symmetry, the QBO wind regimes are in thermal wind balance. Tropical QBO temperatures are, therefore, in thermal wind balance

with the vertical wind shear, so that warm anomalies are associated with westerly vertical shear and cold anomalies are associated with easterly vertical shear, as illustrated in Figure 6. To maintain thermal wind balance against thermal damping requires adiabatic heating in the westerly shear zones provided by the sinking motion within the induced meridional circulation (see Figure 6). Conversely, cooling in the easterly shear zones is provided by rising motion in the induced meridional circulation [*Plumb and Bell*, 1982a, 1982b]. This induced circulation may explain the greater descent rate of the westerly QBO phase noted earlier in section 2. Downward advection of momentum associated with the westerly shear zone enhances the descent of the westerlies, while the upward advection of momentum associated with easterly shear inhibits the descent of easterlies. This has an additional effect of latitudinally narrowing the width of the westerly shear zone while broadening the easterly shear zone.

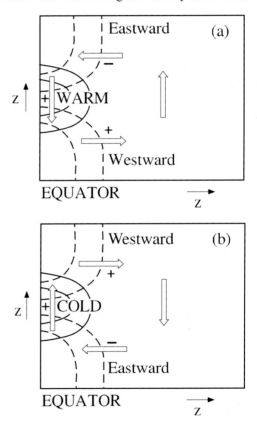

Figure 6. Schematic latitude-height sections showing the mean meridional circulation associated with the equatorial temperature anomaly of the QBO. Solid contours show temperature anomaly isotherms, and dashed contours are zonal wind isopleths. Plus and minus signs designate signs of zonal wind accelerations driven by the mean meridional circulation. (a) Westerly shear zone. (b) Easterly shear zone. After *Plumb and Bell* [1982b]. Copyright Royal Meteorological Society, reprinted with permission.

The greater variability of the easterly shear zone descent is associated with times when the descent appears to "stall" for a short time before continuing its descent. Good examples of this are seen in Figure 1 near 30–50 hPa in 1965 and 1989. In many other years, the descent rate is fairly uniform. A possible explanation for the observed stalling of the descent in some years is that the observed annual cycle in equatorial upwelling has a minimum in NH spring/summer. This enables the easterly shear zone to descend more easily at that time of year (see also Figure 2b), thus creating step-like features in the pattern of descent.

The temperature anomaly associated with the QBO has an amplitude of ±4 K, maximizing near 30–50 hPa. Its relation to the vertical wind shear is demonstrated well in Figure 7. Figure 8 shows the latitudinal structure of the QBO temperature anomaly for NH winter in 1994, during which the lower stratosphere was in an easterly phase. Several features are notable from this plot. First, the negative temperature anomaly over the equator at ~30 hPa, which is associated with the descending easterly phase and, hence, induced upwelling (see arrows), is overlaid by a positive temperature anomaly above ~10 hPa, associated with the next incoming descending westerly QBO wind regime aloft, with its anomalous downwelling and thus adiabatic heating. Second, there are out-of-phase temperature anomalies at subtropical latitudes, consistent with the return arm of the induced circulations shown in Figure 6. Finally, the extratropical temperature anomalies are seasonally synchronized, with much stronger subtropical responses during winter/spring of each hemisphere.

5. IMPACT ON THE EXTRATROPICS

As described in section 2, the QBO is primarily an equatorial phenomenon, as illustrated by Figure 4a. Nevertheless, there appears to be an influence on the extratropical flow, although its identification and characterization is much more difficult because of the increased variability there. Most studies have concentrated on the NH because of the greater

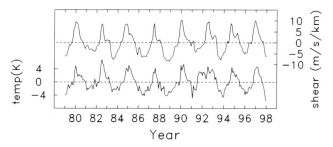

Figure 7. Equatorial temperature anomalies associated with the QBO in the 30- to 50-hPa layer (bottom plot) and vertical wind shear (top plot). Reproduced from *Baldwin et al.* [2001].

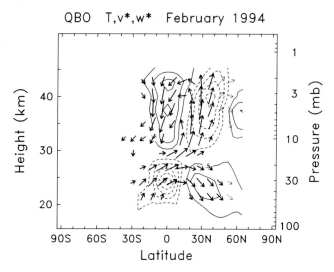

QBO   T,v*,w*   February 1994

Figure 8. Cross-section of QBO anomalies in 1994. Temperature anomalies are contoured (±0.5, 1.0, 1.5 K, etc., with negative anomalies denoted by dashed contours), and components of the residual mean circulation are shown as vectors (scaled by an arbitrary function of altitude). Reprinted from *Randel et al.* [1999]. Copyright American Meteorological Society.

abundance of observations there than in the Southern Hemisphere (SH).

Holton and Tan [1980, 1982] were the first to identify a QBO influence on NH winter flow, using observations of 50 hPa geopotential heights from 1962 to 1977. They found the geopotential heights to be significantly lower during the westerly QBO phase, corresponding to colder, less disturbed winters with a stronger westerly polar vortex. Conversely, during the easterly QBO phase, the winters are warmer and more disturbed by planetary (Rossby) waves, resulting in more stratospheric sudden warmings (SSWs) and a weaker polar vortex. This relationship of a cold/undisturbed vortex under westerly QBO conditions and a warm/disturbed vortex under easterly QBO conditions is referred to as the Holton-Tan relationship. Figure 9 shows a composite of the NH December–January averaged zonal winds for all easterly and westerly QBO phases for the period 1957–2008. The equatorial wind speeds at 44 hPa were used to define the QBO phase. The difference between the two (Figure 9c) shows a statistically significant difference of ~10 m s^{-1} in the strength of the extratropical flow and the polar vortex, in addition to the expected signals at the equator.

The planetary (Rossby) waves that propagate vertically through the background winter westerly flow at midlatitudes and high latitudes and disrupt the polar vortex are generated in the troposphere. They are primarily stationary waves, generated by surface topography and land-sea contrasts [see

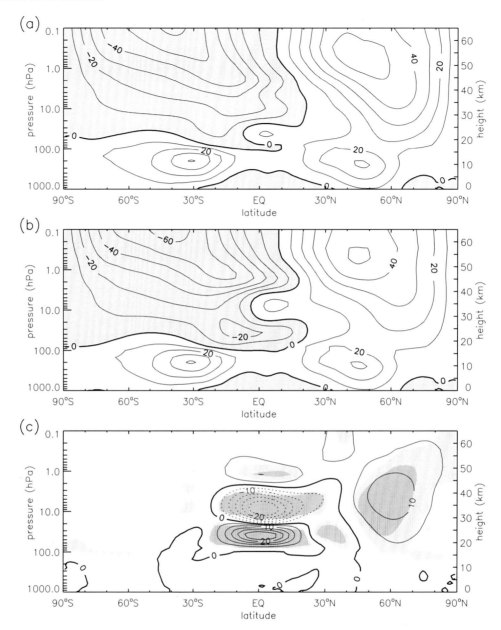

Figure 9. December–January zonal mean zonal wind composites for (a) QBO westerly and (b) QBO easterly years derived for the period 1958–2008. Shaded regions denote easterlies, and contours are 10 m s^{-1} intervals. (c) Westerly minus easterly QBO phase zonal wind difference with t test confidence shading shown at 95% and 99% levels. Contours are 5 m s^{-1}. Updated from *Pascoe et al.* [2005].

Plumb, this volume]. The vertical and meridional propagation of these waves depends upon the background zonal wind field, which can be thought of as refracting the waves as they travel through the stratosphere. In particular, the waves are unable to propagate through background easterly flow. The phase of the QBO is thought to act as a wave guide, since it influences the position of the boundary between the mid-

zlatitude westerlies and easterlies in the tropics/subtropics, which is the critical line for Rossby waves with zero phase speed. In a westerly QBO phase, the waves are able to penetrate into the tropics, while in an easterly QBO phase, the waves encounter a critical line on the winter side of the equator. The effective wave guide for planetary wave propagation in the easterly QBO phase is therefore much narrower, and wave

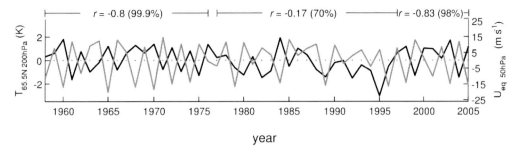

Figure 10. Time series of extratropical zonal-mean temperature (K) at 65.5°N, 200 hPa (black line) plotted against equatorial zonal-mean zonal wind (shaded line) at 50 hPa (m s^{-1}) averaged over the extended winter period (November–March). Correlation coefficient for the entire period 1958–2006 is −0.57 (confidence level 99.7%). Correlation coefficients and confidence levels for subperiods 1958–1976, 1977–1997, and 1998–2006 are given at top of plot. Reproduced from *Lu et al.* [2008].

activity at midlatitudes and high latitudes tends to be stronger, resulting in a more disturbed NH vortex, as suggested by Figure 9.

Although, for convenience, the phase of the QBO is usually denoted by the sign of the equatorial winds at a specified level in the lower stratosphere, e.g., 44 or 50 hPa, it should be stressed that Rossby waves are very deep structures, and their propagation is likely sensitive not only to the lower stratosphere winds but also the QBO in the mid and even possibly the upper equatorial stratosphere [*Gray et al.*, 2001]. Since there are several reversals in QBO phase between the lower and upper stratosphere (see Figure 9), it is unclear at what height the sensitivity lies. It is possible that the sensitivity may not be solely to the sign of the equatorial wind, but also to the vertical extent ("depth") of the QBO wind phases [*Dunkerton et al.*, 1988], the strength/location of vertical gradients in equatorial wind [*Gray et al.*, 2001] or, also related, the timing of the descent of the QBO phases in relation to the annual cycle of Rossby wave forcing [*Anstey and Shepherd*, 2008].

Since Holton and Tan first reported their results, additional analyses have been carried out, as the observational record has increased in length. In Figure 10, we show the temperature anomaly for the extended NH winter (November–March) at 65.5°N, 200 hPa (black line) together with the equatorial zonal wind at 50 hPa (gray line), which acts as a measure of the QBO phase. In the early period before 1977, which is the period studied by Holton and Tan, there is a statistically significant anticorrelation, with warmer temperatures associated with the easterly QBO phase (the so-called Holton-Tan relationship described above). However, as the years of observations accrued, *Labitzke and van Loon* [1988] noted that there were times when this relationship broke down, e.g., around 1967–1970 and around 1977–1979. They noted that these periods coincide with periods of increased solar activity, generally referred to as solar maxima [see *Haigh*, this volume]

and suggested an interaction between the QBO and the 11-year solar cycle and their influence on polar temperatures (see section 6 below).

An examination of the whole time series in Figure 10, however, suggests that the 11-year solar cycle may not be the only additional influence on the behavior of the polar vortex. There is an extended period from 1977 to 1997 when there is very little correlation between equatorial winds and polar temperatures. After 1997, the Holton-Tan anticorrelative behavior reappears. These shifts in behavior appear not to be associated with solar activity and may be more closely connected to regimes shifts in the troposphere that affect the strength of the Rossby wave forcing [e.g., *Lu et al.*, 2008], although more research is required to confirm this. Variability in the strength of Rossby wave generation in the troposphere arises due to internally driven dynamical variability and due to sea surface temperature anomalies such as El Niño. In addition to these possible sources of variability from the troposphere, there is also variability associated with internal stratospheric dynamics, the 11-year solar cycle, and episodic volcanic eruptions, all of which influence polar temperatures and thus make it difficult to extract the QBO signal.

The Holton-Tan relationship appears to be more evident in early winter than in late winter, at least in the midstratosphere, 30–50 hPa (although the period 1977–1997 appears anomalous as described above). This is consistent with increased wave forcing of the polar vortex in the QBO easterly phase, which results in earlier warmings in the easterly phase than the westerly phase. Later in the winter, when the variability of the polar vortex is larger, their influence is less easy to detect, and the signal is also confused by periods of colder than average temperatures that immediately follow a warming, as the vortex is reestablished. On the other hand, at the lower levels, e.g., 200 hPa, the Holton-Tan QBO relationship appears to hold up throughout the winter, as shown in Figure 10. This is possibly due to the longer radiative timescales in the lower stratosphere, which means that an influence from early winter will remain evident for longer.

In the SH, Rossby wave activity is much weaker than in the NH because of weaker land-sea contrast and orography. Observations suggest a modest modulation of midlatitude wave activity, but this is not sufficiently strong to significantly affect the SH vortex until very late in spring, when the vortex amplitude is reduced during the final warming. This observed seasonality of the QBO modulation in the two hemispheres provides further supportive evidence that planetary wave activity is an important mechanism in the QBO modulation of the extratropical flow.

6. INTERACTION WITH THE 11-YEAR SOLAR CYCLE

As described above, *Labitzke* [1987] and *Labitzke and van Loon* [1988] were the first to report an apparent interaction of the QBO signal with the 11-year solar cycle signal in NH winter temperatures. Their early observations have stood the

test of time: Figure 11 shows the results of an extended analysis covering the period 1942–2010, in which the 30-hPa geopotential heights at the North Pole have been plotted as a function of both the QBO phase and the solar flux. Warmer temperatures are found in the QBO easterly phase (i.e., the Holton-Tan relationship) under low solar flux conditions (solar min), but warmer temperatures are found in the westerly QBO phase under high solar flux conditions (solar max), suggesting a reversal of the Holton-Tan relationship in solar maximum. Holton and Tan's analysis had included two solar minima and only one solar maximum, and had therefore been biased toward solar minimum conditions.

Although there is still much uncertainty over the veracity of the solar cycle/QBO interaction because of the short data period available, a number of possible mechanisms for the interaction have been suggested. As described by *Haigh* [this volume], the solar cycle influence on the stratosphere is believed to be associated with increased irradiance during solar maximum periods and the associated increases in ozone, both of which result in a heating of the equatorial upper stratosphere

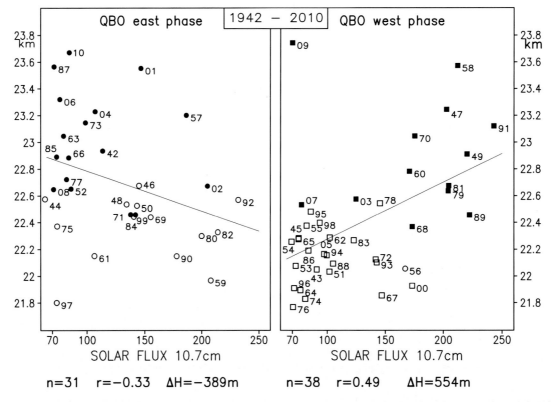

Figure 11. Scatter diagrams of the monthly mean 30 hPa geopotential heights (geopot. km) in February at the North Pole (1942 to 2010), plotted against the 10.7-cm solar flux in solar flux units (1 sfu = 10^{-22} W m^{-2} Hz^{-1}). (left) Years in the east phase of the QBO ($n = 31$). (right) Years in the west phase ($n = 38$). The numbers indicate the respective years; solid symbols indicate major midwinter warmings; r = correlation coefficient; ΔH gives the mean difference of the heights (geopot. m) between solar maxima and minima (minima are defined by solar flux values below 100). Updated from *Labitzke et al.* [2006]. Copyright E. Schweizerbart Science publishers, http://www.borntraeger-cramer.de.

during solar maximum periods. The resulting temperature changes alter the winds in the subtropical upper stratosphere, and these may then influence the propagation of Rossby waves [*Kodera and Kuroda*, 2002] and hence affect the polar vortex. The source of the solar cycle/QBO interaction in this scenario is thought to lie in the upper stratosphere, where there are wind anomalies associated with both the solar cycle and the QBO [*Gray et al.*, 2004].

In addition to this "polar" mechanism route, there is also the possibility of a direct "equatorial" route, where temperature and wind changes associated with the solar cycle in the upper equatorial stratosphere directly affect the descent rate of the QBO. As described in section 3, the wave momentum transfer that gives rise to the QBO at any given level depends only on the background winds at that level and below; one possible equatorial mechanism is that the solar cycle influences the QBO phases in the upper equatorial stratosphere through changes in the background winds there [*Pascoe et al.*, 2005]. There will also be induced secondary meridional circulations associated with the adjustment of the winds to the solar cycle temperature changes, and these can penetrate and influence the levels below the temperature anomaly.

It is not clear which of these various solar cycle/QBO interaction mechanisms is primarily responsible for the observed interactions. If the "polar" route mechanism is acting, 11-year variations in wave momentum transfer at high latitudes will change the strength of the Brewer-Dobson circulation, which will then influence the equatorial upwelling in the lower stratosphere and thus the ability of the QBO phases to descend. Any variations in the vertical structure of the equatorial winds will also have the potential for influencing the propagation of Rossby waves and hence feed back onto the polar vortex strength in winter. Thus, the polar and equatorial routes are intimately entwined, so it may not be possible to distinguish between them from observations alone. There is also an added complication resulting from the interaction with ozone, which has a sufficiently long lifetime in the lower stratosphere that anomalous circulations due to the QBO and solar cycle will result in anomalous ozone distributions, which provide an additional feedback mechanism through the resulting heating.

Figure 11 and other observations suggest that the solar/QBO interaction is evident at polar latitudes. There is also some observational evidence that the equatorial mechanism may be acting. *Salby and Callaghan* [2006] have suggested a solar modulation of the duration of the westerly QBO phase, which would be consistent with a solar modulation of the equatorial upwelling and possibly a modulation of the frequency of "stalling" events (see section 4), but there is still much debate about this apparent signal, and longer data sets are required before it can be verified.

7. IMPACT ON TRACER TRANSPORT

The induced meridional circulations associated with both the QBO and SAO are also evident in the distribution of chemical species and aerosols. Many species that have a relatively short chemical lifetime display QBO modulations because they respond quickly to QBO-induced changes in temperature or to QBO-induced transport that changes the abundance of reactive species. On the other hand, those species that have relatively long chemical lifetimes can be

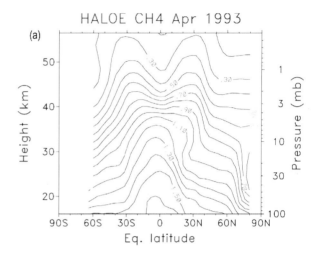

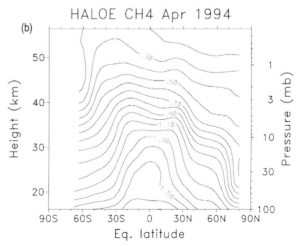

Figure 12. Halogen Occultation Experiment observations of methane in April for (a) 1993 and (b) 1994, illustrating a double peak in the upper stratosphere caused by the SAO and its modulation by the QBO below. Note the strong double peak in 1993 and indication of strong upwelling in the region 25–35 km associated with the descending easterly QBO phase compared with the reduced-amplitude double peak and reduced upwelling near 35 km in 1994, associated with the descending westerly phase. Adapted from *Baldwin et al.* [2001].

thought of as "tracers" of the flow. If they have vertical or horizontal gradients in background abundances in the locality of induced circulations, their observed distributions will be sensitive to transport processes across those gradients.

This was illustrated well by *Trepte and Hitchmann* [1992], who studied the distribution of volcanic aerosol following the major eruption of the El Chichon volcano at equatorial latitudes. When the QBO was in a descending westerly phase, the aerosol distribution reflected the accompanying meridional circulation, with relative descent at the equator (i.e., decreased ascent) and relative ascent in the subtropics, and displayed a "double peak" distribution near 25 km, with local maxima in the subtropics and a local minimum over the equator. In the descending easterly phase, with accompanying increased ascent over the equator, the aerosol distribution was quite different, and had a single maximum over the equator.

Similar induced circulations are also associated with the SAO at higher altitudes. Figure 12a shows satellite observations of methane in April, soon after the descent of the first westerly SAO phase (see Figure 3). The source of methane is in the lower atmosphere, from where it is transported into the stratosphere primarily by upwelling at equatorial regions associated with the upward branches of the Hadley and Brewer-Dobson circulations. It is destroyed in the upper atmosphere, through oxidation, thus forming a vertical gradient. The descending arm of the Brewer-Dobson circulation transports

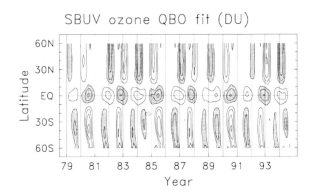

Figure 14. Latitude-time sections of the QBO in column ozone anomalies from the combined SBUV-SBUV/2 data derived using a seasonally varying regression analysis. Contour interval is 3 DU with zero contour omitted and positive values shaded. Vertical lines denote January of each year. Reprinted from *Randel and Wu* [1996]. Copyright American Meteorological Society.

methane-poor air from upper levels down into the extratropics, thus creating horizontal gradients in that region. A very clear double peak structure is evident in the observed methane distribution (Figure 12a) in the region 40–55 km, which coincides with the height region of the SAO, signifying the action of induced relative downwelling at the equator and upwelling in the subtropics associated with the descending westerly phase. A pronounced single peak is present lower down in the height region of the QBO (20–35 km), which was in its easterly descending phase that year. In the following year, however (Figure 12b), there was a descending westerly QBO phase, and this single peak is less pronounced, especially near 35 km where the contours appear flattened. In addition, the SAO double peak is much less evident. This can be explained in terms of a QBO filtering influence on the vertically propagating westerly waves that contribute to the SAO westerly phase, since the QBO-dominated lower atmosphere determines the background flow through which the waves must propagate.

The QBO influence on ozone is rather more complicated to understand, due to its changing chemical lifetime with height. In the upper stratosphere, the ozone lifetime is very short (seconds) and its distribution is controlled by the various chemical reactions that it undergoes. In the lower stratosphere, its chemical lifetime is much longer, and its distribution is controlled by transport processes and is therefore under dynamical control. Figure 13 shows the height distribution of the QBO signal in equatorial ozone. In the lower stratosphere, the positive and negative anomalies are associated with descending westerly and easterly QBO phases, respectively. Remembering that ozone increases with height in the equatorial lower stratosphere, a descending westerly phase at the equator that causes reduced equatorial

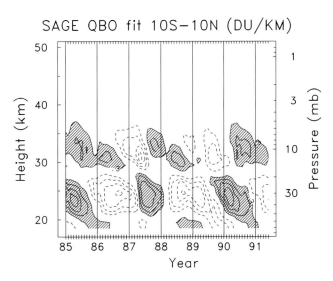

Figure 13. Height-time series of interannual anomalies in ozone density (Dobson unit (DU) km^{-1}, DU = 0.001 atm cm) derived using a regression analysis to isolate the QBO variation. Contour intervals are 0.3 DU km^{-1}, with zero contour omitted and positive values shaded. Reprinted from *Randel and Wu* [1996]. Copyright American Meteorological Society.

ascent via its induced meridional circulation (section 4) will result in anomalously large amounts of ozone. The reverse is true in a descending easterly phase, with increased upwelling transporting low ozone amounts to higher levels. At around 28 km in Figure 13, there is a marked change, with a reversal in sign of the ozone anomaly. This reflects the transition at this level from dynamical to chemical control. The QBO above this level is caused by changes in photochemical sources and sinks of ozone, primarily via transport-induced variations in NO_y [*Chipperfield and Gray*, 1992].

It is evident from Figure 13 that the time-variations of the column (i.e., vertically integrated) ozone amounts will be complicated by the influence from these two different mechanisms, although the lower stratospheric anomalies dominate. Figure 14 shows a time series of column ozone as a function of latitude. The equatorial anomalies have a similar latitudinal width to the QBO, with a reversal in sign in the subtropics. This structure is consistent with anomalous advection of ozone by the induced meridional circulation (Figure 6). A descending westerly phase is associated with a positive anomaly at the equator and a negative anomaly in the subtropics. However, the subtropical anomaly also extends out to higher latitudes, which cannot be explained directly by the QBO-induced meridional circulation, which is confined to low latitudes. Also, the subtropical and high-latitude anomalies are not symmetric about the equator, and the maxima are approximately 6 months apart, coinciding with the local late winter/spring. Occasionally, there is a "missed" subtropical anomaly, as in 1981, 1986, and 1991 in the NH. These unexpected characteristics are due to a modulation of the QBO circulation by the annual cycle, which results in a strengthening of the winter/spring part of the QBO-induced circulation [*Jones et al.*, 1998], as illustrated in Figure 8. The extension of the subtropical signal to high latitudes is not well understood [*Baldwin et al.*, 2001], but is likely to involve an interaction of the QBO with planetary waves, which have a strong annual cycle. Because of this seasonal synchronization, which results in an effective modulation of the QBO period at high latitudes, the QBO may be responsible for a larger component of the high-latitude ozone variability than a simple correlation with the equatorial wind QBO suggests.

8. IMPACT ON THE TROPOSPHERE

There has been increasing evidence over recent years that the state of the lower stratosphere can have an impact on tropospheric flow, especially in wintertime [*Baldwin and Dunkerton*, 2001; see also *Kushner*, this volume]. Stratospheric variability is normally characterized by the Northern and Southern Annular Modes (NAM and SAM), which are the leading EOFs of wintertime geopotential fields. A positive NAM anomaly denotes a strong, cold NH vortex, and a negative NAM index denotes a weak, disturbed vortex that is generally the result of substantial wave activity and the occurrence of a sudden stratospheric warming. The NAM at the Earth's surface is usually referred to as the Arctic Oscillation (AO). Observations suggest that large, sustained variations in the strength of the stratospheric polar vortex in the very lowermost part of the stratosphere propagate downward into the troposphere and remain there for many weeks. Given the evidence that the QBO can influence the strength of the polar vortex, a QBO modulation of the NAM and hence the AO may be expected [e.g., *Barriopedro et al.*, 2008; *Marshall and Scaife*, 2009]. However, the QBO is only one of many factors that influence the strength of the polar vortex, as discussed in section 5, and there is still much uncertainty in the relative contribution of the QBO to the observed surface pressure variations.

Since the QBO originates at equatorial latitudes, there has been sustained interest in whether it may directly influence the underlying tropical troposphere. Observational evidence is hampered by the existence of an approximately 2-year tropospheric oscillation that does not appear to be connected with the stratospheric QBO. Although the QBO itself does not appear to penetrate significantly below the tropopause, and the tropopause temperature QBO is very small, there may be mechanisms such as its influence on static stability, on the vertical wind shear at the upper extent of penetrative storms and on the position of critical levels for tropical easterly waves that might have an influence on the underlying troposphere. In contrast to the stratospheric responses, the spatial scale of these mechanisms is relatively small, and the effect is therefore localized, making detection extremely difficult. Most studies in this area have employed models to explore these mechanisms; while there is some evidence of a QBO impact in the troposphere, there is much uncertainty, possibly related to the models' ability to capture key processes in the tropical troposphere. The only long-standing observational study has been the QBO influence on Atlantic hurricane activity [*Gray et al.*, 1993; S. J. Camargo and A. H. Sobel, Revisiting the influence of the Quasi Biennial Oscillation on tropical cyclone activity, submitted to *Journal of Climate*, 2010]. A further possible impact of the QBO is on the latitudinal width of the Hadley circulation. Studies have shown that this is sensitive to heating anomalies in the equatorial lower stratosphere, such as those associated with the 11-year solar cycle [see *Haigh*, this volume]. Since there are also heating anomalies associated with the QBO in this region, one might expect a similar influence, although there are some differences that may be important; for example, the QBO heating anomalies have a different latitudinal structure to the solar cycle signal, are shorter-lived, and descend with time.

Understanding and quantifying the extent of the stratospheric influence on the underlying troposphere is an important and challenging area that requires much future research.

9. NUMERICAL MODELING OF THE QBO

Significant progress in the representation of the QBO in numerical models has been achieved in recent years. The first simulation in a 2-D (latitude-height) model at equatorial latitudes was that of *Plumb and Bell* [1982b], who parameterized the wave momentum fluxes associated with Kelvin and Rossby-gravity waves. This was extended by *Gray and Pyle* [1989], who achieved the first realistic QBO in a fully global radiative-dynamical-chemical 2-D model. They could, therefore, examine the QBO influence on the extratropics and on ozone and other chemical species distributions.

The use of 3-D (latitude-longitude-height) general circulation models (GCMs) for stratospheric studies was widespread by the early 1990s, albeit at relatively low horizontal and vertical resolution due to the computational costs of including the stratosphere as well as the troposphere. Although they were, therefore, capable of simulating the required wave behavior, none of them achieved a QBO until *Takahashi* [1996], who drastically reduced the vertical grid spacing (from the normal 2–3 km down to 500 m) and also reduced the horizontal diffusion in the model. This allowed the representation of waves with small vertical wavelengths to be represented. However, although many of the widely used stratospheric-resolving GCMs went on to achieve a QBO [e.g., *Hamilton et al.*, 2001; *Scaife et al.*, 2002; *Giorgetta et al.*, 2002; *Shibata and Deushi*, 2005], there is by no means a clear consensus of the factors required to achieve it. Many models now achieve a QBO with the more standard vertical grid spacing. The required momentum comes partly from the resolved waves and partly from parameterized waves such as those generated by nonorographic gravity wave schemes, although the contribution from each varies from model to model. Factors that influence the characteristics of the QBO-like oscillations, such as its period, include both vertical and horizontal resolution, horizontal diffusion, the convection scheme employed in the model, how well the model simulates the background Brewer-Dobson tropical upwelling and also the nature of the local radiative influences in the tropical lower stratosphere, such as the ozone distribution in that region.

It is also noteworthy that the majority of GCMs employed by the Intergovernmental Panel on Climate Change to assess future climate change [*Intergovernmental Panel on Climate Change*, 2007] did not fully resolve stratospheric processes, and none include a representation of the QBO [*Dall'Amico et al.*, 2010a, 2010b]. However, this situation should improve in time for the next IPCC assessment.

REFERENCES

Alexander, M. J. (2010), Gravity waves in the stratosphere, in *The Stratosphere: Dynamics, Transport, and Chemistry, Geophys. Monogr. Ser.*, doi: 10.1029/2009GM000864, this volume.

Andrews, D. G., J. R. Holton, and C. B. Leovy (1987), *Middle Atmosphere Dynamics*, 489 pp., Elsevier, San Diego, Calif.

Baldwin, M. P., and T. J. Dunkerton (2001), Stratospheric harbingers of anomalous weather regimes, *Science*, *294*, 581–584.

Baldwin, M. P., et al. (2001), The quasi-biennial oscillation, *Rev. Geophys.*, *39*, 179–229.

Barriopedro, D., R. Garcia-Herrera, and R. Huth (2008), Solar modulation of Northern Hemisphere winter blocking, *J. Geophys. Res.*, *113*, D14118, doi:10.1029/2008JD009789.

Chipperfield, M. P., and L. J. Gray (1992), Two-dimensional model studies of interannual variability of trace gases in the middle atmosphere, *J. Geophys. Res.*, *97*, 5963–5980.

Dall'Amico, M., L. J. Gray, K. H. Rosenlof, A. A. Scaife, K. P. Shine, and P. A. Stott (2010a), Stratospheric temperature trends: Impact of ozone variability and the QBO, *Clim. Dyn.*, *34*, 381–398.

Dall'Amico, M., P. A. Stott, A. A. Scaife, L. J. Gray, K. H. Rosenlof, and A. Y. Karpechko (2010b), Impact of stratospheric variability on tropospheric climate change, *Clim. Dyn.*, *34*, 399–417.

Dunkerton, T. J. (1997), The role of gravity waves in the quasi-biennial oscillation, *J. Geophys. Res.*, *102*, 26,053–26,076.

Dunkerton, T. J., D. P. Delisi, and M. P. Baldwin (1988), Distribution of major stratospheric warmings in relation to the quasi-biennial oscillation, *Geophys. Res. Lett.*, *15*, 136–139.

Ebdon, R. A. (1960), Notes on the wind flow at 50 mb in tropical and subtropical regions in January 1957 and in 1958, *Q. J. R. Meteorol. Soc.*, *86*, 540–542.

Giorgetta, M. A., E. Manzini, and E. Roeckner (2002), Forcing of the quasi-biennial oscillation from a broad spectrum of atmospheric waves, *Geophys. Res. Lett.*, *29*(8), 1245, doi:10.1029/2002GL014756.

Gray, L. J., and J. A. Pyle (1989), A two-dimensional model of the quasi-biennial oscillation in ozone, *J. Atmos. Sci.*, *46*, 203–220.

Gray, L. J., S. J. Phipps, T. J. Dunkerton, M. P. Baldwin, E. F. Drysdale, and M. R. Allen (2001), A data study of the influence of the upper stratosphere on Northern Hemisphere stratospheric warmings, *Q. J. R. Meteorol. Soc.*, *127*, 1985–2003.

Gray, L. J., S. Crooks, C. Pascoe, S. Sparrow, and M. Palmer (2004), Solar and QBO influences on the timing of stratospheric sudden warmings, *J. Atmos. Sci.*, *61*, 2777–2796.

Gray, W. M., C. W. Landsea, P. W. Mielke Jr., and K. J. Berry (1993), Predicting Atlantic basin seasonal tropical cyclone activity by 1 August, *Weather Forecast.*, *8*, 73–86.

Haigh, J. D. (2010), Solar variability and the stratosphere, in *The Stratosphere: Dynamics, Transport, and Chemistry, Geophys. Monogr. Ser.*, doi: 10.1029/2010GM000937, this volume.

Hamilton, K., R. J. Wilson, and R. S. Hemler (2001), Spontaneous stratospheric QBO-like oscillations simulated by the GFDL SKYHI general circulation model, *J. Atmos. Sci.*, *58*, 3271–3292.

Haynes, P. H. (1998), The latitudinal structure of the quasi-biennial oscillation, *Q. J. R. Meteorol. Soc.*, *124*, 2645–2670.

Holton, J. R., and R. A. Lindzen (1972), An updated theory for the quasi-biennial cycle of the tropical stratosphere, *J. Atmos. Sci.*, *29*, 1076–1080.

Holton, J. R., and H.-C. Tan (1980), The influence of the equatorial quasi-biennial oscillation on the global circulation at 50 mb, *J. Atmos. Sci.*, *37*, 2200–2208.

Holton, J. R., and H.-C. Tan (1982), The quasi-biennial oscillation in the Northern Hemisphere lower stratosphere, *J. Meteorol. Soc. Jpn.*, *60*, 140–148.

Intergovernmental Panel on Climate Change (2007), *Climate Change 2007: The Physical Science Basis: Working Group 1 Contribution to the Fourth Assessment Report of the Intergovernmental Panel on Climate Change*, edited by S. Solomon et al., Cambridge Univ. Press, New York.

Jones, D. B. A., H. R. Schneider, and M. B. McElroy (1998), Effects of the quasi-biennial oscillation on the zonally-averaged transport of tracers, *J. Geophys. Res.*, *103*, 11,235–11,249.

Kodera, K., and Y. Kuroda (2002), Dynamical response to the solar cycle, *J. Geophys. Res.*, *107*(D24), 4749, doi:10.1029/2002JD002224.

Kushner, P. J. (2010) Annular modes of the troposphere and stratosphere, in *The Stratosphere: Dynamics, Transport, and Chemistry, Geophys. Monogr. Ser.*, doi: 10.1029/2009GM000924, this volume.

Labitzke, K. (1987), Sunspots, the QBO and the stratospheric temperature in the north polar region, *Geophys. Res. Lett.*, *14*, 535–537.

Labitzke, K., and H. van Loon (1988), Associations between the 11-year solar cycle, the QBO and the atmosphere, Part I: the troposphere and stratosphere in the northern hemisphere in winter, *J. Atmos. Terr. Phys.*, *50*, 197–206.

Labitzke, K., M. Kunze, and S. Bronnimann (2006), Sunspots, the QBO and the stratosphere in north polar regions—20 years later, *Meteorol. Z.*, *15*, 355–363.

Lindzen, R. S., and J. R. Holton (1968), A theory of the quasi-biennial oscillation, *J. Atmos. Sci.*, *25*, 1095–1107.

Lu, H., M. P. Baldwin, L. J. Gray, and M. J. Jarvis (2008), Decadal-scale changes in the effect of the QBO on the northern stratospheric polar vortex, *J. Geophys. Res.*, *113*, D10114, doi:10.1029/2007JD009647.

Marshall, A. G., and A. A. Scaife (2009), The impact of the QBO on surface winter climate, *J. Geophys. Res.*, *114*, D18110, doi:10.1029/2009JD011737.

Pascoe, C. L., L. J. Gray, S. A. Crooks, M. N. Juckes, and M. P. Baldwin (2005), The quasi-biennial oscillation: Analysis using ERA-40 data, *J. Geophys. Res.*, *110*, D08105, doi:10.1029/2004JD004941.

Plumb, R. A. (1977), The interaction of two internal waves with the mean flow: Implications for the theory of the quasi-biennial oscillation, *J. Atmos. Sci.*, *34*, 1847–1858.

Plumb, R. A. (1984), The quasi-biennial oscillation, in *Dynamics of the Middle Atmosphere*, edited by J. R. Holton and T. Matsuno, pp. 217–251, Terra Sci., Tokyo.

Plumb, R. A. (2010), Planetary waves and the extratropical winter stratosphere, in *The Stratosphere: Dynamics, Transport, and Chemistry, Geophys. Monogr. Ser.*, doi: 10.1029/2009GM000888, this volume.

Plumb, R. A., and R. C. Bell (1982a), Equatorial waves in steady zonal shear flow, *Q. J. R. Meteorol. Soc.*, *108*, 313–334.

Plumb, R. A., and R. C. Bell (1982b), A model of the quasi-biennial oscillation on an equatorial beta-plane, *Q. J. R. Meteorol. Soc.*, *108*, 335–352.

Plumb, R. A., and A. D. McEwan (1978), The instability of a forced standing wave in a viscous stratified fluid: A laboratory analogue of the quasi-biennial oscillation, *J. Atmos. Sci.*, *35*, 1827–1839.

Randel, W. J., and F. Wu (1996), Isolation of the ozone QBO in SAGE II data by singular decomposition, *J. Atmos. Sci.*, *53*, 2546–2559.

Randel, W. J., F. Wu, R. Swinbank, J. Nash, and A. O'Neill (1999), Global QBO circulation derived from UKMO stratospheric analyses, *J. Atmos. Sci.*, *56*, 457–474.

Reed, R. J., W. J. Campbell, L. A. Rasmussen, and D. G. Rogers (1961), Evidence of a downward-propagating annual wind reversal in the equatorial stratosphere, *J. Geophys. Res.*, *66*, 813–818.

Salby, M. L., and P. F. Callaghan (2006), Relationship of the quasi-biennial oscillation to the stratospheric signature of the solar cycle, *J. Geophys. Res.*, *111*, D06110, doi:10.1029/2005JD006012.

Scaife, A. A., N. Butchart, C. D. Warner, and R. Swinbank (2002), Impact of a spectral gravity wave parameterization on the stratosphere in the Met Office Unified Model, *J. Atmos. Sci.*, *59*, 1473–1489.

Scott, R. K., and P. H. Haynes (1998), Internal interannual variability of the extratropical stratospheric circulation: The low latitude flywheel, *Q. J. R. Meteorol. Soc.*, *124*, 2149–2173.

Shibata, K., and M. Deushi (2005), Partitioning between resolved wave forcing and unresolved gravity wave forcing to the quasi-biennial oscillation as revealed with a coupled chemistry-climate model, *Geophys. Res. Lett.*, *32*, L12820, doi:10.1029/2005GL022885.

Takahashi, M. (1996), Simulation of the stratospheric quasi-biennial oscillation using a general circulation model, *Geophys. Res. Lett.*, *23*, 661–664.

Trepte, C. R., and M. H. Hitchman (1992), Tropical stratospheric circulation deduced from satellite aerosol data, *Nature*, *355*, 626–628.

L. J. Gray, Department of Meteorology, University of Reading, Earley Gate, PO Box 243, Reading RG6 6BB, UK. (l.j.gray@reading.ac.uk)

Gravity Waves in the Stratosphere

M. Joan Alexander

Colorado Research Associates Division, North West Research Associates, Boulder, Colorado, USA

This chapter presents a review of some recent research highlighting direct gravity wave effects in the stratosphere. In the last 20 years, our understanding of the range of these effects has grown in tandem with improvements in resolution in both observations and models. The effects include gravity wave driving of the general circulation, temperature structure and related effects on polar ozone chemistry, and effects on ice clouds. Recent observations of gravity waves in the stratosphere that help to quantify these effects are also highlighted. The observations are giving the picture of a collection of events, occurring sporadically in localized wave packets, superimposed on a weaker background spectrum of waves. Finally, new information on the sources of gravity waves gleaned from both the observations and wave-resolving models is also summarized. The improved knowledge of these sources is expected to lead to advances in gravity wave parameterizations in global climate models that will permit more realistic feedback between the waves and future climate change.

1. INTRODUCTION

Prior to 1987 when the work of *Andrews et al.* [1987] was published, we had a clear understanding of the global effects of gravity waves in the mesosphere. A working parameterization of gravity wave forcing effects on the global circulation had been developed for global models [*Lindzen*, 1981; *Holton*, 1982, 1983], and the modeling community had also come to appreciate the effects of mountain wave drag near the tropopause on the general circulation and the importance of this process in both weather forecasting and climate models [*Palmer et al.*, 1986; *McFarlane*, 1987]. A successful theory for the forcing effects of wave dissipation on the mean flow had been developed, and was the cornerstone of these developments [*Andrews and McIntyre*, 1976; *Boyd*, 1976; *Andrews et al.*, 1987]. The transformed Eulerian-mean equations form a set that describes both the direct effects of wave forcing on

the zonal circulation as well as the effects on the meridional transport circulation and the temperature structure of the atmosphere. In their quasigeostrophic form derived on a beta plane, they are a simple set that will be useful for the discussion in this chapter. Reproducing equations 3.5.5 from *Andrews et al.* [1987],

$$\bar{u}_t - f_0\bar{v}^* = \bar{X} + \rho_0^{-1}\nabla\cdot\mathbf{F} \qquad (1)$$

$$\bar{\theta}_t + \bar{w}^*\theta_{0z} = \bar{Q} \qquad (2)$$

$$\bar{v}_y^* + \rho_0^{-1}(\rho_0\bar{w}^*)_z = 0 \qquad (3)$$

The momentum ($\bar{X} + \nabla\cdot\mathbf{F}/\rho_0$) and thermal ($\bar{Q}$) forcing terms are placed on the right-hand sides. Here $(\bar{u},\bar{v})$ is the zonal-mean wind, θ_0 and ρ_0 are reference potential temperature and density that vary with height only, and f_0 the Coriolis parameter defined in the center of the beta plane. These equations describe the temperature and circulation responses to a wave-driven forcing. The momentum equation (1) shows that a driven momentum forcing can lead to both wind accelerations ($\bar{u}_t$) as well as meridional drift through the Coriolis torque ($f_0\bar{v}^*$). Via continuity (equation (3)), the meridional drift is associated with vertical motions ($\bar{w}^*$) that are tied to

The Stratosphere: Dynamics, Transport, and Chemistry
Geophysical Monograph Series 190
Copyright 2010 by the American Geophysical Union.
10.1029/2009GM000864

thermal changes via the thermodynamic equation (2). The meridional transport circulation is approximated here by the residual circulation $(\bar{v}*, \bar{w}*)$ defined by

$$\bar{v}* = \bar{v}_a - \rho_0^{-1}(\rho_0 \overline{v'\theta'}/\theta_{0z})_z,$$

$$\bar{w}* = \bar{w}_a + (\overline{v'\theta'}/\theta_{0z})_y. \qquad (4)$$

In this quasi-geostrophic beta-plane case, the divergence of the Elliassen-Palm flux has only two terms due to eddy momentum and heat fluxes, $\nabla \cdot \mathbf{F} = -(\rho_0 \overline{v'u'})_y + (\rho_0 f_0 \overline{v'\theta'}/\theta_{0z})_z$. The other momentum forcing term could be the forcing due to parameterized gravity waves and be written as the vertical gradient in wave stress or momentum flux $\bar{X} = -\rho_0^{-1}(\rho_0 \overline{u'w'})_z$. In gravity wave parameterization schemes, the momentum flux $\rho_0 \overline{u'w'}$ is specified at some altitude along with other wave propagation properties, and the parameterization determines the force $\bar{X}$. Mountain wave parameterizations treat only stationary waves, while "nonorographic" gravity wave parameterizations treat a broad spectrum of phase speeds.

Holton and Alexander [2000] reviewed the fundamentals of planetary waves and gravity waves, and their roles in driving the transport circulation of the middle atmosphere. In this chapter, we specifically highlight effects of gravity waves at stratospheric levels. Section 2 describes these effects, particularly those that have been discovered in recent decades. Many of these effects have been inferred from global model studies. Knowledge of gravity wave sources and momentum fluxes has been a limitation in quantifying these effects. Recent observations discussed in section 3 show that gravity wave momentum fluxes in the lower stratosphere can vary considerably in individual measurements and can be traced to specific wave sources. The measurements also show seasonal and latitudinal patterns that may begin to describe a climatology. Climate change may result in long-term variations in gravity wave sources, so there is current interest in developing nonorographic source parameterizations for moist convection and jet stream sources that will respond to changing climate in the way mountain wave parameterizations currently do. The nonorographic parameterizations applied in today's chemistry-climate models do not change with changing climate, and this is a limitation in their use for forecasts. We therefore focus in section 4 in gravity wave sources.

2. GRAVITY WAVE EFFECTS IN/ON THE STRATOSPHERE

In the last 20 years, there have been some notable developments in our understanding of gravity wave effects at stratosphere levels in contrast to the previous understanding

of their effects at higher levels in the mesosphere. In the mesosphere, gravity waves have first-order effects. Their dissipation near the mesopause causes complete reversals in the direction of the zonal-mean winds, and the resulting pole-to-pole meridional circulation drives the temperature structure very far from radiative equilibrium. At the polar summer mesopause where the sun shines continuously, but the wave-driven Lagrangian-mean meridional circulation (equation 4) leads to upwelling, temperatures are the coldest found anywhere in the atmosphere. The polar winter mesopause, conversely in complete darkness, is relatively warm because of net downwelling. These upper atmosphere effects were discussed and illustrated by *Andrews et al.* [1987].

Gravity wave effects in the stratosphere, in contrast, are second order. Planetary waves in the extratropics and global-scale equatorial waves account for the majority of the wave-driven circulation effects. However, recent developments have shown the importance of gravity waves. They serve as helpers, with their effects responding to and exaggerating changes in the winds initiated by the global-scale-wave driving. In some situations, gravity waves account for the majority of the wave-driving effects. We next describe several notable examples of the direct effects of gravity waves on the circulation in the stratosphere.

2.1. Extratropical Effects

2.1.1. Wave driving of the Brewer-Dobson circulation. The equator-to-pole Lagrangian-mean meridional transport circulation in the stratosphere is called the Brewer-Dobson circulation for researchers who first proposed it [*Brewer*, 1949; *Dobson*, 1956]. The circulation is largely driven by planetary wave drag [*Yulaeva et al.*, 1994], but the summer hemisphere branch and the seasonal variation in the strength of the circulation is linked to forcing from dissipation of smaller-scale gravity waves that are not resolved in most global models and instead are treated via parameterization. *Alexander and Rosenlof* [2003] used observations to derive the seasonal cycle of net extratropical wave forcing across the 90.7-hPa surface in the lower stratosphere of each hemisphere using the solution for the residual circulation [*Rosenlof*, 1996] and the net wave-driven force F required to balance the transformed Eulerian mean momentum equation. They then computed the resolved wave contribution to EP-flux divergence from a global analysis and by subtraction, the gravity wave (unresolved) forcing. Their results (Figure 1) show that the resolved forcing drives the wintertime maximum in the Brewer-Dobson circulation in each hemisphere, but that smaller-scale gravity waves dominate the wave forcing in the spring-to-summer transition season in each hemisphere.

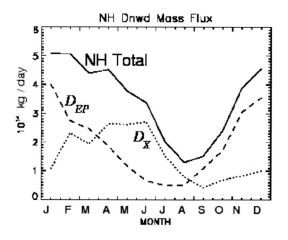
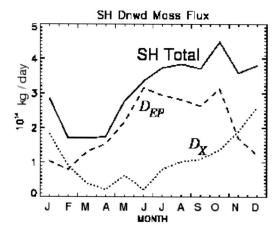

Figure 1. Seasonal variations in downward mass flux at 90 hPa in the northern (NH) and southern (SH) hemispheres. The dashed and dotted curves show contributions from resolved planetary waves (D_{EP}) and unresolved gravity waves (D_X) derived from the two components of the wave-driven force and downward control at the latitude marking the poleward edge of tropical upwelling in each hemisphere. After *Alexander and Rosenlof* [2003].

2.1.2. Transition to summer easterlies. Global circulation models that do not include parameterized nonorographic gravity waves tend to have trouble in describing the spring-to-summer transition of extratropical winds and temperatures in the stratosphere. The transition from winter eastward winds to summer westward winds tends to occur approximately a month too late. *Scaife et al.* [2002] demonstrated that this problem can be resolved using a parameterization of nonorographic gravity waves (Figure 2). A spectrum of westward propagating gravity waves with nonzero phase speeds is required to accelerate the winds and transition from eastward to westward. Orographic waves can only drag the winds toward zero but not accelerate them to westward values. The westward gravity wave forcing responsible for this will not only improve the timing of the transition from eastward to westward winds, but will also help to drive the summer-hemisphere cell of the Brewer-Dobson circulation.

2.1.3. The cold-pole problem and ozone chemistry. Global models without parameterized small-scale wave drag have stratospheric polar winter temperatures that are far colder than observed [*Palmer et al.*, 1986; *McFarlane*, 1987]. Resolved planetary wave drag is insufficient to drive the full strength of the winter residual circulation downwelling. Gravity wave drag at both stratospheric and mesospheric levels is important to accurate modeling of stratospheric temperatures [*Garcia and Boville*, 1994]. Some of the required drag is provided via mountain wave parameterizations [*Boville*, 1991]. Spectral gravity wave parameterizations provide additional drag at winter polar latitudes that is particularly important in the Southern Hemisphere (SH) winter. Climate simulations designed to predict future ozone

changes, so-called "chemistry-climate models," require some form of gravity wave parameterization, which has the effect of adjusting stratospheric polar winter temperatures to more realistic values, before temperature-sensitive chemical reactions can be accurately modeled [*Austin et al.*, 2003; *Eyring et al.*, 2007].

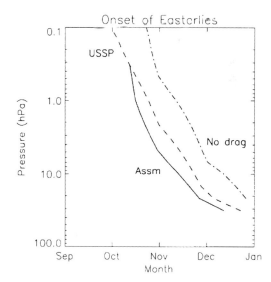

Figure 2. Timing of the onset of westward winds near 60°S latitude as a function of pressure through the spring-to-summer season in the UK Met Office model. "Assm" shows the result after assimilation of observations, and "No drag" is the free-running model without the parameterization of gravity wave forcing. "USSP" shows the improvement in the model timing with parameterized gravity wave drag. After *Scaife et al.* [2002]. Copyright American Meteorological Society.

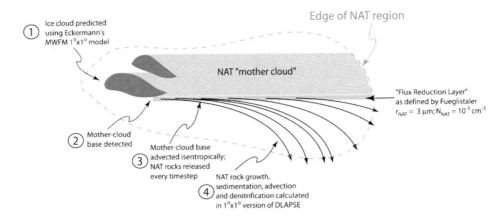

Figure 3. Schematic showing the role of mountain wave temperature anomalies in denitrification of stratospheric air. The mountain waves (via a Mountain Wave Forecast Model) initiate formation of nitric acid trihydrate (NAT) clouds, followed by NAT particle growth and sedimentation of NAT "rocks" (larger particles) downstream. After *Mann et al.* [2005].

2.1.4. Polar stratospheric clouds and ozone chemistry. Polar stratospheric clouds (PSCs) form as winter stratospheric temperatures drop below 195 K. Chemical reactions on the surfaces of these PSCs convert reservoir chlorine species into chemically active forms that destroy ozone when sunlight returns in spring. The clouds may also denitrify polar air, inhibiting reactions that convert the active chlorine back to the reservoir forms [*Solomon and Schoeberl*, 1988; *Tolbert and Toon*, 2001]. Northern Hemisphere (NH) winter polar temperatures are warmer than those in the SH, and correspondingly PSCs are more widespread and persistent in the south. These differences in PSC occurrence are a primary reason for hemispheric differences in seasonal ozone loss. High-latitude mountain waves cause temperature perturbations, and PSCs have been observed to form in the cold phases of the waves in conditions that are otherwise too warm [*Dörnbrack et al.*, 2002]. The chemistry occurring on the surfaces of these clouds is not reversed when the air parcel subsequently passes through the adjacent warm phase of the wave. So net chlorine activation results, and the absence of these small-scale wave features may account for underprediction of Arctic ozone depletion in some models [*Carslaw et al.*, 1998]. Wave-induced clouds may also affect ozone abundance in the Arctic and Antarctic through denitrification following the formation and sedimentation of PSCs [*Mann et al.*, 2005; *Höpfner et al.*, 2006; *Eckermann et al.*, 2009] (see Figure 3).

2.2. Tropical Effects

2.2.1. Gravity wave forcing of the QBO. Twenty years ago, the quasi-biennial oscillation (QBO) in stratospheric zonal winds was understood to be driven by dissipation of tropical waves, but a quantitative working model of the QBO

remained elusive. The primary wave forcing was believed to come from eastward propagating Kelvin waves and westward propagating mixed Rossby-gravity waves [*Holton and Lindzen*, 1972], and the mechanism further elucidated by *Plumb* [1977]. These equatorial wave mechanisms supplanted an earlier theory for gravity wave driving [*Lindzen and Holton*, 1968]. It is now known that a broad spectrum of waves in the tropics contribute to the forcing of the QBO, including both equatorial wave modes and higher-frequency gravity waves [*Dunkerton*, 1997; *Sato and Dunkerton*, 1997]. Reproducing the phenomenon in a three-dimensional (3-D) global model requires high vertical resolution (0.5–0.75 km) to simulate the resolved wave and mean-flow interaction [*Baldwin et al.*, 2001; *Hamilton*, 2008]. Realistic simulations at zonal resolutions near 4° have additionally required parameterization of the effects of small-scale gravity waves [*Giorgetta et al.*, 2002; *Scaife et al.*, 2002] as well as use of a convective parameterization that reproduces high-frequency variability in convection, which results in a broad spectrum of resolved waves [*Takahashi*, 1996; *Ricciardulli and Garcia*, 2000; *Horinouchi et al.*, 2003]. Recent model studies suggest that small-scale gravity waves provide roughly half of the eastward propagating wave flux and half or possibly much more of the westward propagating wave flux needed to drive the QBO [*Giorgetta et al.*, 2002; *Scaife et al.*, 2000; *Kawatani et al.*, 2010], with the gravity waves contributing relatively more in the upper stratosphere and the resolved waves more in the lower stratosphere.

2.2.2. Gravity wave forcing of the stratopause semiannual oscillation. The semiannual oscillation (SAO) in zonal winds near the stratopause is driven by a combination of processes. The westward phases are, in part, due to the advection of summer westward winds across the equator by the residual circulation. Additional forcing comes from wave dissipation, likely including Kelvin, Rossby, Rossby-gravity, and gravity waves [*Hitchman and Leovy*, 1988; *Sassi and Garcia*, 1997; *Garcia et al.*, 1997; *Ray et al.*, 1998].

Small-scale gravity waves may play a relatively important role in descent of the eastward phases [*Garcia et al.*, 1997; *Ray et al.*, 1998]. This conclusion follows from two things: (1) the observation that the eastward phase descent is asymmetric about the equator indicating a lesser role for planetary-scale Kelvin waves, which are symmetric and (2) the fact that underlying QBO eastward winds are much weaker than their westward counterparts and will therefore filter a smaller fraction of the upward propagating wave spectrum that reaches the upper stratosphere to drive the SAO.

2.2.3. Tropical cirrus. Waves of all scales can modulate or initiate the formation of cirrus clouds near the tropical tropopause in the cold phases of the waves. The wave amplitude can determine ice-cloud formation when air is otherwise too warm, and the cooling rate (related to the wave intrinsic frequency) will influence ice particle sizes, number densities, and sedimentation. When ice particles grow large enough to fall, this further affects water vapor concentrations. If the waves affect water vapor near the tropical tropopause in this way, they may thus play a role in the dehydration of air entering the tropical stratosphere, which is subsequently transported globally via the stratospheric residual circulation. Waves may thus influence stratospheric dehydration, cirrus cloud occurrence frequencies, cirrus optical depth, and cloud radiative properties.

Observations have shown that Kelvin waves influence cirrus cloud formation [*Boehm and Verlinde*, 2000; *Fujiwara et al.*, 2009]. The effects of gravity waves in the above processes are poorly understood at present, but model studies indicate a potentially important role for gravity waves in determining cirrus cloud occurrence frequencies [*Jensen and Pfister*, 2004] and particle sizes [*Jensen et al.*, 2009].

2.3. Role of Gravity Waves in Model Responses to Climate Change

2.3.1. Mountain wave parameterization effects on the climate response to increasing CO_2. *Sigmond et al.* [2008] examined differences in the atmospheric circulation response to CO_2 doubling in two models: one designed for climate forecasts, and the other designed for chemistry-climate forecasts. The latter is distinguished by a well-resolved stratosphere, but the two models also include different tunings due to the different purposes for which they were designed. The authors show substantially different NH winter surface pressure response patterns in the two models and

investigate possible causes of the differences. The responses were not very sensitive to the differences in stratospheric resolution or the level of the model top. Instead, differences in the orographic gravity wave parameterizations explained most of the differences in the surface pressure response pattern. Specifically, a parameter that describes the momentum flux of the orographic waves when given identical settings in the two models brought the models into much closer agreement. Larger flux settings cause larger wave drag and weaker winds in the lowermost stratosphere, which in turn affect planetary wave propagation into the stratosphere that leads to changes in surface pressure patterns. The result underscores how uncertainty in mountain wave parameterization settings can influence climate response patterns.

2.3.2. Trends in the Brewer-Dobson circulation. Chemistry-climate models require a well-resolved middle atmosphere. Recent model intercomparisons describe increased upwelling across the tropical tropopause and increases in the Brewer-Dobson transport circulation as robust features of the model responses to CO_2 increases [*Butchart et al.*, 2006]. The cause of the transport circulation increases must be related to changes in stratospheric wave forcing because stratospheric wave dissipation drives the transport circulation. However, the changes in wave dissipation need not be related to changes in wave fluxes from the troposphere. Instead, the wave dissipation may simply change due to changes in winds and stability in the stratosphere. The relative contributions of these two factors are not yet known.

Chemistry-climate models include realistic scenarios for increasing greenhouse gases in the 21st century. All climate models show that increasing greenhouse gases result in both tropospheric warming as well as stratospheric cooling. These changes are, in turn, associated with an increased latitudinal temperature gradient in the subtropical upper troposphere/lower stratosphere, which in turn is related to increases in the subtropical jet strength at upper levels. The wind response to changes in the temperature gradient is easily understood from geostrophic balance.

Several recent studies address the question of which waves are responsible for the future trends in the tropical upwelling in chemistry-climate models. Planetary waves, orographic gravity waves, and equatorial waves likely all contribute to the changes to some degree, but different model analyses have come to slightly different conclusions. Two recent studies found that changes in parameterized orographic wave drag were important in explaining the upwelling trends [*Li et al.*, 2008; *McLandress and Shepherd*, 2009]. Subtropical orographic wave drag shifts to higher altitudes in the future climate in response to increases in the winds in the lowermost stratosphere in these models. Planetary-scale waves are also

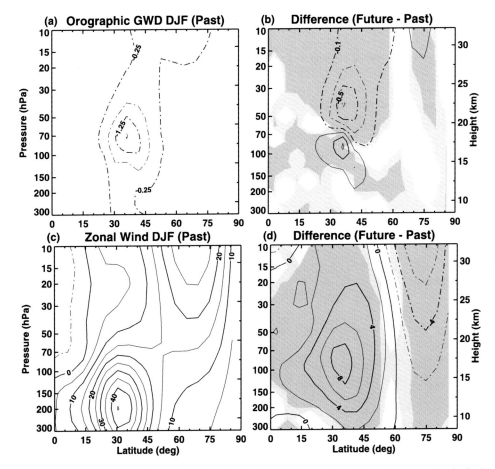

Figure 4. Analysis of NH chemistry-climate model results showing (a) distribution of mountain wave drag in the "past," (b) change in the mountain wave drag in the "future," (c) zonal mean winds in the "past," and (d) change in the zonal mean winds in the "future." "Past" here is the average in years 1960–1979, and "future" is the average in 2080–2099. After *McLandress and Shepherd* [2009]. Shading denotes significance.

important to explaining the trends, but the small- and large-scale waves may be most important at somewhat higher and lower altitudes, respectively [*McLandress and Shepherd*, 2009] (see Figure 4). *McLandress and Shepherd* [2009] further showed that these conclusions can be very sensitive to the range of tropical latitudes included in the diagnosis of the trends in tropical upwelling and that this sensitivity may account for apparently different conclusions about the type of waves responsible drawn by *Li et al.* [2008] and *Garcia and Randel* [2008].

3. OBSERVATIONS OF GRAVITY WAVE MOMENTUM FLUX

Many of the issues described above are treated via parameterization of gravity wave mean-flow forcing effects. These parameterizations require specification of the wave stress or momentum flux at the locations of wave sources (for mountain waves) or as a function of latitude at some arbitrary level near tropopause (for nonorographic gravity waves). Observational constraints on these fluxes have been lagging behind the application of the parameterizations in global models. Mountain wave parameterizations were employed in most of the climate models that participated in the last IPCC AR4 report [*Intergovernmental Panel on Climate Change*, 2007]. The parameterizations are also widely used in global weather forecasting assimilation systems particularly the widely used products of the European (ECMWF) and US (NCEP) operational forecast centers. Nonorographic gravity wave parameterizations have become an important component of models used for ozone chemistry-climate forecasts for the last ozone assessment (see WMO/UNEP [2007] and more in-depth analyses described by *Eyring et al.* [2007]). As mentioned in section 2.1.3, these non-orographic wave parameterizations were needed to tune the

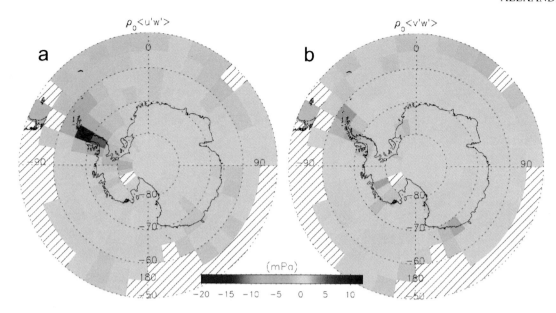

Plate 1. (a) Zonal and (b) meridional components of vector momentum flux derived from long-duration balloon flights in the southern lower stratosphere during September 2005 to February 2006. Large values over the Antarctic Peninsula are associated with mountain waves. Too few measurements were made in hatched regions for a determination. After *Hertzog et al.* [2008]. Copyright American Meteorological Society.

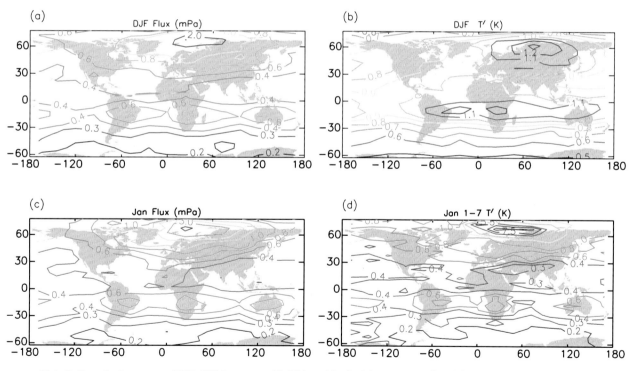

Plate 2. Boreal winter maps, 2005–2006 season at 20–30 km altitude: (a) momentum flux (Flux) averaged over 3 months; (b) temperature amplitude (T') averaged over 3 months; (c) flux averaged over 1 month; and (d) flux averaged over 1 week. Comparing Plates 2a, 2c, and 2d, maximum fluxes increase (3, 6.5, 11 mPa) as the length of averaging time decreases (3 months, 1 month, 1 week) because of wave intermittency. On individual days, maximum fluxes are ~50 mPa, but wave events have limited duration and move from place to place. Averaging such intermittent events decreases the maximum values (derived from High-Resolution Dynamics Limb Sounder temperature data with analysis methods described by *Alexander et al.* [2008]).

polar circulation and subsequent effects on polar winter temperatures in order to describe realistic chemical changes in the models [*Austin et al.*, 2003]. Without the gravity wave parameterizations, the polar winter temperatures are far too cold, and this leads to serious errors in the temperature-sensitive ozone chemistry in these models. The modelers could tune the gravity wave schemes to adjust these winter polar temperatures to reasonable values that allowed realistic hindcasts and forecasts of ozone changes and recovery.

The key uncertain parameter in these gravity wave schemes is the wave stress or momentum flux. The research community looked to observations to constrain these fluxes and to give guidance on how they might vary with latitude and season. Global data sets were needed. Satellites could provide the needed coverage, and improvements in resolution of satellite measurements meant that they began to resolve gravity waves in the 1990s [*Fetzer and Gille*, 1994; *Wu and Waters*, 1996]. However, these measurements provided only temperature variance and potential energy, whereas the gravity wave schemes require constraints on momentum flux. Early theories of the wave spectrum predicted that the potential energy distributions could be used as a proxy for momentum flux [*Fritts and VanZandt*, 1993]; however, mounting observational evidence called these assumptions in the spectral theory into question. Patterns in satellite maps of gravity wave potential energy were further noted to be highly dependent on the portion of the wave spectrum visible to each satellite instrument [*Alexander*, 1998; *Preusse et al.*, 2008].

Long-duration quasi-Lagrangian balloon flights in the lower stratosphere have provided some of the most accurate measurements of gravity wave momentum fluxes and their direction. Vector momentum fluxes from one of these balloon campaigns are shown in Plate 1. Regional variations tied to mountain wave sources are apparent. A high degree of intermittency in the waves was also observed and quantified from these data [*Hertzog et al.*, 2008]. This intermittency means that long-term averaging of momentum fluxes results in much smaller values than short-term averaging (see Plate 2).

Recent high-resolution satellite observations of gravity wave temperature fluctuations have been used to produce global maps of momentum flux computed directly from the observations at altitudes in the lower stratosphere [*Ern et al.*, 2004; *Alexander et al.*, 2008]. These momentum flux calculations require simultaneous measurement of the 3-D wave structure (both horizontal and vertical wavelength) to convert the observed wave temperature amplitudes to momentum fluxes. Plates 1a and 1b show global maps of momentum flux and temperature amplitude, respectively. These show similarities, but the flux is not directly proportional to temperature amplitude or potential energy. Global momentum fluxes so far determined this way have lacked information on the wave propagation direction. This limitation renders the estimates of horizontal wavelength uncertain, and it leaves the estimates of momentum flux uncertain to the same degree. Despite these uncertainties, the five following patterns emerge: (1) Average horizontal wavelengths tend to be much longer in the equatorial region than at higher latitudes; this result agrees with analyses of radiosonde data [*Wang et al.*, 2005]. (2) Because of the horizontal wavelength trend, potential energy maps show weaker latitudinal gradients than momentum flux maps in side-by-side comparisons, a result that confirms that wave potential energy measurements alone cannot be used quantitatively to constrain momentum flux. (3) Wave momentum fluxes are generally much larger in winter seasons than in summer. (4) Large-amplitude mountain waves have been identified, and large average fluxes appear where notable mountain wave sources occur, such as over the Andes and Antarctic peninsula. (5) A high degree of day-to-day variability has been observed (also called intermittency), a result also quantified in analyses of superpressure balloon data [*Hertzog et al.*, 2008].

Some cautionary statements are needed along with the presentation of these general results:

1. Although the satellite measurements used to estimate momentum fluxes are sensitive to a broad range of vertical wavelengths, some observational filtering associated with the analysis methods are likely to still affect the global patterns observed to some degree.

2. The noise and other limits in sensitivity of the measurements will mean that weak waves in the stratosphere will not be observed, and these results therefore focus on larger-amplitude events. This could limit the applicability of these observational constraints on gravity wave forcing in the mesosphere and lower thermosphere because very weak waves occurring in the stratosphere can still grow to large amplitudes before reaching the mesopause. Therefore, weak waves that are either not observed or not emphasized in the existing satellite results might still have profound effects near the mesopause.

3. The satellite analyses assume that the waves observed are propagating upward from sources below. Wave-breaking events can induce secondary wave emission [*Holton and Alexander*, 1999; *Satomura and Sato*, 1999; *Vadas et al.*, 2003], generating waves that propagate both upward and downward. The prevalence or rarity of this process is not currently known. Because wave amplitudes increase with height due to the exponential decrease in density with height, these secondary waves may be far more prominent at higher altitudes, one indication that nonlinear processes become more important in shaping the gravity wave spectrum at mesospheric altitudes. Approximately linear propagation from identifiable tropospheric sources is more common at stratospheric altitudes.

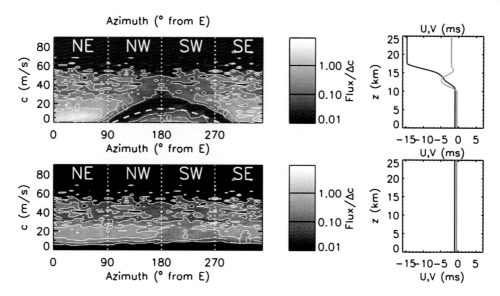

Figure 5. (left) Momentum flux versus phase speed and propagation direction computed for waves in the stratosphere above convection with (right) two different wind profiles. (top) The case with strong winds near the top of the convective heating. A peak in momentum flux occurs at phase speeds 0–10 m s^{-1} with northeastward (NE) propagation directions. (bottom) The case without shear shows the absence of this peak. These NE propagating waves match the motions of latent heating cells within the storm, marking the signature of the "obstacle effect" wave generation mechanism. After *Alexander et al.* [2006]. Copyright American Meteorological Society.

4. SOURCES AND WAVE-GENERATION MECHANISMS

Section 2 summarized some of the ways that gravity wave processes can affect climate forecast models. These processes are treated via parameterization in global climate and chemistry-climate models. The computational demands for such forecasts across century or longer timescales will likely keep model vertical and horizontal resolution too coarse into the near future to model gravity wave process directly, so the parameterizations will continue to be needed. Current operational versions of the climate models include mountain wave parameterizations that will respond to some degree to climate change. Changes in surface wind and stability will change the waves generated by orography, and thus mountain wave source parameterizations have some prognostic capability.

If considered at all, gravity waves from other sources are usually given properties that do not vary with time and may vary only gradually with latitude in operational models. These nonorographic wave parameterizations are a particularly important component of chemistry-climate models, and the sources for the waves modeled in this way are meant to include convection, various processes active in the jet stream like frontal development and jet imbalance, and others (see, e.g., *Fritts and Alexander* [2003]). Wave source parameterizations for fronts [*Charron and Manzini*, 2002] and convection [*Chun et al.*, 2004; *Beres et al.*, 2005] have been

proposed and implemented in research versions of climate models, but the parameterizations are poorly validated against observations at present. High-resolution theoretical model studies have examined waves emanating from these sources and also from regions of jet imbalance. Below, we briefly summarize some recent results from high-resolution model studies of convection and jet sources.

4.1. Waves Generated by Convection

Latent heating within convection is a source for waves because it is localized and time-dependent. In the absence of wind shear, the duration and scale of the heating roughly describe lower limits on the wave period and wavelength that can be generated. So, intense, short-duration, small-scale heating events are efficient wave sources over a broad range of gravity wave properties.

Convection can, in general, generate a broad spectrum of wave phase speeds, with peaks in the stratospheric spectrum above the storm related to the depth of the heating within the storm [*Alexander et al.*, 1995; *Piani et al.*, 2000; *Beres et al.*, 2002; *Holton et al.*, 2001].

A separate peak in the spectrum at a phase speed approximately matching the motions of the heating cells within the storm also appears when there is upper-level shear or when the wind near the top of the heating cell is at least ~5 m s^{-1} relative to the heating cell motion [*Pfister et al.*,

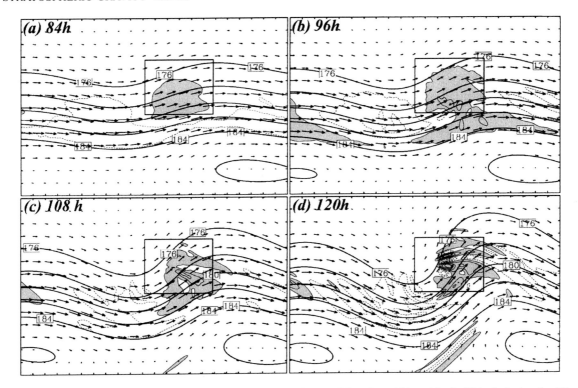

Figure 6. A time series of waves emitted by a jet-front system in a model study at 13 km altitude. Thin dashed and solid contours show wind divergence, and thick contours show isobars. Wind vectors are also shown. After *Zhang* [2004].

1993; *Beres et al.*, 2004; *Alexander et al.*, 2006; *Kuester et al.*, 2008] (see Figure 5). This wind-sensitive peak is associated with waves generated by the so-called "obstacle effect," a mechanism analogous to orographic wave generation [*Chun and Baik*, 1998].

Additional spectral peaks may be associated with oscillations within the storm at specific frequencies [*Fovell et al.*, 1992; *Alexander et al.*, 1995]. *Lane and Reeder* [2001] proposed that the "oscillator" could be described as moist air parcels rising to their level of neutral buoyancy and oscillating there at the local moist buoyancy frequency. This implies waves from this source would be associated with very high frequencies, those with frequencies characterized by buoyancy frequencies that occur in the upper troposphere.

One recent model study of waves generated by convection has been validated by comparison to observation from the AIRS satellite [*Grimsdell et al.*, 2010].

4.2. Waves From Jet Sources

Models of waves generated by jet sources are apparently sensitive to model resolution [*O'Sullivan and Dunkerton*, 1995; *Zhang*, 2004] because such models require both high-resolution and large domain sizes (see Figure 6). Analysis of the waves in these models is complicated by the

complexities of the wind field in the vicinity of the wave source [*Plougonven and Snyder*, 2005; *Bühler and McIntyre*, 2005]. Strong horizontal and vertical gradients in the winds lead to dramatic changes in the observed wave spectrum through processes like wave refraction, critical level filtering, trapping, and preferential propagation. These can all have dramatic effects on the spectrum, making it very difficult to isolate the spectrum emanating from the source and difficult to develop a parameterization for waves from this source. Validation of a model of waves from jet sources by comparison to satellite observation was shown by *Wu and Zhang* [2004].

5. SUMMARY

In the last 20 years, our understanding of the range of gravity wave effects in the stratosphere has grown in tandem with improved resolution in both observations and models. The direct effects of gravity waves in the stratosphere are numerous and significant, but generally smaller in magnitude than the effects of planetary-scale waves. Treatment of gravity wave effects on the circulation via parameterization has become standard practice in global climate and weather forecasting models.

A lack of observational constraints for parameterizations of gravity wave circulation effects in global models has limited our quantitative understanding of these effects. Recent satellite observations have sufficient resolution to provide some global constraints. Global constraints on momentum fluxes and wave propagation properties are needed. The observations give a picture of the waves as a collection of events in the stratosphere, sporadically occurring in very localized wave packets, probably superimposed upon a weaker background wavefield.

High-resolution models with observational validation are giving us a clearer understanding of gravity wave sources and mechanisms, which is expected to lead to improved wave source parameterizations for global models. Existing parameterizations for mountain wave sources and future parameterizations for convection and jet sources will improve the prognostic capability of parameterizations of gravity wave circulation effects. These will permit more realistic feedback between future changes in climate and gravity waves in global models.

Acknowledgments. This work was supported by NASA Earth Science Division contract NNH08CD37C and NSF Physical and Dynamic Meteorology Program grant ATM-0632378.

REFERENCES

Alexander, M. J. (1998), Interpretations of observed climatological patterns in stratospheric gravity wave variance, *J. Geophys. Res.*, *103*, 8627–8640.

Alexander, M. J., and K. H. Rosenlof (2003), Gravity-wave forcing in the stratosphere: Observational constraints from Upper Atmosphere Research Satellite and implications for parameterization in global models, *J. Geophys. Res.*, *108*(D19), 4597, doi:10.1029/2003JD003373.

Alexander, M. J., J. R. Holton, and D. R. Durran (1995), The gravity wave response above deep convection in a squall line simulation, *J. Atmos. Sci.*, *52*, 2212–2226.

Alexander, M. J., J. H. Richter, and B. R. Sutherland (2006), Generation and trapping of gravity waves from convection, with comparison to parameterization, *J. Atmos. Sci.*, *63*(11), 2963–2977.

Alexander, M. J., et al. (2008), Global estimates of gravity wave momentum flux from High Resolution Dynamics Limb Sounder observations, *J. Geophys. Res.*, *113*, D15S18, doi:10.1029/2007JD008807.

Andrews, D. G., and M. E. Mc Intyre (1976), Planetary waves in horizontal and vertical shear: The generalized Eliassen-Palm relation and the mean zonal acceleration, *J. Atmos. Sci.*, *33*, 2031–2048.

Andrews, D. G., J. Holton, and C. Leovy (1987), in *Middle Atmosphere Dynamics*, 489 pp., Elsevier, Orlando, Fla.

Austin, J., et al. (2003), Uncertainties and assessments of chemistry-climate models of the stratosphere, *Atmos. Chem. Phys.*, *3*, 1–27.

Baldwin, M. P., et al. (2001), The quasi-biennial oscillation, *Rev. Geophys.*, *39*, 179–229.

Beres, J. H., M. J. Alexander, and J. R. Holton (2002), Effects of tropospheric wind shear on the spectrum of convectively generated gravity waves, *J. Atmos. Sci.*, *59*, 1805–1824.

Beres, J. H., M. J. Alexander, and J. R. Holton (2004), A method of specifying the gravity wave spectrum above convection based on latent heating properties and background wind, *J. Atmos. Sci.*, *61*, 324–337.

Beres, J. H., R. R. Garcia, B. A. Boville, and F. Sassi (2005), Implementation of a gravity wave source spectrum parameterization dependent on the properties of convection in the Whole Atmosphere Community Climate Model (WACCM), *J. Geophys. Res.*, *110*, D10108, doi:10.1029/2004JD005504.

Boehm, M. T., and J. Verlinde (2000), Stratospheric influence on upper tropospheric tropical cirrus, *Geophys. Res. Lett.*, *27*(19), 3209–3212.

Boville, B. A. (1991), Sensitivity of simulated climate to model resolution, *J. Clim.*, *4*, 469–485.

Boyd, J. P. (1976), The noninteraction of waves with the zonally averaged flow on a spherical Earth and the interrelationships of eddy fluxes of energy, heat and momentum, *J. Atmos. Sci.*, *33*, 2285–2291.

Brewer, A. (1949), Evidence for a world circulation provided by the measurements of helium and water vapor distribution in the stratosphere, *Q. J. R. Meteorol. Soc.*, *75*, 351–363.

Bühler, O., and M. McIntyre (2005), Wave capture and wave-vortex duality, *J. Fluid Mech.*, *534*, 67–95.

Butchart, N., et al. (2006), Simulations of anthropogenic change in the strength of the Brewer-Dobson circulation, *Clim. Dyn.*, *27*, 727–741.

Carslaw, K. S., et al. (1998), Increased stratospheric ozone depletion due to mountain-induced atmospheric waves, *Nature*, *391*, 675–678.

Charron, M., and E. Manzini (2002), Gravity waves from fronts: Parameterization and middle atmosphere response in a general circulation model, *J. Atmos. Sci.*, *59*, 923–941.

Chun, H.-Y., and J.-J. Baik (1998), Momentum flux by thermally induced internal gravity waves and its approximation for large-scale models, *J. Atmos. Sci.*, *55*, 3299–3310.

Chun, H.-Y., I.-S. Song, J.-J. Baik, and Y.-J. Kim (2004), Impact of a convectively forced gravity wave parameterization in NCAR CCM3, *J. Clim.*, *17*(18), 3530–3547.

Dobson, G. M. B. (1956), Origin and distribution of the polyatomic molecules in the atmosphere, *Proc. R. Soc. London, Ser. A*, *236*, 187–193.

Dörnbrack, A., T. Birner, A. Fix, H. Flentje, A. Meister, H. Schmid, E. V. Browell, and M. J. Mahoney (2002), Evidence for inertia-gravity waves forming polar stratospheric clouds over Scandinavia, *J. Geophys. Res.*, *107*(D20), 8287, doi:10.1029/2001JD000452.

Dunkerton, T. J. (1997), The role of gravity waves in the quasi-biennial oscillation, *J. Geophys. Res.*, *102*, 26,053–26,076.

Eckermann, S. D., L. Hoffmann, M. Höpfner, D. L. Wu, and M. J. Alexander (2009), Antarctic NAT PSC belt of June 2003: Observational validation of the mountain wave seeding hypothesis, *Geophys. Res. Lett.*, *36*, L02807, doi:10.1029/2008GL036629.

Ern, M., P. Preusse, M. J. Alexander, and C. D. Warner (2004), Absolute values of gravity wave momentum flux derived from satellite data, *J. Geophys. Res.*, *109*, D20103, doi:10.1029/2004JD004752.

Eyring, V., et al. (2007), Multimodel projections of stratospheric ozone in the 21st century, *J. Geophys. Res.*, *112*, D16303, doi:10.1029/2006JD008332.

Fetzer, E. J., and J. C. Gille (1994), Gravity wave variance in LIMS temperatures. Part I: Variability and comparison with background winds, *J. Atmos. Sci.*, *51*, 2461–2483.

Fovell, R., D. Durran, and J. R. Holton (1992), Numerical simulations of convectively generated stratospheric gravity waves, *J. Atmos. Sci.*, *49*, 1427–1442.

Fritts, D. C., and M. J. Alexander (2003), Gravity wave dynamics and effects on the middle atmosphere, *Rev. Geophys.*, *41*(1), 1003, doi:10.1029/2001RG000106.

Fritts, D. C., and T. E. VanZandt (1993), Spectral estimates of gravity wave energy and momentum fluxes, I: Energy dissipation, acceleration, and constraints, *J. Atmos. Sci.*, *50*, 3685–3694.

Fujiwara, M., et al. (2009), Cirrus observations in the tropical tropopause layer over the western Pacific, *J. Geophys. Res.*, *114*, D09304, doi:10.1029/2008JD011040.

Garcia, R. R., and B. A. Boville (1994), "Downward control" of the mean meridional circulation and temperature distribution of the polar winter stratosphere, *J. Atmos. Sci.*, *51*, 2238–2245.

Garcia, R. R., and W. J. Randel (2008), Acceleration of the Brewer-Dobson circulation due to increases in greenhouse gases, *J. Atmos. Sci.*, *65*, 2731–2739.

Garcia, R. R., T. J. Dunkerton, R. S. Lieberman, and R. A. Vincent (1997), Climatology of the semiannual oscillation of the tropical middle atmosphere, *J. Geophys. Res.*, *102*, 26,019–26,032.

Giorgetta, M. A., E. Manzini, and E. Roeckner (2002), Forcing of the quasi-biennial oscillation from a broad spectrum of atmospheric waves, *Geophys. Res. Lett.*, *29*(8), 1245, doi:10.1029/2002GL014756.

Grimsdell, A. W., M. J. Alexander, P. May, and L. Hoffmann (2010), Model study of waves generated by convection with direct validation via satellite, *J. Atmos. Sci.*, *67*, 1617–1631.

Hamilton, K. (2008), Numerical resolution and modeling of the global atmospheric circulation: A review of our current understanding and outstanding issues, in *High Resolution Numerical Modelling of the Atmosphere and Ocean*, chap. 1, pp. 8–27, Springer, New York.

Hertzog, A., G. Boccara, R. A. Vincent, F. Vial, and P. Cocquerez (2008), Estimation of gravity wave momentum flux and phase speeds from quasi-Lagrangian stratospheric balloon flights. Part II: Results from the Vorcore campaign in Antarctica, *J. Atmos. Sci.*, *65*, 3056–3070, doi:10.1175/2008JAS2710.1.

Hitchman, M. H., and C. B. Leovy (1988), Estimation of the Kelvin wave contribution to the semiannual oscillation, *J. Atmos. Sci.*, *45*, 1462–1475.

Holton, J. R. (1982), The role of gravity wave induced drag and diffusion in the momentum budget of the mesosphere, *J. Atmos. Sci.*, *39*, 791–799.

Holton, J. R. (1983), The influence of gravity wave breaking on the general circulation of the middle atmosphere, *J. Atmos. Sci.*, *40*, 2497–2507.

Holton, J. R., and M. Alexander (1999), Gravity waves in the mesosphere generated by tropospheric convection, *Tellus, Ser. B*, *51*, 45–58.

Holton, J. R., and M. J. Alexander (2000), The role of waves in the transport circulation of the middle atmosphere, in *Atmospheric Science Across the Stratopause, Geophys. Monogr. Ser.*, vol. 123, edited by D. E. Siskind, S. D. Eckermann, and M. E. Summers, pp. 21–35, AGU, Washington, D. C.

Holton, J. R., and R. S. Lindzen (1972), An updated theory for the quasi-biennial cycle of the tropical stratosphere, *J. Atmos. Sci.*, *29*, 1076–1080.

Holton, J. R., M. J. Alexander, and M. T. Boehm (2001), Evidence for short vertical wavelength Kelvin waves in the Department of Energy-Atmospheric Radiation Measurement Nauru99 radiosonde data, *J. Geophys. Res.*, *106*, 20,125–20,129.

Höpfner, M., et al. (2006), MIPAS detects Antarctic stratospheric belt of NAT PSCs caused by mountain waves, *Atmos. Chem. Phys.*, *6*, 1221–1230.

Horinouchi, T., et al. (2003), Tropical cumulus convection and upward-propagating waves in middle-atmospheric GCMs, *J. Atmos. Sci.*, *60*, 2765–2782.

Intergovernmental Panel on Climate Change (2007), *Climate Change 2007: The Physical Science Basis: Working Group 1 Contribution to the Fourth Assessment Report of the Intergovernmental Panel on Climate Change*, edited by S. Solomon et al., 996 pp., Cambridge Univ. Press, New York.

Jensen, E., and L. Pfister (2004), Transport and freeze-drying in the tropical tropopause layer, *J. Geophys. Res.*, *109*, D02207, doi:10.1029/2003JD004022.

Jensen, E. J., et al. (2009), On the importance of small ice crystals in tropical anvil cirrus, *Atmos. Chem. Phys.*, *9*, 5519–5537.

Kawatani, Y., S. Watanabe, K. Sato, T. J. Dunkerton, S. Miyahara, and M. Takahashi (2010), The roles of equatorial trapped waves and internal inertia-gravity in driving the quasi-biennial oscillation. Part I: Zonal mean wave forcing, *J. Atmos. Sci.*, *67*, 963–980.

Kuester, M., M. Alexander, and E. Ray (2008), A model study of gravity waves over hurricane Humberto (2001), *J. Atmos. Sci.*, *65*, 3231–3246.

Lane, T., and M. Reeder (2001), Modelling the generation of gravity waves by a Maritime Continent thunderstorm, *Q. J. R. Meteorol. Soc.*, *127*, 2705–2724.

Li, F., J. Austin, and J. Wilson (2008), The strength of the Brewer-Dobson circulation in a changing climate: Coupled chemistry-climate model simulations, *J. Clim.*, *21*, 40–57, doi:10.1175/2007JCLI1663.1.

Lindzen, R. S. (1981), Turbulence and stress owing to gravity wave and tidal breakdown, *J. Geophys. Res.*, *86*, 9707–9714.

Lindzen, R. S., and J. R. Holton (1968), A theory of the quasi-biennial oscillation, *J. Atmos. Sci.*, *25*, 1095–1107.

Mann, G. W., K. S. Carslaw, M. P. Chipperfield, S. Davies, and S. D. Eckermann (2005), Large nitric acid trihydrate particles and denitrification caused by mountain waves in the Arctic stratosphere, *J. Geophys. Res.*, *110*, D08202, doi:10.1029/2004JD005271.

McFarlane, N. A. (1987), The effect of orographically excited gravity wave drag on the general circulation of the lower stratosphere and troposphere, *J. Atmos. Sci.*, *44*, 1775–1800.

McLandress, C., and T. G. Shepherd (2009), Simulated anthropogenic changes in the Brewer-Dobson circulation, including its extension to high latitudes, *J. Clim.*, *22*, 1516–1540, doi:10.1175/2008JCLI26791.

O'Sullivan, D., and T. Dunkerton (1995), Generation of inertia-gravity waves in a simulated life cycle of baroclinic instability, *J. Atmos. Sci.*, *52*(21), 3695–3716.

Palmer, T. N., G. J. Shutts, and R. Swinbank (1986), Alleviation of a systematic westerly bias in general circulation and numerical weather prediction models through an orographic gravity wave drag parameterization, *Q. J. R. Meteorol. Soc.*, *112*, 1001–1039.

Pfister, L., K. R. Chan, T. P. Bui, S. Bowen, M. Legg, B. Gary, K. Kelly, M. Proffitt, and W. Starr (1993), Gravity waves generated by a tropical cyclone during the STEP Tropical Field Program: A case study, *J. Geophys. Res.*, *98*, 8611–8638.

Piani, C., D. Durran, M. Alexander, and J. Holton (2000), A numerical study of three-dimensional gravity waves triggered by deep tropical convection and their role in the dynamics of the QBO, *J. Atmos. Sci.*, *57*(22), 3689–3702.

Plougonven, R., and C. Snyder (2005), Gravity waves excited by jets: Propagation versus generation, *Geophys. Res. Lett.*, *32*, L18802, doi:10.1029/2005GL023730.

Plumb, R. A. (1977), The interaction of two internal waves with the mean flow: Implications for the theory of the quasi-biennial oscillation, *J. Atmos. Sci.*, *34*, 1847–1858.

Preusse, P., S. D. Eckermann, and M. Ern (2008), Transparency of the atmosphere to short horizontal wavelength gravity waves, *J. Geophys. Res.*, *113*, D24104, doi:10.1029/2007JD009682.

Ray, E. A., M. J. Alexander, and J. R. Holton (1998), An analysis of the structure and forcing of the equatorial semiannual oscillation in zonal wind, *J. Geophys. Res.*, *103*, 1759–1774.

Ricciardulli, L., and R. Garcia (2000), The excitation of equatorial waves by deep convection in the NCAR Community Climate Model (CCM3), *J. Atmos. Sci.*, *57*(21), 3461–3487.

Rosenlof, K. H. (1996), Summer hemisphere differences in temperature and transport in the lower stratosphere, *J. Geophys. Res.*, *101*, 19,129–19,136.

Sassi, F., and R. Garcia (1997), The role of equatorial waves forced by convection in the tropical semiannual oscillation, *J. Atmos. Sci.*, *54*, 1925–1942.

Sato, K., and T. J. Dunkerton (1997), Estimates of momentum flux associated with equatorial Kelvin and gravity waves, *J. Geophys. Res.*, *102*, 26,247–26,261.

Satomura, T., and K. Sato (1999), Secondary generation of gravity waves associated with the breaking of mountain waves, *J. Atmos. Sci.*, *56*(22), 3847–3858.

Scaife, A. A., N. Butchart, C. D. Warner, D. Stainforth, and W. Norton (2000), Realistic quasi-biennial oscillations in a simulation of the global climate, *Geophys. Res. Lett.*, *27*(21), 3481–3484.

Scaife, A., N. Butchart, C. Warner, and R. Swinbank (2002), Impact of a spectral gravity wave parameterization on the stratosphere in the Met Office Unified Model, *J. Atmos. Sci.*, *59*, 1473–1489.

Sigmond, M., J. F. Scinocca, and P. J. Kushner (2008), Impact of the stratosphere on tropospheric climate change, *Geophys. Res. Lett.*, *35*, L12706, doi:10.1029/2008GL033573.

Solomon, S., and M. R. Schoeberl (1988), Overview of the polar ozone issue, *Geophys. Res. Lett.*, *15*(8), 845–846.

Takahashi, M. (1996), Simulation of the stratospheric quasi-biennial oscillation using a general circulation model, *Geophys. Res. Lett.*, *23*(6), 661–664.

Tolbert, M. A., and O. B. Toon (2001), Solving the PSC mystery, *Science*, *292*(5514), 61–63.

Vadas, S., D. Fritts, and M. Alexander (2003), Mechanism for the generation of secondary waves in wave breaking regions, *J. Atmos. Sci.*, *60*, 194–214.

Wang, L., M. A. Geller, and M. J. Alexander (2005), Spatial and temporal variations of gravity wave parameters. Part 1: Intrinsic frequency, wavelength, and vertical propagation direction, *J. Atmos. Sci.*, *62*, 125–142.

Wu, D. L., and J. W. Waters (1996), Gravity-wave-scale temperature fluctuations seen by the UARS MLS, *Geophys. Res. Lett.*, *23*(23), 3289–3292.

Wu, D. L., and F. Zhang (2004), A study of mesoscale gravity waves over the North Atlantic with satellite observations and a mesoscale model, *J. Geophys. Res.*, *109*, D22104, doi:10.1029/2004JD005090.

Yulaeva, E., J. R. Holton, and J. M. Wallace (1994), On the cause of the annual cycle in the tropical lower stratospheric temperature, *J. Atmos. Sci.*, *51*, 169–174.

Zhang, F. (2004), Generation of mesoscale gravity waves in upper-tropospheric jet-front systems, *J. Atmos. Sci.*, *61*(4), 440–457.

M. J. Alexander, Colorado Research Associates Division, NWRA, 3380 Mitchell Lane, Boulder, CO 80301, USA. (alexand@cora.nwra.com)

Variability and Trends in Stratospheric Temperature and Water Vapor

William J. Randel

National Center for Atmospheric Research, Boulder, Colorado, USA

Long-term variability and trends in global stratospheric temperature are described, based on radiosonde observations since the 1960s and satellite measurements since 1979. New radiosonde-based data sets are available that include adjustments for instrumental inhomogeneities, and these data show good agreement with satellite measurements in the lower stratosphere. The stratosphere exhibits well-known transient warming linked to large volcanic eruptions, plus long-term cooling with magnitudes ~-0.5 K/decade in the lower stratosphere to ~-1.2 K/decade in the upper stratosphere. Observations of stratospheric water vapor are also analyzed, based on satellite measurements for 1993–2010. Observed interannual variability is dominated by the quasi-biennial oscillation, plus a step-like drop after 2001. For the observed satellite record, variability in stratospheric water vapor is closely tied to temperature anomalies near the equatorial cold point tropopause.

1. INTRODUCTION

The stratosphere is well-recognized as a key component of the climate system and exhibits coupling to the troposphere across synoptic to decadal time scales. Temperature changes and trends in the stratosphere are an important aspect of global change and are crucial for interpreting and understanding anthropogenic climate change and stratospheric ozone trends (including predictions of future changes). The observed variability and trends in temperatures provides a fingerprint of key processes that influence the stratosphere and are fundamental diagnostics for evaluating model simulations [e.g. *Garcia et al.*, 2007; *Intergovernmental Panel on Climate Change*, 2007; *Austin et al.*, 2009; *SPARC*, 2010].

The historical observational record of global stratospheric temperature is relatively short compared to surface climate records, beginning in the late 1950s for the lower stratosphere (from balloon measurements) and in 1979 for the middle and upper stratosphere (based on satellite data). Furthermore, these historical data were intended for use in operational weather analysis and forecasting and not for producing high-quality climate records, and the measurement record is plagued by artificial effects linked to changes in instrumentation or observational practices (such as improvements in radiosonde instruments or changes in operational satellites) [e.g. *Gaffen*, 1994; *Lanzante et al.*, 2003a]. Such artificial effects are important to take into account when trying to quantify relatively small climate signals. This problem is now well-recognized, and several research groups have developed techniques to make adjustments in historical data and produce climate data sets (for both radiosondes and satellites). Because the artificial changes can be subtle and difficult to identify in the presence of natural variability, it is valuable to have independent analyses of the data sets (and these then provide one measure of uncertainty in the final products). One objective of this paper is to give an overview of stratospheric temperature variability and trends from the historical record and briefly discuss current understanding of this variability. Six separate stratospheric temperature data sets have recently been reviewed by *Randel et al.* [2009a], and the results here focus on overall behavior from a few of those data sets.

Stratospheric water vapor is important because of its radiative effects on stratospheric temperature and surface climate [e.g., *Forster and Shine*, 1999; *Solomon et al.*, 2010],

The Stratosphere: Dynamics, Transport, and Chemistry
Geophysical Monograph Series 190
10.1029/2009GM000870

and chemical effects on ozone [*Dvortsov and Solomon*, 2001]. Air enters the stratosphere primarily in the tropics and is dehydrated on passing the cold tropical tropopause, and this accounts for the overall extreme dryness of the stratosphere [*Brewer*, 1949]. The annual cycle of tropical tropopause temperatures (8 K maximum to minimum) furthermore imparts a strong seasonal cycle (approximately 3.0 to 4.5 ppmv) to stratospheric water vapor near the tropopause, which then propagates coherently throughout the stratosphere [*Mote et al.*, 1996]. *Fueglistaler et al.* [2005] showed that the observed seasonal cycle can be accurately simulated using Lagrangian trajectory calculations (based on analyzed large-scale temperature and wind fields), confirming freeze-out near the cold point as a simple explanation of the large seasonal variation. Interest in stratospheric water vapor increased substantially following the observations of large positive trends (~10% per decade) from a long record of balloon measurements by *Oltmans and Hoffman* [1995]. More recently, satellites have provided global observations and increasingly long records of stratospheric water vapor, and substantial effort has focused on quantifying and understanding interannual variability in both satellite and balloon data [e.g., *Stratospheric Processes and Their Role in Climate* (*SPARC*), 2000; *Randel et al.*, 2004; *Fueglistaler and Haynes*, 2005; *Scherer et al.*, 2008]. Analyses of seasonal and interannual changes in water vapor are now a standard diagnostic for stratospheric model simulations [e.g., *Eyring et al.*, 2006; *Garcia et al.*, 2007; *Oman et al.*, 2008; *SPARC*, 2010]. Here an update of the satellite-based record covering 1993–2010 is presented, including comparisons with tropical tropopause temperatures over this period, and inferences regarding the control of interannual changes in stratospheric water vapor are discussed.

2. TEMPERATURE DATA

Historical observations of stratospheric temperature are primarily derived from radiosonde (balloon) and satellite measurements. Radiosonde measurements in the lower stratosphere extend back to the late 1950s, although regular global coverage above 100 hPa did not occur until after approximately 1965. A key aspect of the radiosonde record is that there have been changes in instrumentation and observational practice over the 50-year period, so that the raw radiosonde record contains substantial inhomogeneities that particularly influence the stratosphere [e.g., *Gaffen*, 1994; *Lanzante et al.*, 2003a]. One important problem is that the older measurements often have systematic warm biases in the stratosphere (related to radiation effects on the temperature sensor), so that time series can include artificial cooling biases. These trend biases can be as large as or larger

than the true climate signal [*Lanzante et al.*, 2003b], and these artificial effects require correction before reliable stratospheric trends can be estimated. This problem is well recognized in the research community, and several radiosonde-based data sets have been developed during the last several years, which employ a variety of techniques to isolate and correct data inhomogeneities. *Randel et al.* [2009a] discuss and compare results from six such data sets, and the overall variability is similar among these data (although trend estimates can vary substantially in some regions). Here we focus on results from two of these data sets, which show good overall agreement with satellite measurements. The Radiosonde Atmospheric Temperature Products for Assessing Climate (RATPAC) [*Free et al.*, 2005] is a data set based on an 85-station network, whose data are adjusted using the approach described by *Lanzante et al.* [2003a, 2003b]. The so-called RATPAC-lite data set is a 47-station subset of the RATPAC data, where satellite-radiosonde comparisons have been used to isolate and remove stations with remaining inhomogeneities [*Randel and Wu*, 2006; *Randel et al.*, 2009a]. The RATPAC-lite data cover the period beginning in 1979. We also use the Radiosonde Innovation Composite Homogenization (RICH) data set [*Haimberger et al.*, 2008], which uses meteorological reanalyses to identify artificial break points in radiosonde time series, which are then adjusted using neighboring radiosonde comparisons. While the RICH data extend back to 1958, we note that there are substantial uncertainties in all the radiosonde data sets for the presatellite era, especially for the data sparse tropics and Southern Hemisphere (SH) [*Randel et al.*, 2009a].

Near-global satellite observations of stratospheric temperatures started in the early 1970s, with the first continuous series of observations beginning in the late 1970s with the NOAA operational satellites. These instruments include the Microwave Sounding Unit (MSU) and the Stratospheric Sounding Unit (SSU), which provide ~10–15 km thick layer-mean temperatures for a number of layers spanning the lower to upper stratosphere [see *Randel et al.*, 2009a]. A key point is that individual satellite instruments are relatively short-lived, so that data from 13 different satellites have been used since 1979. This presents challenges for creating climate quality data sets, as each instrument has slightly different calibration characteristics, the orbits differ between satellites and drift for individual satellites (which can alias stratospheric diurnal tides into trends), and the overlap period between different satellites is sometimes small. For the MSU Channel 4 (hereafter MSU4) data (covering ~13–22 km), there are two separate analyses that are routinely updated for the long-term record, from Remote Sensing Systems [*Mears and Wentz*, 2009] and from the University of Alabama at Huntsville

(UAH) [*Christy et al.*, 2003]. There are relatively small differences between these MSU4 data sets, and we focus here on the UAH analyses; note the last MSU instrument ceased operation in 2005, and the time series have been extended to present using data from a very similar channel on the Advanced Microwave Sounding Unit since 1998. SSU measurements are available from 1979 to 2005 and are the only near-global source of temperature measurements above the lower stratosphere over this period. The SSU data include measurements for three nadir-viewing channels, plus several synthetic so-called x-channels, which combine nadir and off-nadir measurements [*Nash and Forrester*, 1986]. The time series shown here combine measurements for seven separate SSU instruments and are an extension of the time series derived by *Nash and Forrester* [1986] and *Nash* [1988]; one key uncertainty is that this is the only analysis of the combined SSU data to date. The SSU data have been corrected to account for effects of increasing atmospheric CO_2 on the measurements, which result in systematically raising the altitude of the SSU weighting functions and positively biasing resulting temperature trends [*Shine et al.*, 2008].

3. TEMPERATURE OBSERVATIONS

An overview of lower stratospheric temperature variability for the period 1960–2008 is shown in Figure 1, which shows

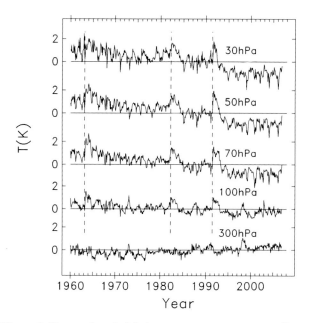

Figure 1. Time series of global average temperature anomalies at pressure levels spanning the upper troposphere to lower stratosphere, derived from the RICH data. The dashed lines denote the volcanic eruptions of Agung (March 1963), El Chichon (April 1982), and Mount Pinatubo (June 1991).

deseasonalized global-mean temperature anomalies at individual pressure levels 300–30 hPa, based on the RICH data. In the stratosphere, the primary components of global variability are the transient warming events linked to the volcanic eruptions of Agung (March 1963), El Chichon (April 1982), and Mount Pinatubo (July 1991), together with long-term net cooling changes of ~2 K. The 300-hPa time series in Figure 1 is included to contrast upper tropospheric temperature behavior, which shows long-term warming, and episodic variations tied to the El Niño–Southern Oscillation (ENSO). The spatial structure of the linear trend and ENSO variations in the zonal mean RICH temperature data for the period 1970–2006 are shown in Figure 2, derived from a standard multivariate linear regression analysis (described in Appendix A). The linear trends in Figure 2a show warming in the troposphere (largest in the NH extratropics and in the tropical upper troposphere) and cooling in the stratosphere. Although the time series in Figure 1 shows that the long-term stratospheric changes are not linear (possibly more step-like) [e.g., *Seidel and Lanzante*, 2004], the linear trends provide a concise measure of net long-term changes. There are substantial latitudinal gradients in trends over 10–15 km linked to the sloping tropopause in Figure 2a, which will lead via the thermal wind relation to increased subtropical jets above 10 km. In the stratosphere above 100 hPa, the trends in Figure 2a show a relatively flat latitudinal structure outside of polar regions. The RICH trends show somewhat larger cooling near the equator and over the SH at 30 and 50 hPa in Figure 2a, but this detail may be suspect because of data uncertainties in these regions.

Zonal mean temperature variations associated with ENSO (Figure 2b) show coherent variations throughout the tropical troposphere, which are approximately in-phase (a 1-month lag) with the multivariate ENSO index, which we use to statistically model ENSO variability (see Appendix A). Note that while Figure 2b shows the zonal mean signature, there is also strong longitudinal (planetary wave) structure to the ENSO temperature response [e.g., *Yulaeva and Wallace*, 1994; *Calvo Fernandez et al.*, 2004]. The ENSO pattern in Figure 2b shows an out-of-phase response in the tropical lower stratosphere (near 70 hPa) that is of similar magnitude to the tropospheric signal and is associated with local temperature variations of ±1 K. A time series of stratospheric temperatures in this region is shown in Figure 3, together with the components associated with separate terms in the regression fit. The ENSO and quasi-biennial oscillation (QBO) components are of similar amplitude at this location, and the net response often depends on the relative phasing of the two signals (for example, there is a near cancellation of these signals during the large ENSO event of 1997–1998, whereas an in-phase behavior in 2000 results in a relatively large net

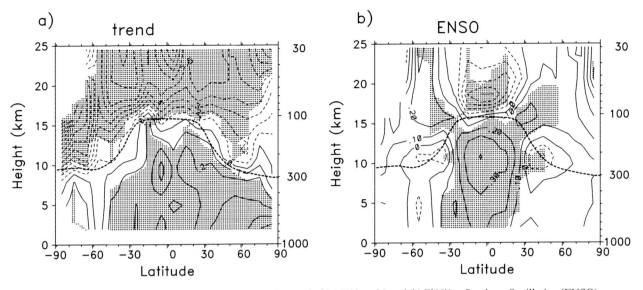

Figure 2. Cross sections of (a) linear trends (contour interval of 0.1 K/decade) and (b) El Niño–Southern Oscillation (ENSO) temperature variations (contour interval of 0.1 K/multivariate ENSO index) in zonally averaged RICH data, for the period 1970–2006. In both plots, solid and dashed lines denote positive and negative values, and shaded regions denote the statistical fits are significant at the 2-sigma level. The dark dashed line denotes the tropopause.

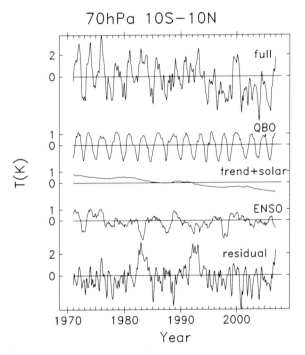

Figure 3. Top curve shows time series of zonal mean temperature at 70 hPa over 10°N–10°S from the RICH data during 1970–2006. The lower curve shows components of variability derived from the multivariate regression fit, together with the residual (bottom curve). Note that the volcanic warming signals of El Chichon (1982) and Pinatubo (1991) are clearly seen in the residual time series, although they are not evident amid the other variability in the full time series.

anomaly). The importance of both the ENSO and QBO variations in this region was previously discussed by *Reid* [1994]. *Free and Seidel* [2009] furthermore show a large ENSO signal in the Arctic stratosphere that maximizes during winter (not evident in the annual mean results shown in Figure 2b).

Figure 4 shows the spatial structure of the temperature anomalies associated with the El Chichon and Pinatubo volcanic eruptions. These are calculated based on the residuals to the full regression fit (see Appendix A and Figure 3), taking the difference between the temperature for 1 year following each eruption and the previous 3 years. Both eruptions result in warm temperatures of 2–3 K centered in the tropical stratosphere (somewhat larger and situated higher for Pinatubo). These temperature anomalies persist for 1–2 years after each eruption (Figure 1). The volcanic periods are also linked to significant cooling in the troposphere, and these volcanic signals provide sensitive tests of tropospheric climate feedback process [*Soden et al.*, 2002]. These volcanic temperature variations have been discussed in more detail by *Free and Angell* [2002], who also analyze the patterns associated with Agung (which shows more asymmetry, with stratospheric warming shifted toward the SH).

Figure 5 shows time series of deseasonalized global-average MSU4 satellite temperature anomalies, together with equivalent results from the RICH and RATPAC-lite radiosonde data (integrated with the MSU4 weighting functions). This compares the direct global satellite measurements with

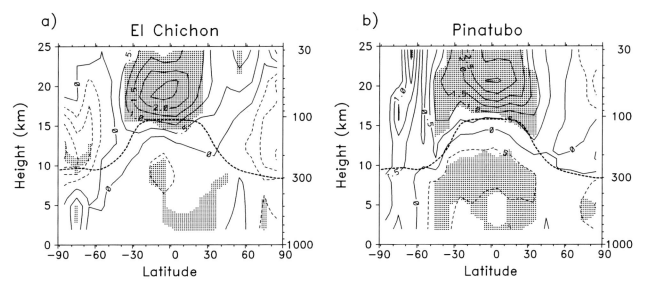

Figure 4. Cross sections of temperature anomalies associated with the (a) El Chichon and (b) Mount Pinatubo volcanic eruptions. These anomalies are estimated from the residuals to the multivariate regression fit (see Appendix A), taking the difference between the 1-year average after each eruption and the previous 3 years. Shading denotes regions where the anomalies are greater than twice the standard deviation of annual mean temperature anomalies at each location. The dark dashed line indicates the tropopause.

equivalent radiosonde-based data sets. The overall behavior of MSU4 is very similar to the 70- to 50-hPa time series in Figure 1, with volcanic effects and step-like temperature decreases (and relatively constant temperatures since ~1995). There is excellent agreement in detail between the satellite measurements and the integrated (homogeneity-adjusted) radiosonde

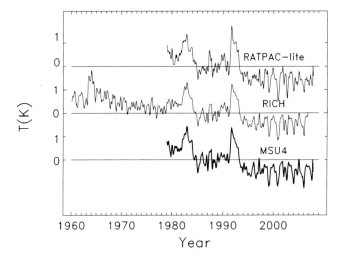

Figure 5. Time series of global mean temperature anomalies from MSU4 satellite data, together with corresponding time series derived from the RICH and RATPAC-lite data sets (vertically integrated using the MSU4 weighting function). Each time series has been normalized to zero for the period 1985–1990.

data, and this agreement is a substantial improvement over similar comparisons using unhomogenized data [*Randel and Wu*, 2006]. The longer record from RICH data provides a longer perspective on the recent record, including the clear signature of the Agung volcanic eruption in 1963.

The vertical profile of linear trends over 1979–2006 in the RICH and RATPAC-lite data are shown in Plate 1, for tropical and extratropical latitude bands, together with corresponding trends derived from the UAH MSU4 data. Trends calculated from the two radiosonde data sets agree well, with the RICH data showing somewhat larger cooling at uppermost levels and slightly different vertical structure in the tropics (including the altitude of the crossover from tropospheric warming to stratospheric cooling). The MSU4 satellite trends show reasonable agreement with the radiosonde results, and overall, the trends show a relatively flat (approximately constant) latitudinal structure over 60°N-S.

Stratospheric temperature changes at polar latitudes deserve separate attention because of the high level of (natural) year-to-year variability during winter and spring. This well-known behavior [e.g., *Labitzke and van Loon*, 1999; *Yoden et al.*, 2002] is illustrated in Figure 6, which shows 70 hPa polar temperature anomalies for seasonal averages (December-January-February (DJF), March-April-May (MAM), etc.) for both the Arctic (60°N–90°N) and Antarctic (60°S–90°S). Large year-to-year variability is observed in the Antarctic during spring (September-October-November (SON)) and in the Arctic during winter (DJF). Long-term cooling trends are evident in the SH during spring (SON) and summer (DJF), and these are associated with development of the Antarctic ozone hole after 1980 [*Randel and Wu*, 1999]. Plate 2b shows the vertical profile of seasonal temperature trends in the Antarctic (for 1970–2006), highlighting cooling throughout the lower stratosphere during spring and summer. Arctic time series (Figure 6a) and trends (Plate 2a) show cooling during summer (June-July-August (JJA)), which is statistically significant because of low natural variability during this season.

Temperature observations in the middle and upper stratosphere derived from the SSU data are available from 1979 to 2005, and Figure 7 shows time series of near-global (60°N–60°S) anomalies for several SSU channels (along with MSU4 for comparison). Time series show overall cooling throughout the stratosphere, with largest net changes (~3 K) in the upper stratosphere. The changes are not monotonic, however, with the transient warming of the El Chichon and Pinatubo eruptions evident from the lower through the middle stratosphere (in SSU channel 26). The upper stratosphere (SSU channels 27 and 36x) shows the influence of long-term cooling superimposed on the 11-year solar cycle (with maxima centered near 1980, 1991, and 2002), which results in a stair-step structure. As with the lower stratosphere, temperatures were relatively constant in the middle and upper stratosphere during 1995–2005.

The vertical structure of near-global mean temperature trends throughout the stratosphere during 1979–2005 is shown in Plate 3, combining results for the SSU and MSU satellites, plus radiosonde data. The overall pattern shows trends increasing with altitude from the lower (~−0.5 K/ decade) to upper stratosphere (~−1.2 K/decade). There is good agreement between the radiosonde and satellite-derived trends for the region where they overlap. Unfortunately, there are no independent measurements of upper stratospheric temperatures on a global scale to compare with the SSU trend results. Long-term measurements of temperatures over 30–80 km from lidar measurements are available from a few stations [*Keckhut et al.*, 2004]. *Randel et al.* [2009a] show there is reasonable overall agreement between the SSU satellite data and lidar measurements from three stations with the longest records (i.e., the statistical trend uncertainties overlap), although there are large differences in sampling that preclude constraining trend uncertainties in either the satellite or lidar data sets.

4. STRATOSPHERIC WATER VAPOR

Observations of stratospheric water vapor have been made by balloon, aircraft, and satellite measurements (as reviewed by *SPARC* [2000]). Estimates of long-term variability and trends derived from combining different data sets are

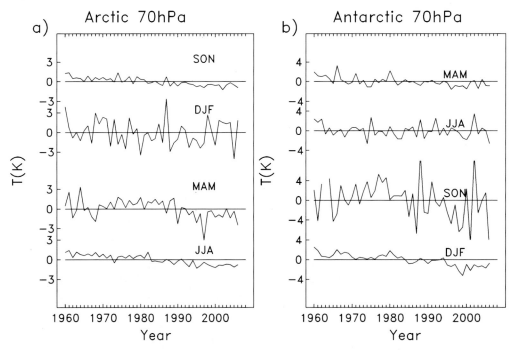

Figure 6. Time series of 70 hPa temperature anomalies in the (a) Arctic (60°N–90°N) and (b) Antarctic (60°S–90°S), calculated from RICH data for each season (December–January–February, DJF, etc.).

problematic because of the relatively large uncertainties and biases (~10%–20%) among different data and measurement techniques [*SPARC*, 2000]. The longest time series of observations from a single location are from balloon measurements from Boulder, Colorado, which began in the early 1980s, with the sampling of individual profiles approximately once per month (or less). These data have been examined in a number of analyses [*Oltmans and Hoffman*, 1995; *Oltmans et al.*, 2000; *Scherer et al.*, 2008; *Solomon et al.*, 2010] and show an overall increase of water vapor since 1980 but with significant variability associated with individual (snapshot) profile measurements.

Global satellite observations allow vastly improved space-time sampling of stratospheric water vapor, so that large-scale coherent variability can be examined in detail, but these are limited in terms of long-term measurements. Here we examine satellite data from the Halogen Occultation Experiment (HALOE) covering January 1992 to August 2005 (using retrieval version v19), combined with Aura Microwave Limb Sounder (MLS) for the period June 2004 to May 2010 (v2.2), and produce a single time series by adjusting the data using the overlap period during 2004–2005. HALOE is based on solar occultation measurements [*Russell et al.*, 1993], which have high vertical resolution (~2 km) but limited spatial sampling (requiring approximately 1 month to sample the region 60°N–60°S). The MLS data [*Read et al.*, 2007] have somewhat lower vertical resolution (~ 3 km), but much denser spatial sampling, with near-global coverage everyday.

While both HALOE and MLS provide high-quality measurements, there are systematic differences of order 10% between the data (related to vertical resolution and

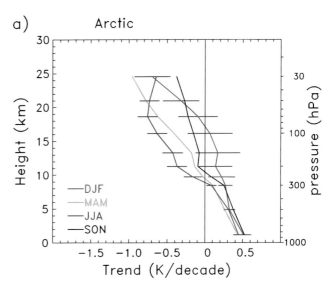

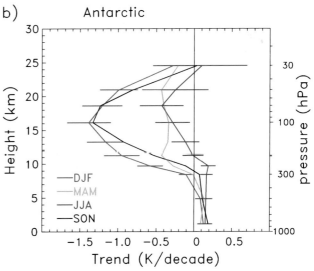

Plate 2. Vertical profile of temperature trends for 1979–2006 over the (a) Arctic (60°N–90°N) and (b) Antarctic (60°S–90°S), based on RICH radiosonde data. Trends are calculated for each season, and error bars denote 1-sigma uncertainties (for clarity, shown only for DJF and JJA statistics).

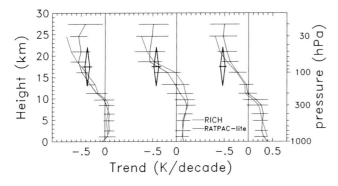

Plate 1. Vertical profile of temperature trends during 1979–2007 derived from RICH and RATPAC-lite data, for latitude bands 30°S–60°S, 30°N–30° S, and 30°N–60°N (left to right). The diamonds denote corresponding trends derived from the UAH MSU4 data, and the height of the diamond corresponds to the MSU4 weighting function. Error bars denote the 2-sigma statistical trend uncertainties.

retrieval details) that require adjustment to produce a single continuous data set. Here we simply deseasonalize both the HALOE and MLS data sets individually (which removes the systematic bias) and then adjust the MLS anomalies to match the HALOE data for the overlap period June 2004 to August 2005. The results are illustrated in Plate 4a, which shows near-global (50°N–50°S) anomalies for the HALOE and MLS data at 82 hPa over 1993–2008. The overlap period is highlighted in Plate 4b, showing reasonable agreement between interannual anomalies derived from both data sets (the

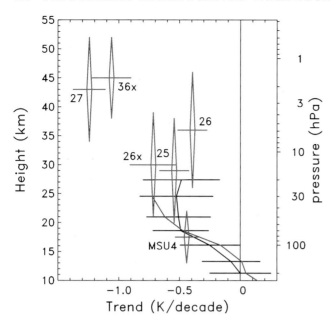

Plate 3. Vertical profile of near-global (60°N-S) temperature trends over 1979–2005 derived from satellite and radiosonde data sets. The lines in the lower stratosphere indicate trends from the RICH (red) and RATPAC-lite (black) data. Blue diamonds indicate trends from MSU4 and SSU satellite data, with the height of the diamond representing the respective weighting function. Error bars denote 2-sigma statistical trend uncertainties.

confidence for using the MLS data to extend the HALOE record.

The overall behavior of water vapor interannual changes in the lower stratosphere (Plate 4a) show variations with an approximate 2-year periodicity, related to the QBO influence on tropical tropopause temperature, combined with a significant drop in water vapor (~0.4 ppmv) after ~2001 and a suggestion of recent increasing values. These interannual variations in water vapor originate near the tropical tropopause and propagate to higher latitudes of both hemisphere in the lower stratosphere and also to higher altitudes in the tropics [*Randel et al.*, 2004]. Plate 5 shows a height-time section of near-global (50°N-S) average water vapor anomalies from 1993–2010, showing this vertical propagation and highlighting the tropopause as a source region for the global anomalies.

Brewer [1949] proposed a simple mechanism by which tropical tropopause temperatures control stratospheric water vapor, and observations [*Randel et al.*, 2004] and trajectory calculations [*Fueglistaler et al.*, 2005; *Fueglistaler and Haynes*, 2005] have confirmed this for both the annual cycle and interannual changes. This behavior is demonstrated for the time series over 1993–2010 in Plate 6, which shows the 82-hPa global water vapor fluctuations together with anomalies in tropical tropopause (cold point) temperatures. This latter time series is derived from a small group of near-equatorial radiosonde stations, chosen based on wide spatial sampling and consideration of data quality (via comparison with MSU satellite data, as described in the work of *Randel and Wu* [2006]). These stations include Nairobi (1°S, 37°E), Majuro (7°N, 171°E), and Manaus (3°S, 60°W), and the time series for each station is shown in Figure 8, showing overall

anomalies for this period are primarily related to the QBO). Note the month-to-month variability in Plate 3 is somewhat smoother in the MLS data, probably due to the denser space-time sampling compared to HALOE. This approximate agreement in variability during the overlap period provides

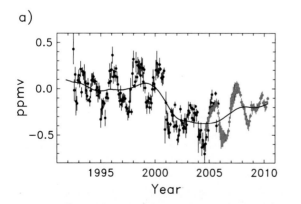

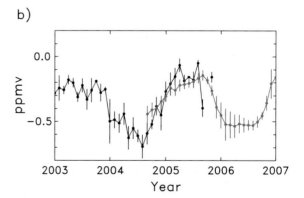

Plate 4. (a) Time series of deseasonalized near-global (50°N–50°S) water vapor anomalies in the lower stratosphere (82 hPa) during 1993–2008. Black points show data derived from the Halogen Occultation Experiment (HALOE) measurements, and blue points from Aura Microwave Limb Sounder (MLS). The black line is a smooth fit to the data, using a Gaussian smoother with a half-width of 1 year. (b) A highlight of the overlap period during 2004–2005, illustrating how the MLS anomalies are adjusted to match the HALOE data to construct a continuous record. The vertical bars denote the standard deviation of the near-global anomalies for each month.

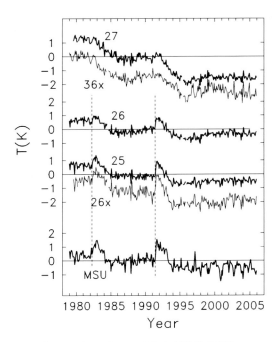

Figure 7. Time series of near-global (60°N–60°S) temperature anomalies from satellite data covering the lower to upper stratosphere, for the period 1979–2005. The lower curve shows results for MSU4, and the upper curves show results for separate SSU channels spanning the middle to upper stratosphere (with approximate altitudes indicated in Plate 3). The SSU data series ends in 2005. The dashed lines indicate the El Chichon and Pinatubo volcanic eruptions.

coherent behavior for anomalies in cold point temperature among the stations. Note that the cold point tropopause is typically near 90–105 hPa, not at a standard pressure level, and thus, cold point anomalies are not available in the homogenized (standard pressure level) radiosonde data sets such as RICH or RATPAC. There is a high level of agreement between the tropopause temperature and water vapor anomalies in Plate 6, with a correlation of 0.76 (with water vapor lagging temperature by 2 months). Both the QBO variations and the decrease after 2001 are observed in both time series. The relationship in Plate 6 suggests a water vapor-temperature sensitivity of ~0.5 ppmv K^{-1}, and this value is consistent with results derived from the trajectory calculations of *Fueglistaler and Haynes* [2005], which are based on large-scale meteorological analyses and assumption of 100% saturation of water vapor with respect to ice. This observed correlation suggests a relatively simple control of global stratospheric water vapor by freeze drying near the tropical tropopause, at least for the period 1993–2010, when global-scale measurements of water vapor from satellites are available.

The main component of interannual variability for water vapor in Plate 6 is due to the QBO, and that is why the near-

equatorial radiosonde measurements (stations within ±10° of the equator) show strongest correlations. Figure 8 also shows time series of temperature anomalies at 100 and 70 hPa (slightly below and above the cold point). While there is coherence among the temperature variations over these nearby levels, the strongest correlation to stratospheric water vapor anomalies is found for the cold point. Also, the relatively abrupt decrease in temperature after 2001 (echoed in stratospheric water vapor) is most evident at the cold point, and this behavior reinforces the relatively simple picture of water vapor control by freeze-out near the equatorial cold point. *Rosenlof and Reid* [2008] also discussed correlations of stratospheric water vapor with temperatures near the tropical tropopause, noting the strong changes after 2001, although *Lanzante* [2009] pointed out large potential biases in the associated radiosonde data, due to unadjusted instrumental changes.

5. SUMMARY AND DISCUSSION

Interannual variability in stratospheric temperature is linked to forcing associated with large volcanic events, long-term changes (trends) in radiative gases, and solar variability, in addition to dynamical variability linked to the QBO and ENSO, and natural year-to-year variations (which are largest in the winter-spring polar regions). Each of these forced

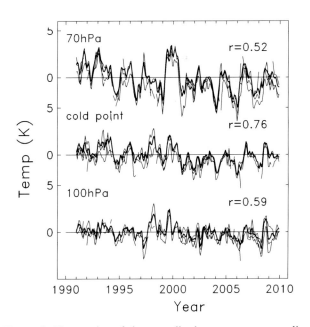

Figure 8. Time series of deseasonalized temperature anomalies derived from several near-equatorial radiosonde stations at 100 hPa, the cold-point tropopause, and 70 hPa. The thin lines show results at each of three stations (Nairobi, Majuro, and Manaus), and the thick line is the average. Correlations with lower stratospheric water vapor (time series in Plate 6 are indicated for each level).

H2O anomaly 50S–50N

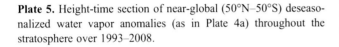

Plate 5. Height-time section of near-global (50°N–50°S) deseasonalized water vapor anomalies (as in Plate 4a) throughout the stratosphere over 1993–2008.

signals is relatively well understood and simulated to some degree in current stratospheric chemistry-climate models [*SPARC*, 2010]. In the global mean, the changes do not appear monotonic, but rather step-like; this behavior has been examined and discussed by *Seidel and Lanzante* [2004] and *Ramaswamy et al.* [2006]. The long-term cooling of the stratosphere is linked to increases in greenhouse gases and decreases in stratospheric ozone, with ozone losses dominating in the lower stratosphere, and more-or-less equal contributions in the upper stratosphere (for the period 1979–1999) [*Shine et al.*, 2003]. The recent flattening of trends throughout the stratosphere over the last decade seen in Figures 1 and 7 (with near constant temperatures after 1995) is most interesting, given the continued increases in CO_2, combined with relatively small changes in stratospheric ozone over this period [*World Meteorological Organization*, 2006].

The analyses here have not included detailed discussions of the QBO variations in stratospheric temperatures, which have magnitudes up to ±4 K and span the tropics to middle latitudes [e.g., *Crooks and Gray*, 2005]. The QBO is relatively easy to isolate statistically, as there are over 10 complete cycles in the satellite observational record. We have also not discussed the 11-year solar cycle variations in stratospheric temperature, which have been recently discussed in the work of *Randel et al.* [2009a]; both the radiosonde and satellite data sets show coherent solar variations throughout the stratosphere in low latitudes (~30°N–30°S), with amplitudes ranging from 0.5 K in the lower stratosphere to 1.0 K in the upper stratosphere. The ENSO effects on zonal mean temperature in the lower stratosphere (Figure 2b) are also an important component of interannual variability in this region (Figure 3). We note that similar behavior is observed in stratospheric ozone

observations and that such temperature and ozone variations are found in a recent chemistry-climate model simulation that incorporates observed sea surface temperature forcing [*Randel et al.*, 2009b]. *Marsh and Garcia* [2007] suggest that there can be confusion of the ENSO and solar signal components in short data records, and neglecting this effect may result in overestimating the ozone solar signal in the tropical lower stratosphere.

Global satellite measurements of stratospheric water vapor are available for 1993–2010, and these data allow accurate mapping of the seasonal cycle and interannual variability over this period. Interannual changes in water vapor show strong coherence throughout the stratosphere, with anomalies originating near the tropical tropopause and propagating latitudinally in the lower stratosphere and vertically in the tropics (advected by the Brewer-Dobson circulation). The observed water vapor anomalies are highly correlated with temperatures near the equatorial cold point tropopause, and the observations for 1993–2010 are consistent with simple dehydration of air entering the stratosphere across the cold point (as simulated in Lagrangian trajectory calculations of *Fueglistaler and Haynes* [2005]). There was an observed drop in stratospheric water vapor (~0.4 ppmv) and cold point temperature (~1 K) after 2001, which has continued to the present, albeit modulated by the QBO with a suggestion of

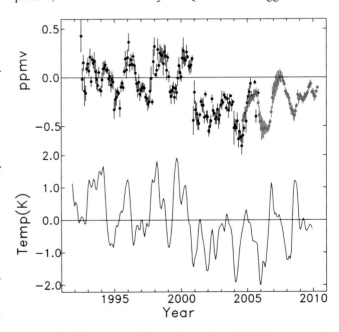

Plate 6. (top) Time series of lower stratosphere (82 hPa) water vapor anomalies from HALOE + MLS data, as in Plate 5a. (bottom) Time series of tropical cold-point tropopause temperature anomalies, derived from several radiosonde stations, as described in the text. The correlation coefficient is 0.75, with water vapor lagging the temperatures by 2 months.

recent increasing values. The cooling associated with the change after 2001 is largest in a narrow vertical layer centered near the cold point and may be associated with a corresponding increase in tropical upwelling [*Randel et al.*, 2006]. However, given the short observational record, it is difficult to link this step-like change to any decadal-scale trends in water vapor, tropopause temperature, or upwelling.

As noted above, the longest record of stratospheric water vapor comes from the balloon measurements at Boulder, Colorado, beginning in 1980 (as recently reviewed in the work of *Scherer et al.* [2008]). These data show positive trends over the period 1980–2006, which seems at odds with the near-zero or cooling trends near the tropical tropopause over this period (Plate 1) [see also *Seidel et al.*, 2001]. Interpretation of the Boulder record is also hampered by disagreement with trends derived from the HALOE record for the overlap period 1992–2005 [*Scherer et al.*, 2008]. *Fueglistaler and Haynes* [2005] also show that the Boulder trends for 1980–2004 are difficult to reconcile with Lagrangian trajectory results. Thus, while the 1993–2010 satellite record suggests a relatively simple interpretation of stratospheric vapor changes linked to equatorial tropopause temperatures, interpretation of the longer record of Boulder balloon measurements remains a topic of ongoing research.

APPENDIX A: LINEAR REGRESSION ANALYSIS

Statistical climate signals in the temperature data are derived using a multivariate linear regression analysis, as in the work of *Ramaswamy et al.* [2001]. The statistical model includes terms to account for linear trends, solar cycle (using the solar F10.7 radio flux as a proxy), ENSO (using the Multivariate ENSO Index from the NOAA Climate Diagnostics Center, http://www.cdc.noaa.gov/people/klaus. wolter/MEI/, with atmospheric temperatures lagged by 1 month), plus two orthogonal time series to model the QBO [*Wallace et al.*, 1993]. We omit 2 years after each of the large volcanic eruptions (El Chichon in April 1982 and Mount Pinatubo in June 1991) from the regression analysis, to avoid influence from the associated large transient warming events. Uncertainty estimates for the statistical fits are calculated using a bootstrap resampling technique [*Efron and Tibshirani*, 1993], which includes the effects of serial autocorrelation.

Acknowledgments. We thank Rolando Garcia and Eric Jensen for comments that helped improve the manuscript and appreciate a constructive review provided by Dian Seidel. Fei Wu provided assistance with data analysis and graphics. This work was partially supported by the NASA ACMAP Program. The National Center for Atmospheric Research is operated by the University Corporation for Atmospheric Research, under sponsorship of the National Science Foundation.

REFERENCES

Austin, J., et al. (2009), Coupled chemistry climate model simulations of stratospheric temperatures and their trends for the recent past, *Geophys. Res. Lett.*, *36*, L13809, doi:10.1029/2009GL038462.

Brewer, A. W. (1949), Evidence for a world circulation provided by measurements of helium and water vapor distribution in the stratosphere, *Q. J. R. Meteorol. Soc.*, *75*, 351–363.

Calvo Fernandez, N., et al. (2004), Analysis of the ENSO signal in tropospheric and stratospheric temperatures observed by MSU, 1979–2000, *J. Clim.*, *17*, 3934–3946.

Christy, J. R., R. W. Spencer, W. B. Norris, W. D. Braswell, and D. E. Parker (2003), Error estimates of version 5.0 of MSU-AMSU bulk atmospheric temperatures, *J. Atmos. Oceanic Technol.*, *20*, 613–629.

Crooks, S. A., and L. J. Gray (2005), Characterization of the 11-year solar signal using a multiple regression analysis of the ERA-40 dataset, *J. Clim.*, *18*, 996–1015.

Dvortsov, V. L., and S. Solomon (2001), Response of the stratospheric temperatures and ozone to past and future increases in stratospheric humidity, *J. Geophys. Res.*, *106*, 7505–7514.

Efron, B., and R. J. Tibshirani (1993), *An Introduction to the Bootstrap*, 436 pp., Chapman and Hall, London.

Eyring, V., et al. (2006), Assessment of temperature, trace species, and ozone in chemistry-climate model simulations of the recent past, *J. Geophys. Res.*, *111*, D22308, doi:10.1029/2006JD007327.

Forster, P. M. d. F., and K. P. Shine (1999), Stratospheric water vapour changes as a possible contributor to observed stratospheric cooling, *Geophys. Res. Lett.*, *26*, 3309–3312.

Free, M., and J. K. Angell (2002), Effect of volcanoes on the vertical temperature profile in radiosonde data, *J. Geophys. Res.*, *107* (D10), 4101, doi:10.1029/2001JD001128.

Free, M., and D. J. Seidel (2009), Observed El Niño–Southern Oscillation temperature signal in the stratosphere, *J. Geophys. Res.*, *114*, D23108, doi:10.1029/2009JD012420.

Free, M., D. J. Seidel, J. K. Angell, J. Lanzante, I. Durre, and T. C. Peterson (2005), Radiosonde Atmospheric Temperature Products for Assessing Climate (RATPAC): A new data set of large-area anomaly time series, *J. Geophys. Res.*, *110*, D22101, doi:10.1029/2005JD006169.

Fueglistaler, S., and P. H. Haynes (2005), Control of interannual and longer-term variability of stratospheric water vapor, *J. Geophys. Res.*, *110*, D24108, doi:10.1029/2005JD006019.

Fueglistaler, S., M. Bonazzola, P. H. Haynes, and T. Peter (2005), Stratospheric water vapor predicted from the Lagrangian temperature history of air entering the stratosphere in the tropics, *J. Geophys. Res.*, *110*, D08107, doi:10.1029/2004JD005516.

Gaffen, D. J. (1994), Temporal inhomogeneities in radiosonde temperature records, *J. Geophys. Res.*, *99*, 3667–3676.

Garcia, R. R., D. R. Marsh, D. E. Kinnison, B. A. Boville, and F. Sassi (2007), Simulations of secular trends in the middle atmosphere, 1950–2003, *J. Geophys. Res.*, *112*, D09301, doi:10.1029/2006JD007485.

Haimberger, L., C. Tavolato, and S. Sperka (2008), Towards elimination of the warm bias in historic radiosonde temperature records—Some new results from a comprehensive intercomparison of upper air data, *J. Clim.*, *21*, 4587–4606.

Intergovernmental Panel on Climate Change (2007), Climate Change 2007: The Physical Science Basis. Contribution of Working Group I to the Fourth Assessment Report of the Intergovernmental Panel on Climate Change, edited by S. Solomon, et al., 996 pp., Cambridge Univ. Press, New York.

Keckhut, P., et al. (2004), Review of ozone and temperature lidar validations performed within the framework of the Network for the Detection of Stratospheric Change, *J. Environ. Monit.*, *6*, 721–733.

Labitzke, K. G., and H. van Loon (1999), *The Stratosphere: Phenomena, History and Relevance*, 179 pp., Springer, New York.

Lanzante, J., S. Klein, and D. J. Seidel (2003a), Temporal homogenization of monthly radiosonde temperature data. Part I: Methodology, *J. Clim.*, *16*, 224–240.

Lanzante, J., S. Klein, and D. J. Seidel (2003b), Temporal homogenization of monthly radiosonde temperature data. Part II: Trends, sensitivities and MSU comparisons, *J. Clim.*, *16*, 241–262.

Lanzante, J. R. (2009), Comment on "Trends in the temperature and water vapor content of the tropical lower stratosphere: Sea surface connection" by Karen H. Rosenlof and George C. Reid, *J. Geophys. Res.*, *114*, D12104, doi:10.1029/2008JD010542.

Marsh, D. R., and R. R. Garcia (2007), Attribution of decadal variability in lower-stratospheric tropical ozone, *Geophys. Res. Lett.*, *34*, L21807, doi:10.1029/2007GL030935.

Mears, C. A., and F. J. Wentz (2009), Construction of the Remote Sensing Systems V3.2 atmospheric temperature records from the MSU and AMSU microwave sounders, *J. Atmos. Oceanic Technol.*, *26*, 1040–1056.

Mote, P. W., K. H. Rosenlof, M. E. McIntyre, E. S. Carr, J. C. Gille, J. R. Holton, J. S. Kinnersley, H. C. Pumphrey, J. M. Russell III, and J. W. Waters (1996), An atmospheric tape recorder: The imprint of tropical tropopause temperatures on stratospheric water vapor, *J. Geophys. Res.*, *101*, 3989–4006.

Nash, J. (1988), Extension of explicit radiance observations by the Stratospheric Sounding Unit into the lower stratosphere and lower mesosphere, *Q. J. R. Meteorol. Soc.*, *114*, 1153–1171.

Nash, J., and G. F. Forrester (1986), Long-term monitoring of stratospheric temperature trends using radiance measurements obtained by the TIROS-N series of NOAA spacecraft, *Adv. Space Res.*, *6*, 37–44.

Oltmans, S. J., and D. J. Hofmann (1995), Increase in lower-stratospheric water vapour at a mid-latitude Northern Hemisphere site from 1981 to 1994, *Nature*, *374*, 146–149.

Oltmans, S. J., H. Vömel, D. J. Hofmann, K. H. Rosenlof, and D. Kley (2000), The increase in stratospheric water vapor from balloonborne, frostpoint hygrometer measurements at Washington, D.C., and Boulder, Colorado, *Geophys. Res. Lett.*, *27*(21), 3453–3456.

Oman, L., D. W. Waugh, S. Pawson, R. S. Stolarski, and J. E. Nielsen (2008), Understanding the changes of stratospheric water vapor in coupled chemistry-climate model simulations, *J. Atmos. Sci.*, *65*, 3278–3291.

Ramaswamy, V., et al. (2001), Stratospheric temperature trends: Observations and model simulations, *Rev. Geophys.*, *39*, 71–122.

Ramaswamy, V., M. Schwarzkopf, W. J. Randel, B. D. Santer, B. J. Soden, and G. L. Stenchikov (2006), Anthropogenic and natural influences in the evolution of lower stratospheric cooling, *Science*, *311*, 1138–1141.

Randel, W. J., and F. Wu (1999), Cooling of the Arctic and Antarctic polar stratospheres due to ozone depletion, *J. Clim.*, *12*, 1467–1479.

Randel, W. J., and F. Wu (2006), Biases in stratospheric and tropospheric temperature trends derived from historical radiosonde data, *J. Clim.*, *19*, 2094–2104.

Randel, W. J., F. Wu, S. Oltmans, K. Rosenlof, and G. Nedoluha (2004), Interannual changes of stratospheric water vapor and correlations with tropical tropopause temperatures, *J. Atmos. Sci.*, *61*, 2133–2148.

Randel, W. J., F. Wu, H. Vömel, G. Nedoluha, and P. Forster (2006), Decreases in stratospheric water vapor since 2001: Links to changes in the tropical tropopause and the Brewer-Dobson circulation, *J. Geophys. Res.*, *111*, D12312, doi:10.1029/2005JD006744.

Randel, W. J., et al. (2009a), An update of observed stratospheric temperature trends, *J. Geophys. Res.*, *114*, D02107, doi:10.1029/2008JD010421.

Randel, W. J., R. R. Garcia, N. Calvo, and D. Marsh (2009b), ENSO influence on zonal mean temperature and ozone in the tropical lower stratosphere, *Geophys. Res. Lett.*, *36*, L15822, doi:10.1029/2009GL039343.

Read, W. G., et al. (2007), Aura Microwave Limb Sounder upper tropospheric and lower stratospheric H_2O and relative humidity with respect to ice validation, *J. Geophys. Res.*, *112*, D24S35, doi:10.1029/2007JD008752.

Reid, G. C. (1994), Seasonal and interannual temperature variations in the tropical stratosphere, *J. Geophys. Res.*, *99*, 18,923–18,932.

Rosenlof, K. H., and G. C. Reid (2008), Trends in the temperature and water vapor content of the tropical lower stratosphere: Sea surface connection, *J. Geophys. Res.*, *113*, D06107, doi:10.1029/2007JD009109.

Russell, J. M., III, L. L. Gordley III, J. H. Park, S. R. Drayson, W. D. Hesketh, R. J. Cicerone, A. F. Tuck, J. E. Frederick, J. E. Harries, and P. J. Crutzen (1993), The Halogen Occultation Experiment, *J. Geophys. Res.*, *98*, 10,777–10,797.

Scherer, M., H. Vömel, S. Fueglistaler, S. J. Oltmans, and J. Staehelin (2008), Trends and variability of midlatitude stratospheric water vapour deduced from the re-evaluated Boulder balloon series and HALOE, *Atmos. Chem. Phys.*, *8*, 1391–1402.

Seidel, D. J., and J. R. Lanzante (2004), An assessment of three alternatives to linear trends for characterizing global atmospheric temperature changes, *J. Geophys. Res.*, *109*, D14108, doi:10.1029/2003JD004414.

Seidel, D. J., R. J. Ross, J. K. Angell, and G. C. Reid (2001), Climatological characteristics of the tropical tropopause as revealed by radiosondes, *J. Geophys. Res.*, *106* (D8), 7857–7878.

Shine, K. P., et al. (2003), A comparison of model-simulated trends in stratospheric temperatures, *Q. J. R. Meteorol. Soc.*, *129*, 1565–1588.

Shine, K. P., J. J. Barnett, and W. J. Randel (2008), Temperature trends derived from Stratospheric Sounding Unit radiances: The effect of increasing CO_2 on the weighting function, *Geophys. Res. Lett.*, *35*, L02710, doi:10.1029/2007GL032218.

Soden, B. J., R. T. Wetherald, G. L. Stenchikov, and A. Robock (2002), Global cooling after the eruption of Mount Pinatubo: A test of climate feedback by water vapor, *Science*, *296*, 727–730, doi:10.1126/science.296.5568.727.

Solomon, S., et al. (2010), Contributions of stratospheric water vapor to decadal changes in the rate of global warming, *Science*, *327*, 1219–1223, doi:10.1126/science.1182488.

SPARC CCMVal (2010), SPARC report on the evaluation of chemistry-climate models, edited by V. Eyring, T. G. Shepherd, and D. W. Waugh, *WMO/TD-No. 1526*, World Meteorol. Organ., Geneva, Switzerland. (Available at http://www.atmosp.physics.utoronto.ca/SPARC)

Stratospheric Processes and Their Role in Climate (SPARC) (2000), SPARC Assessment of Upper Tropospheric and Stratospheric Water Vapor, edited by D. Kley, J. M. Russell III, and C. Phillips, *SPARC Rep. 2*, 312 pp., World Meteorol. Organ., Geneva, Switzerland.

Wallace, J. M., R. L. Panetta, and J. Estberg (1993), Representation of the equatorial quasi-biennial oscillation in EOF phase space, *J. Atmos. Sci.*, *50*, 1751–1762.

World Meteorological Organization (2006), Scientific assessment of ozone depletion: 2006, *Rep. 47*, Global Ozone Res. and Monit. Proj., Geneva, Switzerland.

Yoden, S., M. Taguchi, and Y. Naito (2002), Numerical studies on time variations of the troposphere-stratosphere coupled system, *J. Meteorol. Soc. Jpn.*, *80*, 811–830.

Yulaeva, E., and J. M. Wallace (1994), The signature of ENSO in global temperature and precipitation fields derived from the microwave sounding unit, *J. Clim.*, *7*, 1719–1736.

W. J. Randel, Atmospheric Chemistry Division, National Center for Atmospheric Research, PO Box 3000 Boulder, CO 80307-3000, Boulder, CO 80307, USA. (randel@ucar.edu)

Trace Gas Transport in the Stratosphere: Diagnostic Tools and Techniques

Mark R. Schoeberl[1] and Anne R. Douglass

NASA Goddard Space Flight Center, Greenbelt, Maryland, USA

The study of stratospheric trace gases over the last 20 years has spurred the development of new methods and techniques for analyzing observations that have resulted in a comprehensive understanding of stratospheric chemistry and dynamics and a significant improvement in numerical methods. These analysis techniques and numerical methods are somewhat scattered throughout the literature and are gathered together here and placed in context. To do this, we first provide a brief overview of stratospheric dynamics and transport as seen through the distribution of trace gases. We then provide a summary of some of the more popular methods used to analyze observations including trajectory models, potential vorticity-based analyses, and the age-of-air diagnostic. We then describe the class of conservative numerical schemes now used to model trace gas distributions.

1. INTRODUCTION

Observations of constituents that have long photochemical lifetimes show the fingerprints of the stratospheric circulation. In fact, the stratospheric Brewer-Dobson circulation, the slow movement of air from the tropics to the middle latitudes, was originally diagnosed through observations of water vapor and ozone [*Brewer*, 1949; *Dobson*, 1956]. In this article, we describe some of the important concepts and tools developed during the last 20 years of research on the distribution stratospheric trace gases (tracers). These concepts and tools have contributed significantly to the explanation of satellite and aircraft observations, have led to the observations being used in new ways, and have led to the improvement of models used to predict changes in the ozone layer and the response of the stratosphere to climate change. For those starting on this subject, the text book on stratospheric dynamics by *Andrews*

et al. [1987] is a necessity; however, the reader is also referred to two excellent reviews by *Plumb* [2002] and *Shepherd* [2007] that update much of the research that has taken place since the publication of *Andrews et al.* [1987]

This article is divided into three sections. In the next section (section 2), we show some of the key observations that have guided our thinking. We also discuss the theory behind tracer transport in the stratosphere and relevant aspects of dynamics. Other articles in this volume provide more comprehensive reviews of the dynamics of the stratosphere. In section 3, we discuss some of the tools and concepts used by researchers to analyze observations. In these analyses, researchers have discovered important relationships that can be used to test numerical models. These tests have spurred the development of new classes of conservative transport schemes. Use of these new schemes in numerical models has subsequently improved the results from these models. The last section contains a brief summary.

This article does not cover statistical analysis methods such as the analysis of time series, data fitting techniques, empirical orthogonal functions, or wavelet analysis. Many of these topics are covered in textbooks including the popular "Numerical Recipes" books.

2. OBSERVATIONS AND THEORY

In this section, we briefly discuss observations and the modern theory of stratospheric dynamics.

---

[1]Now at Science and Technology Corporation, Columbia, Maryland, USA.

The Stratosphere: Dynamics, Transport, and Chemistry
Geophysical Monograph Series 190
This paper is not subject to U.S. copyright.
Published in 2010 by the American Geophysical Union.
10.1029/2009GM000855

2.1. Observations

The Limb Infrared Monitor of the Stratosphere and the Stratospheric and Mesospheric Sounder on the Nimbus 7 satellite, launched in the late 1978, made the first global high vertical resolution observations of ozone and a few other stratospheric trace gases [*Gille and Russell*, 1984; *Wale and Peskett*, 1984]. The Upper Atmosphere Research Satellite (UARS, launched in 1991) [*Reber et al.*, 1993], the Environmental Satellite (launched in 2002) [see *Gottwald et al.*, 2006, chapter 2], and Aura (launched in 2004) [*Schoeberl et al.*, 2006a] expanded these measurements. In 1987, the first measurements of stratospheric trace gases from highly instrumented high-altitude aircraft began. Complementing the satellite global coverage, the aircraft measurements provided stratospheric trace gas data at very high spatial resolution. Balloon-borne instruments also have provided simultaneous profiles of tracers and many other constituents and have played an important role in validating the satellite measurements.

Plate 1 shows zonal mean measurements of the long-lived tracer N_2O made by the Microwave Limb Sounder (MLS) on the Earth Observing System satellite Aura. The measurements of N_2O and other long-lived tracers that experience photochemical loss in the upper stratosphere show similar features: a tropical upward bulge with lower values at extratropical latitudes. A cartoon of the circulation, along with "barriers" showing the zones where steep gradients occur, is overlaid on top of the measurements. If only the mean circulation were acting on the N_2O trace gas distribution, and the sources and sinks can be neglected, then the tracer contours will approximately follow the zonal mean circulation streamlines. This is not the case. In the winter hemisphere, the N_2O distribution is somewhat flat in the middle latitudes, with shaper gradients at the edges of the polar and tropical regions. We now know that the sharp meridional gradients in N_2O form at the edges of regions where rapid mixing occurs. In the stratosphere, this mixing is produced by large-scale waves propagating up from the troposphere. These waves or eddies normally do not penetrate into the tropical regions where the Coriolis force weakens and are generally blocked by the strong circumpolar wind jet that forms at higher altitudes and latitudes in winter [*Charney and Drazin*, 1961]. The stratospheric circumpolar wind jet is called the winter polar vortex, and it creates a transport barrier that gives rise to the sharp trace gas gradients. The tropics and the polar vortex are not completely isolated from the middle latitude stratosphere but are isolated sufficiently that steep gradients can be maintained. The barriers to transport show up clearly in the trace gas distributions [e.g., *Sparling*, 2000] and in age-of-air tracers discussed below.

The action of the Brewer-Dobson overturning circulation on the trace gases is to steepen the N_2O gradient with respect to latitude. Opposing this effect is the tendency of the eddy mixing to flatten the N_2O isopleths along potential temperature surfaces. In steady state, the slope of the tracer gradient with respect to latitude is proportional to the meridional mixing rate [*Mahlman et al.*, 1986; *Holton*, 1986] as is the vertical flux of tracer.

The upward bulge of high N_2O in the tropics indicates transport from the entry point for trace gases in the stratosphere, consistent with the Brewer-Dobson circulation. Steep gradients in the subtropics suggest that there is little horizontal mixing between the middle latitudes and tropics. Just as in turbulent systems, tracer gradients form at the edges of the mixing regions. Plate 1b shows the temperature and zonal wind structure from assimilated observations. The tropopause is shown as the dash-dot line. Key features in Plate 1b are the cold tropical upper troposphere and the cold winter polar stratosphere. The thermal wind equation [*Holton*, 1992, Eq 3.30] relates strong meridional temperature gradients to the vertical wind shear. The observed zonal mean wind shear is consistent with the observed temperature field.

The net diabatic zonal mean heating rate (minus the global mean) is shown in Plate 1c. In the upper tropical troposphere, the temperatures are so cold that there is net infrared absorption leading to a net heating. These cold temperatures result from the induced upwelling circulation due to waves dissipating at higher altitudes. The tropical heating rate gets stronger with altitude with the addition of stratospheric ozone heating. Although the stratospheric ozone heating rate decreases at higher latitudes, adiabatic downwelling maintains the stratosphere above radiative equilibrium temperatures, so there is net cooling. The downwelling is the other branch of the tropical upwelling circulation in the lower stratosphere. The pattern of diabatic heating and cooling is consistent with the circulation shown in Plate 1a, upwelling in the tropics and descent at middle and higher latitudes. As discussed, the large-scale circulation is produced by the combination of dissipation of waves propagating upward from the troposphere and the net heating due to radiatively active gases, mostly heating due to absorption of UV radiation by ozone and IR cooling from CO_2, ozone, and water vapor.

The upwelling tropical lower stratosphere shows the imprint of trace gas seasonal variability at the tropopause. The phenomenon is popularly called the "tape recorder" because the signal imprinted at the tropopause is carried aloft like a tape moving away from the tape head. Tape recorder signals were first discovered from satellite observations of water vapor [*Mote et al.*, 1996]. Not surprisingly, tape recorders have been discovered in other trace gases such as

(a)

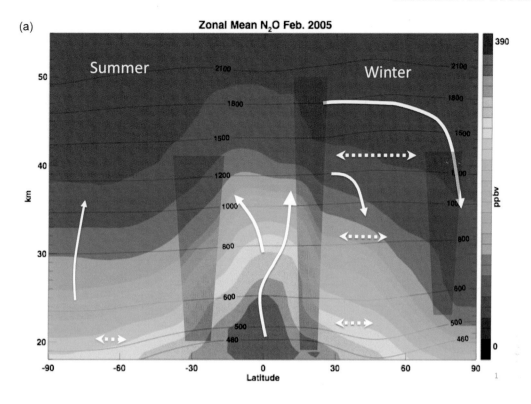

(b)

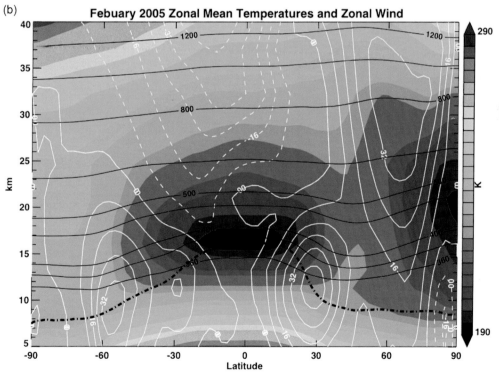

Plate 1. (a) Zonal mean, monthly mean, Aura Microwave Limb Sounder (MLS) measurements of N_2O for February 2005. Solid white arrows indicate the circulation pattern; dashed arrows indicate mixing. Black lines are the potential temperature levels. Blocks show where barriers to transport exist. (b) Zonal mean, monthly mean temperatures (K), zonal mean winds (white contours) in m s^{-1}, and potential temperature K (black contours). Easterly wind contours are dashed.

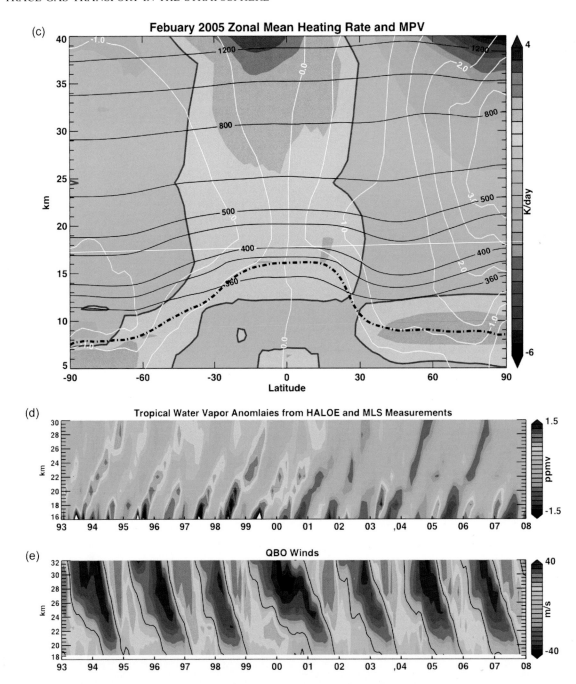

Plate 1. (continued) (c) Zonal mean clear-sky heating rate in K d^{-1}. Black lines are potential temperature contours, while white lines are Lait potential vorticity (PV) contours as described below. The zonal mean tropopause is shown as in Plates 1b and 1c as a dash-dot line. (d) Tropical water vapor anomaly time series from HALOE and MLS measurements updated from *Schoeberl et al.* [2008]. (e) Singapore zonal wind measurements from radiowinsondes revealing the quasi-biennial oscillation (QBO) in the winds.

carbon monoxide [*Schoeberl et al.*, 2006c], HCN [*Pumphrey et al.*, 2008], CO_2 [*Andrews et al.*, 1999], and volcanic aerosols [L. Thomason, personal communication, 2008]. All of these constituents exhibit variable tropical tropopause sources and, as with water vapor, the variations are carried upward by the mean upwelling circulation.

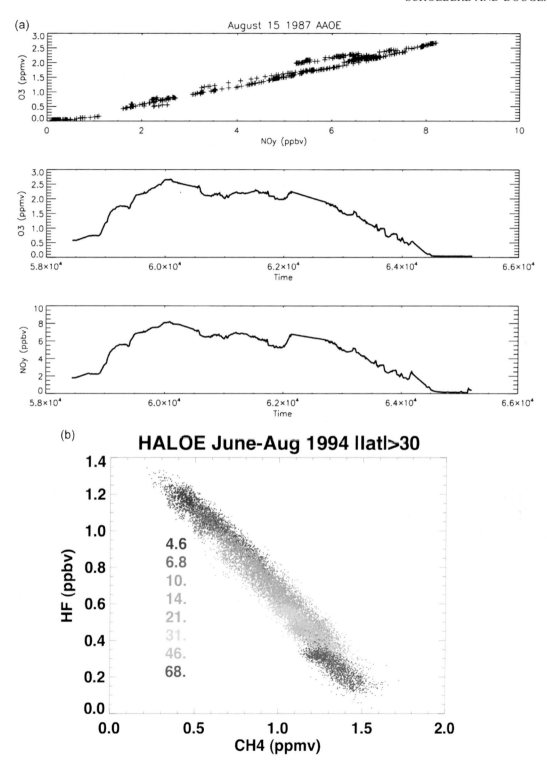

Plate 2. (a) Highly correlated aircraft measurements of ozone and NO_y over Antarctica in 1987. (b) Measurements of HF and methane (CH_4) by the HALOE instrument on Upper Atmosphere Research Satellite (UARS). Colors indicate the pressure range of the measurements in hPa.

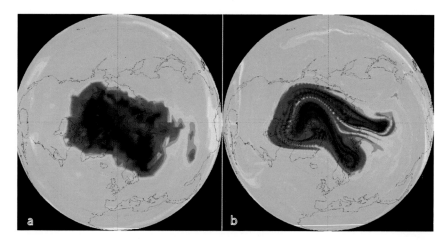

Plate 3. (a) 1 February 2007 MLS observations of N_2O mapped onto a 1° by 1° grid on the 450-K surface. (b) RDF calculation of the N_2O field using observations from 7 days prior. Note that the RDF method has resolved the filament of air injected into the vortex over central Russia. This RDF example was generated by the authors for this publication.

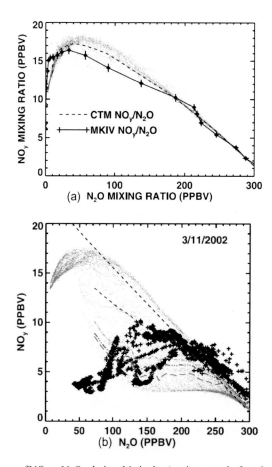

Plate 4. (top) Initial measurements of NO_y to N_2O relationship in the Arctic vortex before the formation of polar stratospheric clouds, November 1999. Crosses are Mark IV balloon measurements, while dashed line and gray dots are chemical transport model simulations. (bottom) Aircraft measurements of NO_y and N_2O in early March 2000. Measurements are shown as crosses. Dashed line shows the relationship in November 1999; colored lines show various model simulations using parameterizations for denitrification by polar stratospheric clouds. Images from *Considine et al.* [2003].

In the lower tropical stratosphere, the zonal mean winds oscillate with fairly regular period. This phenomenon, called the quasi-biennial oscillation (QBO) (see Plate 1c), is caused by upward propagating Kelvin and mixed-Rossby gravity waves as well as higher-frequency gravity waves [*Baldwin et al.*, 2001]. The period of the oscillation varies (~26–28 months) and is partially synchronized with the annual oscillation of the midlatitude circulation. The secondary circulation associated with the QBO, theoretically described by *Plumb and Bell* [1982], also produces identifiable changes in the trace gas distributions such as ozone [*Logan et al.*, 2003] and water [*Schoeberl et al.*, 2008].

Complementing the information obtained from satellites about the large-scale distribution of trace gases, high spatial resolution aircraft measurements have contributed significantly to understanding of the interrelationship between tracers. Plate 2 shows that long-lived tracers are highly correlated and exhibit compact relationships when plotted against each other. This surprising and important feature was discovered in observations made from NASA's ER-2 during the late 1980s [e.g., *Fahey et al.*, 1990] and has been extensively exploited in the study of the physical and chemical processes that lead to the polar ozone loss. Plate 2a shows observations from the ER-2 and the relationship of NO_y (total reactive nitrogen ~ $HNO_3 + NO_2 + NO +$ other less important species) and ozone. The ER-2 measurements are taken at and below 20 km altitude at southern middle latitudes. The compact relationship between tracers is also seen in the satellite data. Plate 2b shows observations of CH_4 and HF from the Halogen Occultation Experiment (HALOE) for pressures between 68 and 4.6 hPa. In addition to facilitating the interpretation of measurements of reactive species, these compact relationships give an important clue to the mixing processes within the stratosphere [*Plumb and Ko*, 1992; *Plumb et al.*, 2000; *Plumb*, 2007]. The compact relationships also place an important constraint on numerical transport algorithms that is discussed further in section 4.

2.2. Tracer Transport Theory

The mixing ratio for a tracer μ is governed by the equation

$$\frac{d\mu}{dt} = S - L, \qquad (1)$$

where d/dt is the Lagrangian derivative following an air parcel. S is the chemical source, and L is the chemical loss. For tracers like N_2O or chlorofluorocarbons, the most important loss process is photolysis. For such tracers, the loss rate is proportional to the concentration and thus can be written as $-\alpha\mu$, where α is called the loss frequency. Assuming that the tracer mixing ratio and the winds can be

written as zonal mean (overbar) and perturbations from the zonal mean (prime), i.e., $\mu = \bar{\mu} + \mu'$, equation (1) can be written in Eulerian form.

$$\frac{\partial\bar{\mu}}{\partial t} + \bar{v}\frac{\partial\bar{\mu}}{\partial y} + \bar{w}\frac{\partial\bar{\mu}}{\partial z} = -\frac{\partial(\overline{v'\mu'})}{\partial y} - \frac{\partial(\overline{w'\mu'})}{\partial z} + \bar{S} - \bar{L}. \qquad (2)$$

The zonal mean transport terms are on the left, while the eddy transport terms are on the right. Using the transformations given by *Andrews et al.* [1987], this expression can be rewritten as

$$\frac{\partial\bar{\mu}}{\partial t} + \bar{v}^*\frac{\partial\bar{\mu}}{\partial y} + \bar{w}^*\frac{\partial\bar{\mu}}{\partial z} = \nabla \cdot M + \bar{S} - \bar{L}. \qquad (3)$$

where $()^*$ indicates the residual circulation. The residual circulation, derived by *Andrews and McIntyre* [1978], combines the zonal mean circulation and the eddy fluxes of heat and momentum. The vector M and the residual circulation terms are defined in *Andrews et al.* [1987]. M contains nonconservative terms, and it is sufficient to know that M vanishes when the motion is adiabatic, linear, and steady.

For long-lived tracers where photochemical production and loss can be neglected (S, L ~ 0), the formulation of the transport equations in isentropic coordinates is given by *Plumb* [2002]:

$$\frac{\partial\bar{\mu}^*}{\partial t} + \bar{v}^*\frac{\partial\bar{\mu}^*}{\partial y} + \bar{\dot{\theta}}\frac{\partial\bar{\mu}^*}{\partial\theta} = -\frac{1}{\sigma}\frac{\partial}{\partial y}(\sigma\overline{v'\mu'}), \qquad (4)$$

where $\sigma = -g^{-1}\frac{\partial p}{\partial\theta}$ is the isentropic density, $\dot{\theta} = Q(p_o/p)^\kappa$ where Q is the net heating rate, p is the pressure, p_o is the surface pressure, and $\kappa = 2/7$, θ is the potential temperature, $\theta = T(p_o/p)^\kappa$. The overbar-star now indicates the mass weighted zonal mean quantities along isentropic surfaces. The primes again indicate perturbations from that mean. Retaining only the horizontal mixing component, the eddy flux term on the right-hand side can be expressed (for small perturbations) as $\overline{v'\mu'} = -K_{yy}\frac{\partial\bar{\mu}}{\partial y}$ where $K_{yy} = \frac{1}{2}\frac{\partial(\overline{\eta'^2})}{\partial t}$, where η' is the meridional displacement of the tracer contour from the zonal mean. K_{yy} is the eddy mixing term expressed as a diffusion coefficient (the flux-gradient approximation).

Ertel potential vorticity (PV), which can be derived solely from wind and temperature measurements, is an important dynamical diagnostic of the flow [e.g., *McIntyre*, 1992]. In the discussion below, we will make use of potential vorticity as a coordinate and as a type of tracer. Following *Andrews et al.* [1987], the Ertel potential vorticity q is defined as

$$q = \zeta/\sigma,$$

where $\zeta = f + \zeta'$, where f is the planetary vorticity, ζ' is the local vorticity. Now from *Andrews et al.* [1987], it can be shown that

$$\frac{\partial q}{\partial t} + u\frac{\partial q}{\partial x} + v\frac{\partial q}{\partial y} + \dot{\theta}\frac{\partial q}{\partial \theta} = q\frac{\partial \dot{\theta}}{\partial \theta} + R, \qquad (5)$$

where R represents the residual forcing terms associated with momentum sources such as breaking gravity waves and isentropic divergence of vorticity. We do not discuss these terms for this article, as they can usually be neglected in the lower stratosphere. Note the similarity in the form of equations (1) and (5). In adiabatic flows, PV behaves much like a tracer, and this explains why PV and tracer fields tend to move together for periods of a week or more as can be seen from maps.

Potential vorticity increases very rapidly with altitude and becomes somewhat awkward to use if a large range of altitudes are to be compared. This complication can be partially remedied by using modified PV (MPV) instead of Ertel PV [*Lait, 1994; Mller and Gunther, 2003*]. MPV relates to PV as follows: $MPV = q(\theta/\theta_o)^{-\varepsilon}$ where θ_o is a constant. MPV has the same conservation properties as Ertel PV. The value of θ_o is usually taken to be 420 K, and ε is $-9/2$ [*Lait, 1994*] or -4 [*Mller and Gunther, 2003*]. Plate 1c shows zonal mean MPV contours along with potential temperature and the heating rates.

We now return to the discussion of the circulation of the stratosphere. In an atmosphere at radiative equilibrium, the net heating and cooling rates would go to zero. The fact that the cooling rate in the polar night region is not zero even in the depth of winter shows that the temperature is largely being maintained against radiative equilibrium by a downwelling circulation. The magnitude of the downwelling can be derived from the heating rate and the change in temperature using the residual thermodynamic equation: $\overline{w}^* = (\overline{Q} - \partial\overline{T}/\partial t)/S$ where S is the static stability defined in the work of *Andrews et al.* [1987], and T is temperature. This relationship between the heating rate and the diabatic circulation can be used to diagnose the mean circulation [e.g., *Rosenlof*, 1995].

Given the residual vertical velocity, the meridional flow can be diagnosed from the residual continuity equation

$$\frac{1}{\rho}\frac{\partial\rho\overline{w}^*}{\partial z} + \frac{\partial\overline{v}^*}{\partial y} = 0. \qquad (6)$$

The simplified zonal mean meridional momentum equation is

$$\frac{\partial\overline{u}}{\partial t} - f\overline{v}^* = F, \qquad (7)$$

where F is the eddy momentum source such as planetary wave or gravity wave dissipation. Substituting for $\overline{v}^*$ into

equation (6) from equation (7), integrating from a level z to infinity, and noting that $\overline{w}^*$ at infinity is zero, we obtain

$$\overline{w}^*(z) = \frac{1}{\rho(z)}\int_z^\infty \rho\frac{\partial}{\partial y}\left[\left(\frac{\partial\overline{u}}{\partial t} - F\right)/f\right]dz, \qquad (8)$$

where ρ is the atmospheric density, and z is the log-pressure height. Equation 8 states that the upwelling circulation at a given level results from the eddy momentum source (F) above that level. This is called the "downward control principle" and was first derived by *Haynes et al.* [1991]. Thus, the circulation shown in Plate 1a is primarily the result of wave dissipation at levels above the upwelling or downwelling regions. In the lower and middle stratosphere, the eddy momentum source is due to the dissipation of large-scale eddies, which also mix the trace gas fields. In the upper stratosphere and lower mesosphere, breaking gravity waves provide an additional force especially in the summer hemisphere because stationary planetary waves cannot propagate into easterly flows [*Charney and Drazin*, 1961].

The transport of tracers by the large-scale zonal mean circulation sets up meridional trace gas gradients shown in Plate 1a. If this were the only transport process, then soon, the trace gas contours would be identical to streamlines of the flow. However, the presence of eddy mixing acts against the steepening of the zonal mean gradients. The slope of the tracer field, with respect to latitude seen in Plate 1a, can be related to the eddy mixing rate by assuming a steady circulation with meridional length L. Then using equation (4), Slope $= \frac{\delta\theta}{\delta y} \sim \frac{L\dot{\theta}}{K_{yy}}$, showing that as the mixing becomes stronger, the slope decreases, or if the circulation becomes stronger (larger values of $\dot{\theta}$), the slope increases.

The conservation of PV and θ in adiabatic and frictionless flow creates barriers to transport. Imagine an air parcel with a given potential temperature and potential vorticity. In order for that air parcel to move from its average location to a different average location, both its potential temperature and potential vorticity must change. The increase in absolute potential vorticity with latitude prevents air from the tropics from easily moving to middle latitudes and air from the exterior of the polar vortex from easily moving to the interior and vice versa. On the other hand, parcels move freely in regions where potential gradients on isentropic surfaces are weak (i.e., well-mixed regions).

The transport of potential vorticity by the diabatic circulation is one way to think about the development of transport barriers during the seasons. The strong eddy activity during winter sets up a downward diabatic flow at high latitudes moving high potential vorticity air to lower altitudes. This process generally maintains the polar night jet against deceleration by dissipating eddies propagating up from the troposphere. (An exception is the case of spectacular sudden warmings, when a wave pulse from the troposphere pulls the vortex off the pole, and circulation around the displaced vortex brings low tropical PV air toward the pole.) In addition, the interior of the vortex becomes a zone of high PV. Midlatitude air, with its lower value of potential vorticity, cannot usually penetrate the polar region. This downward movement of potential vorticity also lowers the tropopause, which can be defined roughly as a fixed potential vorticity contour [*Holton et al.*, 1995; *Mahlman*, 1997]. The reversal of the lower stratospheric polar circulation in the summer tends to lift lower values of potential vorticity into the midstratosphere, decelerating the jet and creating easterly winds. The tropopause is also lifted.

The overall dynamical picture of the transport of tracers by the stratospheric circulation has emerged from the aircraft, balloon, and satellite observations and concerted effort to develop multidimensional models of the stratospheric circulation consistent with those observations. The comparison of models and measurements advanced over the past decades as more measurements have become available, and the quality of models has improved [*Prather and Remsberg*, 1993; *Park et al.*, 1999; *Eyring et al.*, 2005, 2006, 2007; *Waugh and Eyring*, 2008]. The development of advanced numerical modeling tools as well as new data analysis techniques has been core to the improvements in models and achieving an understanding of the stratospheric circulation as seen through the trace gas distribution. In the next sections, we summarize some of the more important techniques and advances in numerical modeling that were developed in the last 20 years.

3. TRACER TRANSPORT DIAGNOSTICS: TOOLS AND TECHNIQUES

From equation (1), it is clear that dynamics and chemical reactions (including photolytic processes) both affect the mixing ratio of a tracer. Separation of these processes is needed to test the chemical mechanisms behind ozone depletion. Indeed, the separation of chemical and dynamical effects has been a core problem in stratospheric research for the last two decades. For example, what is the ozone chemical loss rate within the polar vortex during winter? How does polar vortex exchange air with the middle latitudes? How isolated is the vortex chemical system during winter from the midlatitude system? What is the origin of small-scale chemical anomalies observed by aircraft?

To address these questions and others, researchers have developed Lagrangian models and techniques, and pseudo-Lagrangian coordinate systems, such as potential temperature-equivalent latitude that allow quantitative separation of

dynamical and chemical processes. These Lagrangian models and techniques are described in sections 3.1, 3.4, and 3.5. These methods, with some modifications, also allow us to better quantify mixing processes and production of small-scale features by wave transience as described in section 3.2. Tracer-tracer correlations also provide a practical way to quantify chemical changes observed by aircraft and satellite (section 3.3). Finally, the development of the age-of-air concept has provided important new insights into the overall circulation of the stratosphere and the formation of barriers (section 3.6). Both tracer-tracer correlations and age-of-air estimates from observational data have provided an important link between the models of the stratosphere and observations. Producing compact tracer relationships and a reasonable age spectrum has become a test for the numerical methods described in section 4.

3.1. Trajectory Methods

Atmospheric models fall into two broad categories: Eulerian models, where variables are calculated on a fixed grid as in most general circulation models, and Lagrangian models, where variables are calculated following the motion of a parcel. A trajectory model is a Lagrangian model following equation (1). In a trajectory model, a massless parcel of air is moved by the wind field that is derived from either meteorological analyses or a general circulation model. In the last two decades, the improvements in meteorological analyses have increased the popularity of trajectory models, and now they are used for a variety of problems. Trajectory models track the origin of specific trace gas anomalies, allowing us to identify sources or processes that may have occurred along the parcel path prior to the point that the composition of the air parcel was measured. For example, the fact that air has encountered polar stratospheric clouds, resulting in high levels of active chlorine, can be inferred from parcel temperature history along a back trajectory [*Schoeberl et al.*, 1993; *Kawa et al.*, 1997]. Ozone loss in the polar vortex has been estimated by matching ozone changes computed from ozone profiles obtained by consecutive sonde launches along trajectory paths [e.g., *Rex et al.*, 1999; *Morris et al.*, 2005] and by combining satellite and sonde data [*Schoeberl et al.*, 2002]. Trajectory calculations also allow us to create synoptic maps from asynoptically sampled data (e.g., from satellites or aircraft) [*Morris et al.*, 1995, 2000], much like a very simple assimilation system.

The use of trajectory models for stratospheric work really began with *Danielsen* [1961] who noted that computing trajectories on potential temperature surfaces (isentropic trajectories) would eliminate the high-frequency adiabatic motions that plagued earlier trajectory calculations. Parcel trajectories are easy to compute given the modern analyzed wind and temperature fields. Most researchers use the fourth-order Runge-Kutta numerical integration scheme along with a bicubic interpolation scheme or other advanced numerical methods to obtain the meteorological variables at the parcel position.

Errors associated with trajectory calculations are cumulative and are usually dominated by errors in the analysis wind fields [Schoeberl and Sparling, 1995; Stohl, 1998]. As a rule of thumb, an individual trajectory calculation is unreliable after 5–7 days, but longer integrations may be used to assess statistical properties of the flow. One method often used to determine the reliability of the trajectory calculation is to run multiple trajectories starting in a small ring around the target. If the trajectory paths diverge significantly, the calculation is uncertain. A generalization of this idea is to compute two sets of trajectories at each point with the second trajectory offset small distance from the first. The separation distance between the two trajectories usually will begin to grow. We can express the growth of the separation distance between the two initial points as $\delta D = \delta D_o e^{\lambda t}$ where λ is the finite time Lyapunov coefficient, a concept borrowed from chaos theory. Regions with high values of λ often indicate zones where the flow has a low Richardson number (for a Richardson number less than $\frac{1}{4}$, the flow is unstable). In these regions, there is significant uncertainty in the position of the parcel.

Because a trajectory calculation is basically an integration of equation (1), the mixing ratio is conserved along the trajectory path in the absence of chemical sources or sinks. After a period of integration, the question arises as to whether a parcel's trace gas identity remains intact because subgrid scale processes will mix the parcel with surrounding air. This mixing is not implicitly or explicitly included in the trajectory computation. Treatment of subgrid scale mixing is one of the important differences between Lagrangian methods (like trajectory calculations) and Eulerian fixed-grid methods. For example, using a Lagrangian method, it is much more difficult to estimate trace gas fluxes because the "parcel" cannot be assigned to a volume or mass. Diffusive mixing is also difficult to simulate in Lagrangian models, but there are schemes that seem to do a reasonable job of simulating the mixing that takes place at turbulent scales. The most popular approach to simulating mixing is to have nearby parcels suddenly "mix" when the fluid strain reaches a critical value. In the Chemical Lagrangian Model of the Stratosphere (CLaMS), mixing triggers are related to the finite time Lyapunov coefficient [McKenna et al., 2002; Konopka et al., 2004] mentioned above. Trajectory models with mixing, like CLaMS, generally do a better job matching observations than similar calculations without mixing [Konopka et al., 2005],

and the explicit mixing parameterization gives insight into the properties of the fluid.

3.2. Contour Advection and Reverse Domain Fill

Contour advection and reverse domain fill are applications of trajectory methods that effectively boost the resolution of the tracer field. Contour advection, developed by Dritschel [1988, 1989] uses a series of points to represent contours that are then advected by the wind field using a trajectory code or Lagrangian integrator. New points are added as the contours stretch or are deleted as they shrink. Contours can also be reconnected: contour advection with surgery (CAS). Waugh and Plumb [1994] and Appenzeller et al. [1996], using contour advection, showed that small-scale filamentary features observed by aircraft were generated by the transience of the large-scale flow. Baker and Cunnold [2001] used the CAS method to study vortex evolution, contour lengthening, and filament production.

Reverse domain filling (RDF) is a similar technique to CAS and also based upon trajectory calculations [Sutton et al., 1994; Newman and Schoeberl, 1995]. RDF is also used to forecast or analyze small-scale structures in the trace gas field. RDF is simpler to implement than CAS, since no surgery algorithms are required. RDF techniques have been used to identify inward vortex breaking events seen by aircraft [Plumb et al., 1994; Schoeberl et al., 2006b] and satellites [e.g., Glatthor et al., 2005]. They have also been used to identify the mechanisms for vortex erosion [Schoeberl and Newman, 1995].

There are three steps to the RDF method. First, a regular grid of trajectory starting points is laid down. Second, a backward trajectory calculation is performed for each point on the grid. Finally, the low spatial resolution observations of the target field (for example, N_2O) are interpolated onto the positions of the backward trajectory points, and those values are then displayed on the original grid of points. The result produces fields with filaments and structures characteristic of the type of cascade to small scales seen in fluid mixing. Plate 3 shows an example of results using the RDF method. In this figure, the MLS low-resolution N_2O observations (about $25°$ longitude, $2°$ latitude) show an unusual structure north of Russia. The RDF calculation clearly resolves this structure as part of an inward breaking event that extends from north of Russia across the Arctic and winding up over Greenland.

3.3. Tracer-Tracer Correlations

As noted above, the compact relationships between tracers provide important information on the mixing processes within the stratosphere [Plumb and Ko, 1992; Plumb et al.,

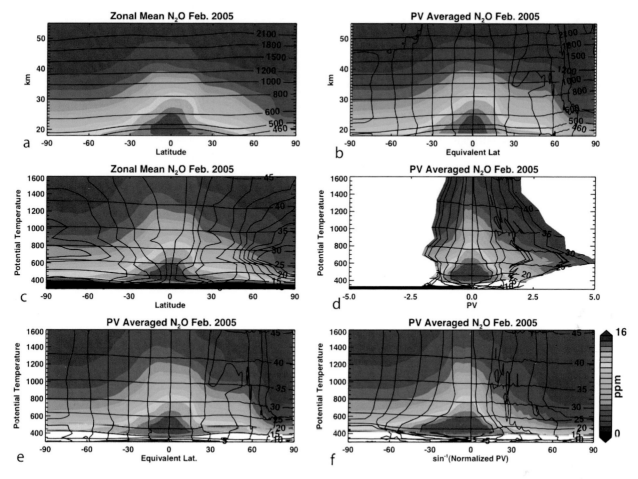

Plate 5. Zonal mean N$_2$O field for February 2005 using different coordinate systems. (a) Zonal mean N$_2$O, potential temperature contours (black) are overlaid. (b) PV averaged N$_2$O in height and equivalent latitude coordinates; contours of latitude (black) are overlaid. (c) Zonal mean N$_2$O in potential temperature coordinates; altitude and equivalent latitude contours are overlaid. (d) PV-potential temperature coordinates; altitude and latitude contours are overlaid. (e) PV-averaged N$_2$O in equivalent latitude coordinates; potential temperature coordinates, height, and latitude contours are overlaid. (f) N$_2$O in normalized PV coordinates; height and latitude contours are overlaid.

2000; *Plumb*, 2007]. As we noted earlier, chemical ozone loss, and physical processes such as denitrification and dehydration can be difficult to quantify in an environment where both dynamical processes and chemistry contribute. The high degree of tracer correlation noted in Plate 2 can be put to use to estimate physical and chemical changes in an air mass.

Fahey et al. [1990] was the first to look at breaks from the tight correlation between two tracers to estimate dehydration and denitrification. Plate 4 from *Considine et al.* [2003] illustrates how tracer-tracer correlations can be used to evaluate physical processes as well as test models. Plate 4 (top) shows observations of the relationship between NO$_y$ and N$_2$O in the Arctic polar vortex for early December, 1999. A chemical transport model simulation of the relationship is also shown. Plate 4 (bottom) shows the same relationship

as measured by aircraft in early March 2000. The changes in NO$_y$ are due to denitrification of the vortex through the formation and sedimentation of polar stratospheric clouds that are composed of ice crystals and the condensate of nitric acid trihydrate. The amount of denitrification is clearly quantifiable from this analysis. The gray dots and colored lines show different model simulations as described in the work of *Considine et al.* [2003].

A caveat is that one must be very careful drawing conclusions from changes in correlations especially in situations where the tracer relationships may not be unique in different air masses [*Plumb et al.*, 2000]. Nonetheless, tracer-tracer correlations are very useful for analysis of certain processes, for example, quantification of chemical ozone destruction under the conditions that a reliable estimate of

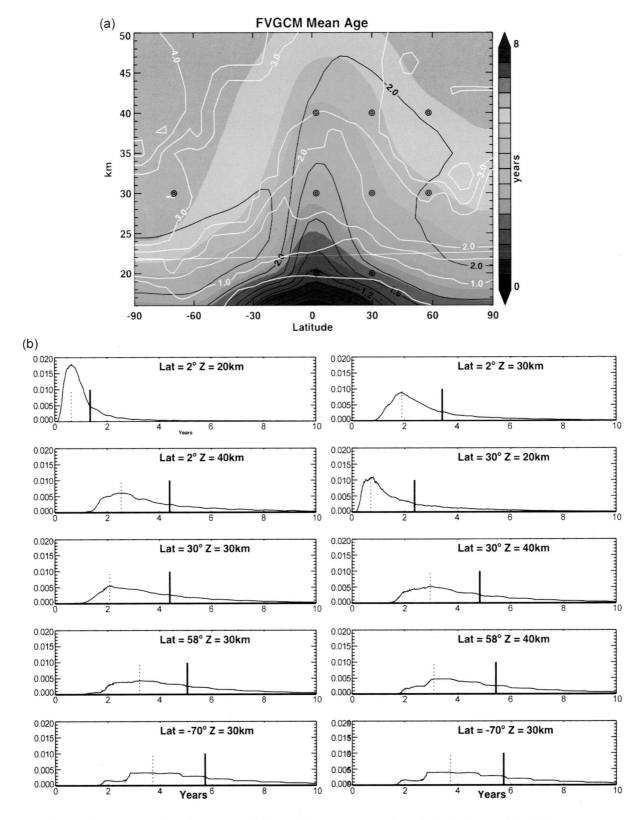

Plate 6. (a) Mean age is shown in color (years). The modal age contours are shown in black. The spectral width is shown in white. (b) Age spectra at the individual points shown as double circles in Plate 6a. The mean age is the thick dark line, while the modal age is the dotted line.

the ozone/tracer correlation is obtained for the early winter and that a transport barrier prevents mixing from midlatitudes into the polar vortex [Mller et al., 2005]. Furthermore, as first noted by Plumb and Ko [1992], the slope of the tracer-tracer correlation curve is proportional to the flux of tracer.

3.4. Potential Vorticity as a Tracer and Potential Vorticity Coordinates

Leovy et al. [1985] were among the first to note that PV and ozone were highly correlated in the high-latitude lower stratosphere. Potential vorticity can be considered a special type of tracer (see the form of equation (5)), and because PV can be computed from the meteorological field, it can be used when other tracers are absent. PV maps can assess mixing events and diagnose wave breaking in the stratosphere [e.g., McIntyre and Palmer, 1983].

The high degree of correlation between PV and the tracer field suggests that PV can be used to greatly reduce the meteorological variability of trace gas measurements. This is accomplished simply by analyzing the data in a PV (q) coordinate [Schoeberl and Lait, 1992]. More formally, using equation (1)

$$\frac{\partial \mu}{\partial t} + \mathbf{V} \cdot \nabla \mu = S - L, \qquad (9)$$

where $\mathbf{V}$ is the flow velocity. Without any loss of generality, we can choose a coordinate system to be s, q, θ where s is the distance along the PV contour. Physically, we can think of q and θ defining a tube of fluid with s being the along-tube dimension. Transforming equation (8),

$$\frac{\partial \mu}{\partial t} + \dot{s}\frac{\partial \mu}{\partial s} + \dot{q}\frac{\partial \mu}{\partial q} + \dot{\theta}\frac{\partial \mu}{\partial \theta} = S - L. \qquad (10)$$

Now consider an adiabatic, frictionless environment where q and θ are conserved. Conservation of q and θ means that the third and fourth terms are zero ($\dot{q} = \dot{\theta} = 0$). We assume that the tracer field is in close alignment with the flow field, consistent with the second stage of an adjustment near equilibrium of a tracer field (the first stage is "shear dispersion," where there is a rapid re-adjustment of the tracer field to align with the streamlines. In the second stage, the tracer field is nearly aligned with the streamlines, and continued adjustment occurs slowly through diffusion of the tracer across the streamlines [Rhines and Young, 1982].) Under such an assumption, $\frac{\partial \mu}{\partial s} \sim 0$. Thus, in this coordinate system $\frac{\partial \mu}{\partial t} = S - L$, which is now identical to equation (1). This means that the (s, q, θ) coordinate system is effectively Lagrangian, and in that system, we are watching chemical processes unfold as we are moving with the fluid tube.

In (s, q, θ) coordinate system, s is "longitude," q is the "latitude," and θ is the "height." Now as noted above, PV

increases rapidly with height, so this coordinate system is highly distorted. However, using MPV for q, the coordinate system becomes a bit more recognizable, as seen in Plate 5d. An alternate approach to reducing the range of PV with altitude is to normalize the PV field (nPV) on each zonal mean potential temperature surface. If the normalization extends between -1 and 1, then PV "latitude" is $\sin^{-1}$(nPV). Another form of modified PV (sPV), introduced by Dunkerton and Delisi [1986], scales PV to a base isentropic density, sPV = $q\sigma_o$, where σ_o is the isentropic density along a fixed potential temperature surface.

A process called "reconstruction" also makes use of (s, q, θ) coordinate systems [Lait et al., 1990]. If trace gases along a (s, q, θ) tube of fluid are well mixed, then the composition of the tube can be defined by a single measurement along the tube. Thus, one can combine occasional trace gas measurements, whether from satellite, balloon, or aircraft or instruments, to fill in the two-dimensional (2-D) (q, θ) domain such as shown in Plate 5d. Once the domain is filled in, the data can be "reconstructed" in physical space using the meteorological analysis of PV on θ surfaces. This technique has been used, for example, by Yang et al. [2007], to produce high-resolution stratospheric ozone maps from which the stratospheric ozone column can be derived.

3.5. Equivalent Latitude and Effective Diffusivity

Another form of potential vorticity coordinates is the popular "equivalent latitude" coordinate developed by Butchart and Remsberg [1986]. Equivalent latitude coordinate transformations tend to preserve the actual tracer gradients seen in nonzonal flows compared to zonal averaging and thus give a somewhat truer picture of the tracer structure. A latitude circle enclosing the same area as that enclosed by the PV contour (or any tracer, in general) is the equivalent latitude. More formally, we can define an area, A, where the value of PV (or mixing ratio of the tracer) is greater than or equal to q_a, then the equivalent latitude is defined as $A = 2\pi r^2 (1 - \sin(\varphi_e))$ where φ_e is the equivalent latitude. Averaging with respect to equivalent latitude is the same as averaging around the contour, so the (φ_e, θ) coordinate system is essentially the same as (s, q, θ) coordinates when the averaging is done along the tube (s). Because equivalent latitude is always normalized to the surface area of the earth, the q values used for averaging the data may change from day to day, thus equivalent latitude is not a conservative coordinate. In practice, this is not a serious problem if high-quality analyses are used, since q will change slowly from day to day.

Plate 5 shows the MLS N_2O field for February 2005 as seen in the various (q, θ) coordinate systems described above. This figure illustrates the kinds of changes in the tracer field structure that appears between the various coordinate systems. During the period shown, the center of the polar vortex is off the pole so that zonal mean averaging tends to wash out the tongue of low N_2O air descending at high northern latitudes. However, this feature is clearly seen in the equivalent latitude $-\theta$ coordinate system. All of the (q, θ) coordinate systems, including the normalized PV coordinate systems, were developed for tracer studies in the presence of highly distorted flows such as those seen at middle and high latitudes. These transformations are usually not needed for tropical analyses.

The concept of effective diffusion and equivalent length arises from equivalent latitude coordinate averaging. *Nakamura* [1996] showed that the mixing ratio in equivalent latitude coordinates satisfies a simple diffusion equation: that the advection terms vanish assuming that the flow is nearly nondivergent. The effective diffusion is given by κ_{eff} and can be written in terms of the equivalent length, L_e. Following *Haynes and Shuckburgh* [2000], the mixing ratio in equivalent latitude is given by

$$\frac{\partial \mu_e}{\partial t} = \frac{1}{a^2 \cos(\phi_e)} \frac{\partial}{\partial \phi_e} \left[\kappa_{eff} \cos(\phi_e) \frac{\partial \mu_e}{\partial \phi_e} \right], \quad (11)$$

where

$$\kappa_{eff} = \frac{\kappa L_{eq}^2}{(2\pi a \, \cos(\phi_e))^2} = \kappa a^2 \frac{\langle |\nabla \mu|^2 \rangle}{\left(\frac{\partial \mu_e}{\partial \phi_e} \right)^2},$$

where κ is the actual atmospheric diffusion rate and the $< >$ brackets indicate average over an area between adjacent tracer contours. The other terms are defined in the work of *Haynes and Shuckburgh* [2000]. The equivalent length is proportional to the length of the tracer contour and is equal to the length of a latitude circle for a pure zonal flow. As the flow becomes less zonal, the equivalent length increases, and the tracer gradient with respect to equivalent latitude weakens. As discussed by *Nakamura* [1996], where highly elongated tracer contours occur, diffusion (κ) is most effective, and mixing occurs across contours. Thus, transport is greatest when the contour lengths are longest as seen in equation (11).

Equivalent length diagnostics appear to work best with artificial tracers that have uniform pole to pole gradients [*Haynes and Shuckburgh*, 2000; *Allen and Nakamura*, 2001]. When artificial tracers are advected by model or analyzed winds and the equivalent length computed, the zones of minimum in equivalent length (or effective diffusivity) clearly show the position transport barriers shown in Plate 1a.

3.6. Age of Air

The concept of age-of-air was developed by *Kida* [1983] and considerably extended by *Hall and Plumb* [1994]. This concept provides important insight into transport process, the origin of air in the stratosphere, and the strength of the circulation. Measurements of the mean age of air using clock-like tracers have been used to diagnose and improve the circulation in numerical models [*Hall et al.*, 1999]. The reader is referred to the review article by *Waugh and Hall* [2002] for details beyond the discussion below.

The concept of age-of-air is particularly useful for the stratosphere, since air mostly enters the stratosphere in a narrow region in the tropics, and thus, the single entry point is the locus of the youngest air. Likewise for deep ocean circulation, the single descent point of cold North Atlantic water near Iceland makes the age concept useful for oceanographers.

The age of air is formally the boundary Green's function solution for the transport operator applied to a tracer with a boundary source. The Green's function, G, is called the age spectrum as shown in Figure 1. The mixing ratio μ inside the domain can be defined in terms of the boundary mixing ratio μ_o using the age spectrum.

$$\mu(t, y, z) = \int_0^t \mu_o(t', y_s, z_s) G(t, t', y, z) dt', \quad (12)$$

where G is normalized so that $\int_0^\infty G dt = 1$. The mean age is defined as $\Gamma = \int_0^t t G dt$, and the spectral width is $\Delta^2 = \frac{1}{2} \int_0^\infty (t - \Gamma)^2 G dt$. These are the first and second moments

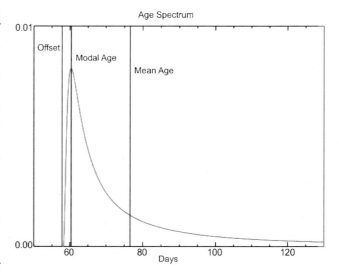

Figure 1. Age spectrum from an analytic model. The mean, modal ages, and spectrum offset are noted.

of G, respectively. The modal age is the peak in the age spectrum (the mode).

Figure 1 shows a typical age spectrum with the mean and mode ages identified.

One can think of a sample of air containing a mix of irreducible parcels with different ages. The age spectrum can then be interpreted as a histogram of the parcel ages within the sample. Age spectra are typically skewed toward older ages, and thus, the modal and mean ages are quite different. This means that even though an air sample may be dominated by young air components (the modal age is young), the mean age may be quite a bit older.

The age spectrum cannot be directly observed, but can be calculated using models. For example, the spectrum for a sample could be calculated using a series of long time back trajectories from a cluster of parcels within the sample domain. The age spectrum can also be computed from a chemical transport model by initiating a pulse of tracer, $\mu_o(t,t_o) = \delta(t - t_o)$, at the boundary and then tracking the tracer mixing ratio at each point. The solution of equation (12), in this case, is $\mu = G$.

Plate 6 shows the chemical transport model generated age spectrum from a tracer pulse experiment. The mean ages are younger in the lower tropical stratosphere and older at high tropical and middle latitudes. Not surprisingly, the shape of the mean age distribution is similar to that of long-lived tracers (Plate 1a) with the change in age following the Brewer-Dobson circulation.

The spectral width expands at higher altitudes and higher latitudes, and this is also seen in the individual age plots shown in Plate 6b. In general, looking at the cross-sections, the age spectrum tends to be monomodal, although there is evidence of seasonal effects that show up as small perturbations in the spectrum. Monomodality means that the tracers at a given location are mostly arriving via the same average pathway [*Schoeberl et al.*, 2005] and that the spread in the spectrum is due to diffusion along that path.

Although the spectrum itself cannot be observed, information about the spectrum can be obtained from tracer observations. For example, the mean age can be estimated from observations using tracers that have a nearly linearly increasing concentration such as SF_6 or CO_2; such estimates have been invaluable in evaluating models as mentioned above [*Hall et al.*, 1999]. Even more information can be extracted from tracers with nonlinear behavior or tracers with chemical sinks. In the simplest example, if the tracer has a loss rate proportional to its time in the stratosphere, then

$$\mu(t,y,z) = \int_{-\infty}^{t} \mu_o(t',y_s,z_s)G(t,t',y,z)e^{-\alpha t'}\,dt'$$

where α is the loss frequency. A radioactive tracer with a source in the troposphere would be an example of such a tracer. Note that if μ_o is constant, then the integral is the Laplace transform of G. Tracers that decay rapidly with time are sensitive to the "young" part of the age spectrum (see Plate 5), while longer-lived tracers are more sensitive to the "old" part of the spectrum. Thus, combining tracer fields or using the variation of these fields with time at the tropopause, allows us to gain more information on the age spectrum. *Andrews et al.* [2001], *Johnson et al.* [1999], and *Schoeberl et al.* [2005] used this approach to create an observationally based age spectra.

4. NUMERICAL TRANSPORT SCHEMES

Many of the important advances in understanding the stratospheric circulation have come from the close interplay between models and observations. Model advances have been spurred on by new observations, and the often disappointing realization that models being used to assess impact of changes in chlorine on the ozone layer were unable to simulate many existing observations. For example, simple model tests such as maintaining tight tracer-trace correlation [*Allen et al.*, 1991] or producing a mean age distribution that matches that deduced from measurements eluded early models [*Hall et al.*, 1999]. Getting the mean age right is especially important, since the mean age distribution is a function of the speed of the Brewer-Dobson overturning circulation, and the speed of the circulation affects the stratospheric lifetime of chlorine-containing source gases.

Improvements to chemistry transport models have come from better laboratory measurements of photochemical reaction rates and developing parameterizations of physical processes such as denitrification by polar stratospheric clouds [*Considine et al.*, 2000]; however, a major step forward has been the development of fast conservative numerical transport schemes. The development of these schemes is the topic of this section.

There are several key requirements for numerical transport schemes used in chemistry transport models [e.g., *Rood*, 1987]: (1) conservation of total tracer mass, (2) no negative mixing ratio (undershoots), and (3) maintaining tracer gradients.

1. Lack of tracer mass conservation clearly creates problems for simulations that attempt to quantify the response of ozone to changes in composition and climate. Assessments needed to compute the impact of chlorofluorcarbons on ozone loss calculate the ozone change for a specified input of chlorine-containing source gases. If the transport scheme does not conserve tracer mass, then the amount of chlorine in the simulation becomes disconnected from the specified input, and the simulated ozone change then depends on both the specified perturbation and the transport numerics.

2. Negative trace gas mixing ratios can be generated by the numerics near sharp gradients. The negative mixing ratios are fatal to the photochemical schemes. Although neighbor filling algorithms can eliminate negative mixing ratios, they can also fail to conserve tracer mass or produce unrealistic diffusion.

3. The numerical transport algorithm must produce and maintain sharp tracer gradients such as observed at the tropopause, at the edge of the polar vortex, and near the subtropical barriers. Many numerical schemes are excessively diffusive because the gradient within a grid box is averaged and is then advected between boxes. Some schemes produce numerical overshoots (create new maxima) and unrealistically sharpen gradients.

4.1. Development of Numerical Schemes

Some early chemistry transport models had success using spectral transport schemes that included a filling algorithm [e.g., *Cunnold et al.*, 1975; *Kurzeja et al.*, 1984; *Rood et al*, 1985], although the schemes could not maintain gradients for long integrations, nor did they conserve mass. Semi-Lagrangian schemes solve the problems of negative mixing ratios and overshoots but require special attention to mass conservation [*Williamson and Rasch*, 1989]. The tracer observations from the Airborne Antarctic Ozone Experiment and Airborne Arctic Stratospheric Expedition field campaigns provided a new constraint for transport algorithms, i.e., the simulated fields of long-lived tracers must be correlated with one another as discussed in section 2. Problems with filling algorithms and mass conservation have an impact on computed constituent distributions, and very few schemes used in atmospheric modeling in the 1980s could produce compact relationships among tracers.

Schemes that compute the constituent change using the winds and the gradient in the direction from which the wind is blowing (called "upwind" schemes) maintain constituent correlations and do not produce negative mixing ratios as long as the Courant-Friedrichs-Lewy (CFL) condition [*Courant et al.*, 1967] is satisfied. However, simple upwind schemes are very diffusive, and the flux from one box to its neighbor must be limited to maintain realistic constituent gradients [*van Leer*, 1974; *Zalasak*, 1979; *Smolarkiewicz*, 1984]. *Russell and Lerner* [1981] developed a scheme that transports the mean and the first moment of each tracer distribution within a grid box; *Prather* [1986] generalized this scheme to also include the second-order moment. The later scheme, referred to as second-order moments (SOM), satisfies the physical criteria for tracer transport, but has huge storage requirements. In a 3-D simulation, besides the mean concentration array, nine additional arrays must be maintained for the moments of each

transported constituent (three arrays for each transport direction times the three directions). The SOM scheme has been used in 2-D models [e.g., *Douglass et al.*, 1989] and 3-D models [*Chipperfield*, 1999]. Nonetheless, the SOM scheme is practical for 2-D and 3-D applications depending on duration and number of transported constituents. SOM can also be practical for multidecadal 3-D applications if the spatial resolution is not too fine. For example, SLIMCAT running at a horizontal resolution of 7.5° longitude by 7.5° latitude produces realistic horizontal transport because the SOM scheme transports the distribution of tracer within a grid box as well as the total amount.

Allen et al. [1991] implemented the upwind scheme with flux limiters described by *van Leer* [1974] to improve the representation of long-lived tracers in the chemistry transport model (CTM) developed at Goddard Space Flight Center. This implementation required smaller time steps at higher latitudes to satisfy the CFL condition and also required the implementation of a small well-mixed polar cap to allow cross-polar flow. Along with improved behavior of the simulated tracers provided by this scheme came improved fidelity of the ozone simulation [*Rood et al.*, 1991].

Despite its successes, the *van Leer* [1974] scheme is not optimal. The simulated gradients were somewhat weaker than observed, and this deficiency was traced to numerical diffusion that is not eliminated by the flux limiters in the transport algorithm. *Lin and Rood* [1996] (LR) developed a piecewise parabolic scheme that allowed transport for more than a single grid cell, thus relaxing the CFL condition. The LR scheme is less diffusive than the van Leer scheme, improving the representation of the constituent gradients. However, even though the LR scheme is more diffusive than SOM, it does not have the large memory requirements of SOM.

4.2. Chemistry Transport Models

An important application of chemistry transport models is to quantify the importance of various anthropogenic emissions to air quality and climate. Usually, the same problem is addressed with different numerical models, which may have different grid resolutions and different transport schemes. Results differ among models, and it is an ongoing effort to evaluate the credibility of various model simulations and to understand the differences among the model results. For example, *Gurney et al.* [2003] analyzed the results from 17 different CTMs that attempted to calculate regional carbon sources and sinks from the same data using a standard approach. They found that the transport was a significant source of uncertainty. Recent studies have attempted to quantify the possible contribution of errors in numerical transport to the transport differences among models. *Wild and*

Prather [2006] performed a series of simulations in a single CTM, increasing the horizontal resolution up to 0.625° longitude by latitude. These simulations showed that the solution to the continuity equation obtained with SOM converged to a "true" solution. *Prather et al.* [2008] compared results from a CTM with the SOM scheme with results from a CTM using the LR scheme for a simulation of CO_2. A constant surface pattern of emissions was specified in both CTMs, and the buildup and dispersal of CO_2 was followed for 10 years. For horizontal resolution of 5° longitude by 4° latitude, the differences attributable to the numerical schemes for the same simulation of CO_2 were as large as the differences reported by *Gurney et al.* [2003]. This is an important finding, since *Gurney et al.* [2003] concluded that the intermodel differences resulted from differences in meteorology not transport methodology. *Prather et al.* [2008] found that the largest differences between LR and SOM simulations were in regions of strong gradients. This is not surprising, since SOM is superior to LR at maintaining strong gradients. The age-of-air can be calculated from the simulated CO_2, and Prather et al. found that the CTM with LR produced age values about 1 year younger than the CTM with SOM at 5° longitude by 4° latitude. The differences in the CO_2 simulation and the derived age distributions were greatly reduced when the resolution for both CTMs was doubled.

To summarize, the choice of an appropriate transport scheme and appropriate spatial resolution depends on the application. For simulations of the evolution of constituents, it is important that the numerical resolution and the transport scheme be suited to the horizontal and vertical gradients that are expected from observations. Correct simulation of tracer-tracer correlation and the components of the age spectrum place stringent constraints on the numerical transport scheme and on assimilation systems as well [*Schoeberl et al.*, 2003; *Douglass et al.*, 2003]. Results of numerical transport can depend on the scheme and the resolution. A scheme that maintains gradients (such as SOM) will provide a more accurate estimate of the mean age of air and the age spectrum for the meteorological fields that are being used compared to a more diffusive scheme (such as LR) if all other factors are equal. As grid resolution is increased, the contribution of numerical errors to sensitive quantities such as the mean age and the age spectrum decrease.

5. SUMMARY

The number of quantitative tools used to understand trace gas observations from satellite and airborne sensors has grown remarkably in the last 20 years. This growth has had its foundation in theoretical developments, improved measurements, improved meteorological analysis, and improved numerical methods. The extensive use of PV-based analyses and trajectory models would not have been possible without high-quality wind and temperature fields from modern meteorological analysis systems. The discovery of the compact tracer-tracer correlations and that linkage between tracer distributions and transport would not have been possible without improved satellite and aircraft observations. Likewise, the development of the age-or-air diagnostic has led to a better understanding of the stratospheric circulation and improvements in models. Although the focus of the article has been on the stratosphere, the alert reader will quickly realize that some of the techniques described here have been independently developed in areas of research unrelated to the stratosphere. Many of the techniques described here are applicable to the troposphere, the ocean, and indeed any planetary fluid problem.

Acknowledgments. This article was submitted as part of Alan Plumb's 60th Birthday Conference in 2008. It is clear from the frequent references to Alan Plumb's works in this paper that he has had a very significant and influential role in the development of our understanding of atmospheric transport processes.

REFERENCES

Allen, D. J., A. R. Douglass, R. B. Rood, and P. D. Guthrie (1991), Application of a monotonic upstream-biased transport scheme to three dimensional constituent transport calculations, *Mon. Weather Rev.*, *119*, 2456–2464.

Allen, D. R., and N. Nakamura (2001), A seasonal climatology of effective diffusivity in the stratosphere, *J. Geophys. Res.*, *106*, 7917–7935.

Andrews, A., K. Boering, S. Daube, S. Wofsy, E. Hintsa, E. Weinstock, and T. Bui (1999), Empirical age spectra for the lower tropical stratosphere from in situ observations of CO_2: Implications for stratospheric transport, *J. Geophys. Res.*, *104*, 26,581–26,595.

Andrews, A. E., K. A. Boering, S. C. Wofsy, B. C. Daube, D. B. Jones, S. Alex, M. Loewenstein, J. R. Podolske, and S. E. Strahan (2001), Empirical age spectra for the midlatitude lower stratosphere from in situ observations of CO_2: Quantitative evidence for a subtropical "barrier" to horizontal transport, *J. Geophys. Res.*, *106*, 10,257–10,274.

Andrews, D. G., and M. E. McIntyre (1978), Generalized Eliassen-Palm and Charney-Drazin theorems for waves in axisymmetric mean flows in compressible atmospheres, *J. Atmos. Sci.*, *35*, 175–185.

Andrews, D. G., J. R. Holton, and C. B. Leovy (1987), in *Middle Atmosphere Dynamics*, Elsevier, New York.

Appenzeller, C., H. C. Davies, and W. A. Norton (1996), Fragmentation of stratospheric intrusions, *J. Geophys. Res.*, *101*, 1435–1456.

Baker, M. N., and D. Cunnold (2001), The uses and limitations of contour advection as a technique for examining arctic vortex dynamics, *J. Atmos. Sci.*, *58*, 2210–2221.

Baldwin, M. P., et al. (2001), The quasi-biennial oscillation, *Rev. Geophys.*, *39*, 179–229.

Brewer, A. W. (1949), Evidence for a world circulation provided by the measurements of helium and water vapor distribution in the stratosphere, *Q. J. R. Meteorol. Soc.*, *75*, 351–363.

Butchart, N., and E. E. Remsberg (1986), The area of the stratospheric polar vortex as a diagnostic for tracer transport on an isentropic surface, *J. Atmos. Sci.*, *43*, 1319–1339.

Charney, J. G., and P. G. Drazin (1961), Propagation of planetary-scale disturbances from the lower into the upper atmosphere, *J. Geophys. Res.*, *66*, 83–109.

Chipperfield, M. P. (1999), Multiannual simulations with a three-dimensional chemical transport model, *J. Geophys. Res.*, *104*, 1781–1805.

Considine, D. B., A. R. Douglass, P. S. Connell, D. E. Kinnison, and D. A. Rotman (2000), A polar stratospheric cloud parameterization for the global modeling initiative three-dimensional model and its response to stratospheric aircraft, *J. Geophys. Res.*, *105*, 3955–3973.

Considine, D. B., S. R. Kawa, M. R. Schoeberl, and A. R. Douglass (2003), N_2O and NO_y observations in the 1999/2000 Arctic polar vortex: Implications for transport processes in a CTM, *J. Geophys. Res.*, *108*(D5), 4170, doi:10.1029/2002JD002525.

Courant, R., K. Friedrichs, and H. Lewy (1967), On the partial difference equations of mathematical physics, *IBM J.*, *11*, 215–234.

Cunnold, D., F. Alyea, N. Phillips, and R. Prinn (1975), Three-dimensional dynamical-chemical model of atmospheric ozone, 2029–2051, *J. Atmos. Sci.*, *32*, 170–194.

Danielsen, E. F. (1961), Trajectories: Isobaric, isentropic and actual, *J. Meteorol.*, *18*, 479–486.

Dobson, G. M. B. (1956), Origin and distribution of the polyatomic molecules in the atmosphere, *Proc. R. Soc. London, Ser. A*, *236*, 187–192.

Douglass, A. R., C. H. Jackman, and R. S. Stolarski (1989), Comparison of model results transporting the odd nitrogen family with results transporting separate nitrogen species, *J. Geophys. Res.*, *94*, 9862–9872.

Douglass, A. R., M. R. Schoeberl, R. B. Rood, and S. Pawson (2003), Evaluation of transport in the lower tropical stratosphere in a global chemistry and transport model, *J. Geophys. Res.*, *108*(D9), 4259, doi:10.1029/2002JD002696.

Dritschel, D. G. (1988), Contour surgery: A topological reconnection scheme for extended integrations using contour dynamics, *J. Comput. Phys.*, *77*, 240–266.

Dritschel, D. G. (1989), Contour dynamics and contour surgery: Numerical algorithms for extended, high-resolution modeling of vortex dynamics in two-dimensional, inviscid, incompressible flows, *Comput. Phys. Rep.*, *10*, 77–146.

Dunkerton, T. J., and D. P. Delisi (1986), Evolution of potential vorticity in the winter stratosphere of January–February 1979, *J. Geophys. Res.*, *91*, 1199–1208.

Eyring, V., et al. (2005), A strategy for process-oriented validation of coupled chemistry-climate models, *Bull. Am. Meteorol. Soc.*, *86*, 1117–1133.

Eyring, V., et al. (2006), Assessment of temperature, trace species and ozone in chemistry-climate model simulations of the recent past, *J. Geophys. Res.*, *111*, D22308, doi:10.1029/2006JD007327.

Eyring, V., et al. (2007), Multimodel projections of stratospheric ozone in the 21st century, *J. Geophys. Res.*, *112*, D16303, doi:10.1029/2006JD008332.

Fahey, D. W., S. Solomon, S. R. Kawa, M. Loewenstein, J. R. Podolske, S. E. Strahan, and K. R. Chan (1990), A diagnostic for denitrification in the winter polar stratospheres, *Nature*, *345*, 698–702.

Gille, J. C., and J. M. Russell III (1984), The Limb Infrared Monitor of the Stratosphere: Experiment description, performance, and results, *J. Geophys. Res.*, *89*, 5125–5140.

Glatthor, N., et al. (2005), Mixing processes during the Antarctic vortex split of September–October 2002 as inferred from source gas and ozone distributions from ENVISAT-MIPAS, *J. Atmos. Sci.*, *62*, 787–800.

Gottwald, M., et al. (2006), in *SCIAMACHY, Monitoring the Changing Earth's Atmosphere*, DLR.

Gurney, K. R., et al. (2003), TransCom3 CO_2 inversion intercomparison: 1. Annual mean control results and sensitivity to transport and prior flux information, *Tellus, Ser. B*, *55*, 555–579.

Hall, T. M., and R. A. Plumb (1994), Age as a diagnostic of stratospheric transport, *J. Geophys. Res.*, *99*, 1059–1070.

Hall, T. M., D. W. Waugh, K. A. Boering, and R. A. Plumb (1999), Evaluation of transport in stratospheric models, *J. Geophys. Res.*, *104*, 18,815–18,839.

Haynes, P., and E. Shuckburgh (2000), Effective diffusivity as a diagnostic of atmospheric transport 1. Stratosphere, *J. Geophys. Res.*, *105*, 22,777–22,794.

Haynes, P. H., C. J. Marks, M. E. McIntyre, T. G. Shepherd, and K. P. Shine (1991), On the downward control of extratropical diabatic circulations by eddy-induced mean zonal forces, *J. Atmos. Sci.*, *48*, 651–678.

Holton, J. R. (1986), A dynamically based transport parameterization for one-dimensional photochemical models of the stratosphere, *J. Geophys. Res.*, *91*, 2681–2686.

Holton, J. R. (1992), in *Dynamic Meteorology*, 3rd ed., 511 pp., Elsevier, San Diego, Calif.

Holton, J. R., P. H. Haynes, M. E. McIntyre, A. R. Douglass, R. B. Rood, and L. Pfister (1995), Stratosphere-troposphere exchange, *Rev. Geophys.*, *33*, 403–439.

Johnson, D. G., K. W. Jucks, W. A. Traub, K. V. Chance, G. C. Toon, J. M. Russell III, and M. P. McCormick (1999), Stratospheric age spectra derived from observations of water vapor and methane, *J. Geophys. Res.*, *104*, 21,595–21,602.

Kawa, S. R., et al. (1997), Activation of chlorine in sulfate aerosol as inferred from aircraft observations, *J. Geophys. Res.*, *102*(D3), 3921–3933.

Kida, H. (1983), General circulation of air parcels and transport characteristics derived from a hemispheric GCM, Part 2, Very long-term motions of air parcels in the troposphere and stratosphere, *J. Meteorol. Soc. Jpn.*, *61*, 510–522.

Konopka, P., et al. (2004), Mixing and ozone loss in the 1999–2000 arctic vortex: Simulations with the three-dimensional Chemical Lagrangian Model of the Stratosphere (CLaMS), *J. Geophys. Res.*, *109*, D02315, doi:10.1029/2003JD003792.

Konopka, P., R. Spang, G. Gunther, R. Muller, D. S. McKenna, D. Offermann, and M. Riese (2005), How homogeneous and isotropic is stratospheric mixing? Comparison of CRISTA-1 observations with transport studies based onthe Chemical Lagrangian Model of the Stratosphere (CLaMS), *Q. J. R. Meteorol. Soc.*, *131*, 565–579.

Kurzeja, R. J., K. V. Haggard, and W. L. Grose (1984), Numerical experiments with a general circulation model concerning the distribution of ozone in the stratosphere, *J. Atmos. Sci.*, *41*, 2029–2051.

Lait, L. R. (1994), An Alternative form for potential vorticity, *J. Atmos. Sci.*, *51*, 1754–1759.

Lait, L. R., et al. (1990), Reconstruction of O_3 and N_2O fields from ER-2, DC-8, and balloon observations, *Geophys. Res. Lett.*, *17*, 521–524.

Leovy, C. B., C. R. Sun, M. H. Hitchman, E. E. Remsberg, J. M. Russell, L. L. Gordley, J. C. Gille, and L. V. Lyjak (1985), The transport of ozone in the middle stratosphere—Evidence for planetary wave breaking, *J. Atmos. Sci.*, *42*, 230–244.

Lin, S.-J., and R. B. Rood (1996), Multidimensional flux form semi-Lagrangian transport schemes, *Mon. Weather Rev.*, *124*, 2046–2070.

Logan, J. A., D. B. A. Jones, I. A. Megretskaia, S. J. Oltmans, B. J. Johnson, H. Vömel, W. J. Randel, W Kimani, and F. J. Schmidlin (2003), Quasibiennial oscillation in tropical ozone as revealed by ozonesonde and satellite data, *J. Geophys. Res.*, *108*(D8), 4244, doi:10.1029/2002JD002170.

Mahlman, J. D. (1997), Dynamics of transport processes in the upper troposphere, *Science*, *276*, 1079–1083.

Mahlman, J. D., H. Levy II, and W. J. Moxim (1986), Three-dimensional simulations of stratospheric N_2O: Predictions for other trace constituents, *J. Geophys. Res.*, *91*, 2687–2707.

McIntyre, M. E. (1992), Atmospheric dynamics: Some fundamentals, with observational implications, in *The Use of EOS for the Study of Atmospheric Physics*, edited by G. Visconti, and J. Gille, *Proc. Int. Sch. Phys. Enrico Fermi*, vol. 115, pp. 313–386, NATO Summer School, North Holland, N. Y.

McIntyre, M. E., and T. N. Palmer (1983), Breaking planetary waves in the stratosphere, *Nature*, *305*, 593–600.

McKenna, D. S., P. Konopka, J.-U. Grooß, G. Günther, R. Müller, R. Spang, D. Offermann, and Y. Orsolini (2002), A new Chemical Lagrangian Model of the Stratosphere (CLaMS) 1. Formulation of advection and mixing, *J. Geophys. Res.*, *107*(D16), 4309, doi:10.1029/2000JD000114.

Morris, G. A., et al. (1995), Trajectory mapping and applications to data from the Upper Atmosphere Research Satellite, *J. Geophys. Res.*, *100*, 16,491–16,505.

Morris, G. A., J. F. Gleason, J. Ziemke, and M. R. Schoeberl (2000), Trajectory mapping: A tool for validation of trace gas observations, *J. Geophys. Res.*, *105*, 17,875–17,894.

Morris, G., B. Bojkiv, L. Lait, and M. Schoeberl (2005), A review of the Match technique as applied to AASE-2/EASOE and SOLVE/THESEO 2000, *Atmos. Chem. Phys.*, *5*, 2571–2592.

Mote, P. W., K. H. Rosenlof, M. E. McIntyre, E. S. Carr, J. C. Gille, J. R. Holton, J. S. Kinnersley, H. C. Pumphrey, J. M. Russell III, and J. W. Waters (1996), An atmospheric tape recorder: The imprint of tropical tropopause temperatures on stratospheric water vapor, *J. Geophys. Res.*, *101*, 3989–4006.

Müller, R., and G. Gunther (2003), A generalized form of Lait's modified potential vorticity, *J. Atmos. Sci.*, *60*, 2229–2237.

Müller, R., S. Tilmes, P. Konopka, J.-U. Groß, and H.-J. Jose (2005), Impact of mixing and chemical change on ozone-tracer relations in the polar vortex, *Atmos. Chem. Phys.*, *5*, 3139–3151.

Nakamura, N. (1996), Two-dimensional mixing, edge formation, and permeability diagnosed in an area coordinate, *J. Atmos. Sci.*, *53*, 1524–1537.

Newman, P. A., and M. R. Schoeberl (1995), A reinterpretation of the data from the NASA stratosphere-troposphere exchange project, *Geophys. Res. Lett.*, *22*, 2501–2504.

Park, J. H., M. K. W. Ko, C. H. Jackman, R. A. Plumb, J. A. Kaye, and K. H. Sage (1999), Models and measurements intercomparison II, *NASA Tech. Memo. 1999–209554*, NASA, Hampton, Va.

Plumb, R. A. (2002), Stratospheric transport, *J. Meteorol. Soc. Jpn.*, *80*, 793–809.

Plumb, R. A. (2007), Tracer interrelationships in the stratosphere, *Rev. Geophys.*, *45*, RG4005, doi:10.1029/2005RG000179.

Plumb, R. A., and R. C. Bell (1982), A model of the quasi-biennial oscillation on an equatorial beta-plane, *Q. J. R. Meteorol. Soc.*, *108*, 335–352.

Plumb, R. A., and M. K. W. Ko (1992), Interrelationships between mixing ratios of long lived stratospheric constituents, *J. Geophys. Res.*, *97*, 10,145–10,156.

Plumb, R. A., D. W. Waugh, R. J. Atkinson, P. A. Newman, L. R. Lait, M. R. Schoeberl, E. V. Browell, A. J. Simmons, and M. Loewenstein (1994), Intrusions into the lower stratospheric Arctic vortex during the winter of 1991–1992, *J. Geophys. Res.*, *99*, 1089–1105.

Plumb, R. A., D. W. Waugh, and M. P. Chipperfield (2000), The effects of mixing on tracer relationships in the polar vortices, *J. Geophys. Res.*, *105*, 10,047–10,062.

Prather, M. J. (1986), Numerical advection by conservation of second-order moments, *J. Geophys. Res.*, *91*, 6671–6681.

Prather, M. J., and E. E. Remsberg (1993), The atmospheric effects of stratospheric aircraft: Report of the 1992 Models and Measurements Workshop, *NASA Ref. Pub. 1292*, NASA, Washington, D. C.

Prather, M. J., Z. Zhu, S. E. Strahan, S. D. Steenrod, and J. M. Rodriguez (2008), Quantifying errors in trace species transport modeling, *Proc. Natl. Acad. Sci. U. S. A.*, *105*, 19,617–19,621.

Pumphrey, H. C., C. Boone, K. A. Walker, P. Bernath, and N. J. Livesey (2008), Tropical tape recorder observed in HCN, *Geophys. Res. Lett.*, *35*, L05801, doi:10.1029/2007GL032137.

Reber, C. A., C. E. Trevathan, R. J. McNeal, and M. R. Luther (1993), The Upper Atmosphere Research Satellite (UARS) mission, *J. Geophys. Res.*, *98*, 10,643–10,647.

Rex, M., et al. (1999), Chemical ozone loss in the Arctic winter 1994/95 as determined by the Match technique, *J. Atmos. Chem.*, *32*, 35–59.

Rhines, P. B., and W. R. Young (1982), Homogenization of potential vorticity in planetary gyres, *J. Fluid Mech.*, *122*, 347–367.

Rood, R. B. (1987), Numerical advection algorithms and their role in atmospheric transport and chemistry models, *Rev. Geophys.*, *25*, 71–100.

Rood, R. B., D. J. Allen, W. E. Baker, D. J. Lamich, and J. A. Kaye (1985), The use of assimilated stratospheric data in constituent transport calculations, *J. Atmos. Sci.*, *46*, 687–701.

Rood, R. B., A. R. Douglass, J. A. Kaye, M. A. Geller, C. Yuechen, D. J. Allen, E. M. Larson, E. R. Nash, and J. E. Nielsen (1991), Three-dimensional simulations of wintertime ozone variability in the lower stratosphere, *J. Geophys. Res.*, *96*, 5055–5071.

Rosenlof, K. H. (1995), Seasonal cycle of the residual mean meridional circulation in the stratosphere, *J. Geophys. Res.*, *100*, 5173–5191.

Russell, G. L., and J. A. Lerner (1981), A new finite-differencing scheme for the tracer transport equation, *J. Appl. Meteorol.*, *20*, 1483–1498.

Schoeberl, M. R., and L. R. Lait (1992), Conservative coordinate transformations for atmospheric measurements, in *The Use of EOS for the Study of Atmospheric Physics*, edited by G. Visconti, and J. Gille, *Proc. Int. Sch. Phys. Enrico Fermi*, vol. 115, pp. 419–430, NATO Summer School, North Holland, N. Y..

Schoeberl, M. R., and P. A. Newman (1995), A multiple-level trajectory analysis of vortex filaments, *J. Geophys. Res.*, *100*, 25,801–25,815.

Schoeberl, M. R., and L. C. Sparling (1995), Trajectory modeling, in *Diagnostic Tools in Atmospheric Physics*, edited by G. Fiocco, and G. Visconti, *Proc. Int. Sch. Phys. Enrico Fermi*, vol. 124, pp. 289–305, IOS Press, Amsterdam.

Schoeberl, M. R., et al. (1993), The evolution of ClO and NO along air parcel trajectories, *Geophys. Res. Lett.*, *20*(22), 2511–2514.

Schoeberl, M. R., et al. (2002), An assessment of the ozone loss during the 1999–2000 SOLVE/THESEO 2000 Arctic campaign, *J. Geophys. Res.*, *107*(D20), 8261, doi:10.1029/2001JD000412.

Schoeberl, M. R., A. R. Douglass, Z. Zhu, and S. Pawson (2003), A comparison of the lower stratospheric age spectra derived from a general circulation model and two data assimilation systems, *J. Geophys. Res.*, *108*(D3), 4113, doi:10.1029/2002JD002652.

Schoeberl, M. R., A. R. Douglass, B. Polansky, C. Boone, K. A. Walker, and P. Bernath (2005), Estimation of stratospheric age spectrum from chemical tracers, *J. Geophys. Res.*, *110*, D21303, doi:10.1029/2005JD006125.

Schoeberl, M. R., et al. (2006a), Overview of the EOS Aura Mission, *IEEE Trans. Geosci. Remote Sens.*, *44*, 1102–1105.

Schoeberl, M. R., et al. (2006b), Chemical observations of a polar vortex intrusion, *J. Geophys. Res.*, *111*, D20306, doi:10.1029/2006JD007134.

Schoeberl, M. R., B. N. Duncan, A. R. Douglass, J. Waters, N. Livesey, W. Read, and M. Filipiak (2006c), The carbon monoxide tape recorder, *Geophys. Res. Lett.*, *33*, L12811, doi:10.1029/2006GL026178.

Schoeberl, M. R., et al. (2008), QBO and annual cycle variations in tropical lower stratosphere trace gases from HALOE and Aura MLS observations, *J. Geophys. Res.*, *113*, D05301, doi:10.1029/2007JD008678.

Shepherd, T. G. (2007), Transport in the middle atmosphere, *J. Meteorol. Soc. Jpn.*, *85B*, 165–191.

Sparling, L. C. (2000), Statistical perspectives on stratospheric transport, *Rev. Geophys.*, *38*, 417–436.

Stohl, A. (1998), Computation, accuracy and applications of trajectories—A review and bibliography, *Atmos. Environ.*, *32*, 947–966.

Smolarkiewicz, P. K. (1984), A fully multidimensional positive definite advection algorithm with small implicit diffusion, *J. Comp. Phys.*, *54*, 325–362.

Sutton, R. T., H. Maclean, R. Swinbank, A. O'Neill, and F. W. Taylor (1994), High resolution stratospheric tracer fields estimated from satellite observations using Lagrangian trajectory calculations, *J. Atmos. Sci.*, *51*, 2995–3005.

van Leer, B. (1974), Towards the ultimate conservative difference scheme, II: Monotonicity and conservation combined in a second-order scheme, *J. Comput. Phys.*, *14*, 361–370.

Wale, M. J., and G. D. Peskett (1984), Some aspects of the design and behavior of the Stratospheric and Mesospheric Sounder, *J. Geophys. Res.*, *89*, 5287–5293.

Waugh, D. W., and V. Eyring (2008), Quantitative performance metrics for stratospheric-resolving chemistry-climate models, *Atmos. Chem. Phys.*, *8*, 5699–5713.

Waugh, D. W., and T. M. Hall (2002), Age of stratospheric air: Theory, observations, and models, *Rev. Geophys.*, *40*(4), 1010, doi:10.1029/2000RG000101.

Waugh, D. W., and R. A. Plumb (1994), Contour advection with surgery: A technique for investigating finescale structure in tracer transport, *J. Atmos. Sci.*, *51*, 530–540.

Wild, O., and M. J. Prather (2006), Global tropospheric ozone modeling: Quantifying errors due to grid resolution, *J. Geophys. Res.*, *111*, D11305, doi:10.1029/2005JD006605.

Williamson, D., and P. Rasch (1989), Two-dimensional semi-Lagrangian transport with shape-preserving interpolation, *Mon. Weather Rev.*, *117*, 102–129.

Yang, Q., D. M. Cunnold, H.-J. Wang, L. Froidevaux, H. Claude, J. Merrill, M. Newchurch, and S. J. Oltmans (2007), Midlatitude tropospheric ozone columns derived from the Aura Ozone Monitoring Instrument and Microwave Limb Sounder measurements, *J. Geophys. Res.*, *112*, D20305, doi:10.1029/2007JD008528.

Zalasak, S. T. (1979), Fully multidimensional flux-corrected transport algorithms for fluids, *J. Comput. Phys.*, *31*, 335–362.

A. R. Douglass, NASA Goddard Space Flight Center, Code 613.3, Greenbelt, MD 20771, USA.

M. A. Schoeberl, Science and Technology Corporation, 9650 Santiago Road, Columbia, MD 21045, USA. (mark.schoeberl@mac.com)

Chemistry and Dynamics of the Antarctic Ozone Hole

Paul A. Newman

NASA Goddard Space Flight Center, Greenbelt, Maryland, USA

The Antarctic ozone hole is caused by human-produced chlorine and bromine compounds. The unique cold conditions of the Antarctic lower stratosphere in winter and spring allow for the development of polar stratospheric clouds. Chemical reactions on these stratospheric cloud particle surfaces (heterogeneous chemistry) release chlorine from reservoir species into highly reactive species that are easily photolyzed. Chlorine and bromine catalytic cycles result in massive ozone depletion during August and September. By early October, ozone is completely destroyed in the lower stratosphere over Antarctica. The amount of total ozone began a downward trend in the 1970s and stopped in the early 1990s. Current models indicate an ozone hole return date around 2067.

1. INTRODUCTION AND BACKGROUND

The ozone hole is an extreme ozone depletion that has been appearing annually since the 1970s over the Antarctic region. Ozone (O_3) amounts in the Antarctic lower stratosphere drop dramatically in the course of a few weeks during the Austral spring (August–October). The ozone hole begins to develop in August and culminates by early October and subsequently disappears by early December. The ozone column amount in this period falls by up to 50% over Antarctica. Observations taken in the Antarctic region from aircraft, the ground, and satellites have confirmed that the ozone hole results from human-produced chlorine and bromine compounds, combined with the peculiar meteorology of the Southern Hemisphere (SH) winter.

Farman et al. [1985] discovered the Antarctic ozone hole using a ground-based instrument called a Dobson spectrophotometer at the Halley Bay station in Antarctica. Plate 1 displays the Halley Bay observations of the total column ozone (black points). *Farman et al.* [1985] originally showed the points from 1956 to 1984. These initial observations

showed a slow decline from 1956 to 1975, followed by a rapid decline to 1984. *Farman et al.* [1985] also suggested that the depletion was a result of the accumulation of man-made CFC in the atmosphere (albeit, their proposed chemical mechanism turned out to be incorrect).

Measurements by the Total Ozone Mapping Spectrometer (TOMS) instrument aboard the Nimbus 7 satellite subsequently showed that the depletion of ozone during the SH spring occurred over the entire Antarctic continent, centered on the South Pole [*Stolarski et al.*, 1986]. Because of the visual appearance of this Antarctic low-ozone region in satellite images, this phenomenon was quickly dubbed the Antarctic "ozone hole" [*Müller*, 2009].

Various theories were initially proposed to account for the existence of the ozone hole. These included the dynamical theory, the nitrogen oxide theory, and the heterogeneous chemistry theory [*Schoeberl and Rodriguez*, 2009]. Ultimately, observations showed that the heterogeneous chemical theory was correct and that human-produced chemicals, CFCs, and bromine compounds (halons), were responsible for the ozone hole.

The heterogeneous chemistry theory proposed that reactions were occurring on the surfaces of polar stratospheric clouds (PSCs) that form in the extremely cold conditions of the Antarctic winter stratosphere [*McElroy et al.*, 1986; *Solomon et al.*, 1986]. The PSCs altered the polar stratospheric chemistry, converting nonreactive chlorine reservoir

The Stratosphere: Dynamics, Transport, and Chemistry
Geophysical Monograph Series 190
Published in 2010 by the American Geophysical Union.
10.1029/2009GM000873

compounds into reactive forms. These reactive chlorine compounds (combined with bromine species) catalytically destroy ozone at an extremely rapid rate in the sunlit atmosphere during the Austral spring [*McElroy et al.*, 1986; *Molina and Molina*, 1987]. Based on the heterogeneous chemistry theory, the rapid increase of CFCs and halons in the 1970 through 1995 period was the primary factor in the rapid decline of Antarctic ozone.

In this chapter, we give a basic description of the Antarctic ozone hole. In section 2, we provide some basic observations of the Antarctic ozone hole. Section 3 describes the dynamics of the SH circulation and temperatures, while section 4 describes the chemistry of the ozone hole. Section 5 delves into the future of the ozone hole and the limitations to our understanding future ozone. The final section provides a summary discussion of the Antarctic ozone hole and the current prognosis for the future.

2. OBSERVATIONS OF THE ANTARCTIC OZONE HOLE

As noted in section 1, the Antarctic ozone hole was discovered by *Farman et al.* [1985], and this was quickly followed by the publication of satellite images of the depleted region [*Stolarski et al.*, 1986]. Plate 2 shows a series of October averages of total ozone from the backscatter ultra-violet instrument on the Nimbus 4 satellite, TOMS, and the Ozone Monitoring Instrument (OMI) on the Aura satellite. Comparing the images from the 1970s with total column ozone from the last decade shows much lower ozone over the Antarctic. The values have decreased from around 280 Dobson units (DU) (green) to less than 150 DU (purple). The satellite images show that this ozone decrease is continental in scale. The minimum Antarctic values of total ozone from these October averages are shown in Plate 1. There is a good agreement between the Halley Bay station and the satellite observations.

Ozone depletion slowly begins in July and continues through the winter and spring to a maximum depletion in September or October [*Gardiner and Shanklin*, 1986]. Plate 3a displays the ozone hole area, and Plate 3b shows the ozone hole minimum value. The ozone hole area is defined by the area enclosed by the 220-DU value (approximately the light blue color in the Plate 2 images). As ozone is destroyed in early August, total ozone values fall below 220 DU somewhere over Antarctica, and the ozone hole area increases. For comparison, the size of the North American continent is 24.7 million km^2. Typically, the maximum size of the ozone hole occurs in middle-to-late September, while the lowest ozone values occur in late September or early October. Return to normal ozone levels begins in early October and is usually

complete by the middle of December [*Bowman*, 1986]. The averages of the area and minimum values are shown as the thick black lines, while 2008 is shown in red. Day-to-day meteorological fluctuations cause both the area and minimum to fluctuate. It should be noted that the ozone hole area rapidly increases, and the ozone minimum rapidly falls as the sun rises over Antarctica. The blue line in Plate 3 shows the solar zenith angle. Angles greater than 90° indicate polar night and are indicated as the shaded region.

As is shown in Plate 3, the decreases of ozone observed during October develop during August and September. In addition, those decreases are largely concentrated in the lower stratosphere between the altitudes of about 12 and 20 km. This temporal and vertical structure was recognized soon after the discovery of the ozone hole using ozone balloon observations [*Chubachi*, 1984; *Hofmann et al.*, 1987]. Plate 4 shows the ozone balloon observations (ozone density) from *Chubachi* [1984]. In the lower stratosphere in 1982, the values fall from about 23 DU km^{-1} to about 9 DU km^{-1} over the August and September period. Values rapidly fluctuate during October through December as the polar vortex alternately moved over and away from Syowa (more information about the polar vortex can be found in the next section and from *Waugh and Polvani* [this volume]). As an aside, *Chubachi* [1984] also pointed out that the 1982 ozone was the smallest since 1966. However, he failed to show the clear secular trend of ozone that was shown by *Farman et al.* [1985]. Current observations of ozone from South Pole ozone sondes show that ozone is virtually zero over the altitude range of 12–20 km by early October [*Solomon et al.*, 2005].

In the last few years, the ozone hole has demonstrated both extreme depletions (e.g., 2000) and rather weak depletions (e.g., 2002). Plate 5a shows ozone values in the lower stratosphere on the 460-K isentropic level from National Oceanographic and Atmospheric Administration (NOAA) ozonesondes launched from the South Pole. Ozone values typically increase over the fall and early winter periods as higher ozone mixing ratios are transported (or advected) poleward and downward (see the next section on the circulation). These mixing ratios peak in midwinter and then precipitously decline in the late August through September period. Values near 460 K typically reach zero by early October. Four of the 5 years shown in Plate 5a show a decline in ozone to a value of zero by early October. The sole exception is 2002, when this ozone decline was sharply arrested in late September by a major stratospheric warming [*Allen et al.*, 2003; *Newman and Nash*, 2005] (see the next section and *Shepherd et al.* [2005] for a description of major warmings). Plate 5b shows the vertical structure of the ozone decreases (in ppmv). The decreases extend approximately from 12 to 24 km (gray shaded region). Furthermore,

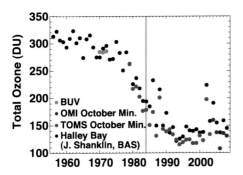

Plate 1. October total column ozone versus year from Halley Bay Station (black points), Total Ozone Mapping Spectrometer (TOMS) (blue points), Ozone Monitoring Instrument (OMI) (red points), and backscatter ultraviolet (BUV) (green points). The satellite points are the minimums observed in the October values over Antarctica. The vertical gray line indicates 1984. See the filled star from the upper left panel of Plate 2 for the location of Halley Bay station.

the values of ozone in the 15- to 20-km layer plunge to near zero.

The Antarctic ozone observations can be summarized in terms of both column amounts and vertical profiles. The downward trend of total ozone began in the 1970s and stopped in the early 1990s. This downward trend is seen in both ground and satellite observations (Plates 1 and 2). An inspection of satellite imagery and a comparison of the various Antarctic ground stations reveal that the ozone depletion is continental in scale (Plate 2). Total ozone satellite, balloon,

and ground observations all show that decreases develop in the August through September period and culminate in October (Plates 3–5). It is also clear that ozone recovers to near normal values during the November through December period. The balloon observations clearly demonstrate that the ozone decreases are mostly confined to the lower stratosphere (Plates 4 and 5). The balloon profiles also reveal that 100% of the ozone is now lost in the 15- to 20-km region. A comparison of the Syowa profiles from 1982 (Plate 4) with the South Pole profiles from 2000 (Plate 5a) shows that the profile decreases became worse between these years in good agreement with the column decreases.

3. DYNAMICS OF THE ANTARCTIC CIRCULATION

The Antarctic stratosphere is a unique region of our atmosphere. The dynamics and meteorology of the SH stratosphere have been described by *Labitzke and van Loon* [1972], *Randel and Newman* [1998], and *Waugh and Polvani* [this volume]. The polar stratospheric regions of both hemispheres are surrounded by a narrow band or stream of fast-moving winds (over 50 m s^{-1}) very high up and blowing from west to east (the polar night jet). Similar to the more familiar upper tropospheric jet stream or subtropical jet, this jet stream develops along a zone of sharp temperature contrasts (or a tight temperature gradient). Plate 6 displays a longitudinally averaged (or zonal-mean) plot of both temperature and winds averaged over 1979–2001 in August.

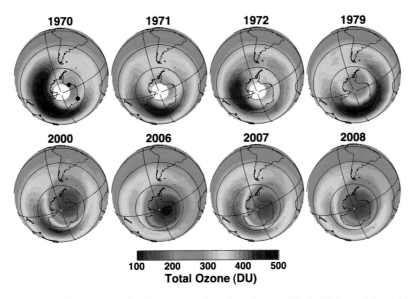

Plate 2. False color images of October total column ozone for selected years. The 1970 through 1972 images are derived from BUV, the 1979 and 2000 images are from TOMS, and the 2006 through 2008 images are from OMI. The filled star and circle on the 1970 (top left) image shows the location of the Halley Bay and Syowa stations, respectively. Observations are in Dobson units.

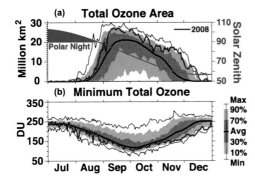

Plate 3. (a) Ozone hole area and (b) minimum SH value of total ozone. The area is defined by the region enclosed by the 220-DU value in the Antarctic region. The daily climatology is determined from the 1979 to 2008 period. The thick black line shows the average for each day of this climatology. The gray shading shows the percentage range of those same values. The red line shows the values for 2008. The blue line in Plate 3a shows the noon solar zenith angle at 80°S. Angles greater than 90° (shaded) indicate that the sun is always below the horizon. The data are from the TOMS series of instruments and from Aura OMI.

The polar night jet extends from the upper troposphere to the mesosphere at about 55°S (the axis of the polar night jet is shown by the gray line in Plate 6).

Plate 6 shows that extremely low temperatures are found over Antarctica during winter, averaging about 183 K at 50 hPa (20 km) in early August. These temperatures are contained inside the polar vortex region. The winds are predominantly westerly, with only a slight north-south meridional com-

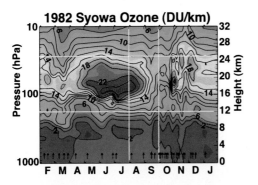

Plate 4. Contour plot of ozone density from balloon observations (Dobson units km^{-1}) from Syowa station during the February 1982 through January 1983 period. See the filled circle from the upper left panel of Plate 2 for the location of Syowa. These observations were first shown by *Chubachi* [1984, Figure 2] in partial pressure units. The vertical bars are set on 1 August and 1 October to show the period of ozone depletion over Syowa station, while the horizontal bars show the altitude range of the depletion. The arrows at the bottom indicate the balloon launch dates.

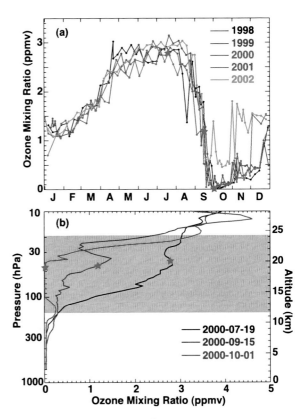

Plate 5. (a) Ozone balloon observations in mixing ratio (ppmv) from South Pole station at the 460-K isentropic level (approximately 18–19 km in altitude and 50 hPa in pressure). The points indicate the balloon launch dates. The three blue stars highlight the values on 19 July 2000, 15 September 2000, and 1 October 2000. (b) Ozone profiles of mixing ratio (ppmv) for the three highlighted days. The three blue stars indicate the 460-K isentropic level from the upper panel. The data are courtesy of the NOAA Global Monitoring Division. See *Hofmann et al.* [1997] for a discussion of these sondes.

ponent. This means that there is little mixing of warmer air from the midlatitudes to the polar region [*Schoeberl and Hartmann*, 1991]. The very low temperatures lead to the formation of PSCs, which require temperatures below approximately 195 K (more about PSCs and the chemical reactions in the next section). The temperature increases toward the north (a positive temperature gradient) by more than 20 K at 60°S, generally confining the PSCs to the Antarctic region.

The low temperatures over Antarctica develop during the April through June period and reach their coldest levels in early August. Plate 7a shows the minimum temperature observed in the Antarctic lower stratosphere region (50°–90°S) on the 50-hPa pressure surface (approximately 20 km). Temperatures decrease in the polar lower stratosphere because of the absence of absorption of UV radiation by ozone. By the middle of May, temperatures are low enough to

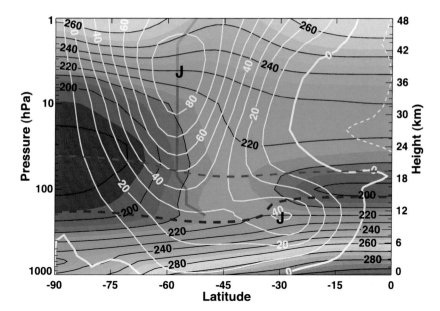

Plate 6. Zonal-mean temperature (colors and solid black isolines) and zonal-mean zonal wind (white isolines) averaged for the August 1979–2001 period. The dark dashed line shows the tropopause, while the dashed red line shows the 460-K isentropic surface. The upper "J" is the polar night jet, while the lower "J" near 30°S and 12 km is the subtropical jet. The vertical gray line between 50°S and 60°S shows the polar night jet axis. The data are from the European Centre for Medium-Range Weather Forecasts ERA-40 reanalyses.

form PSCs (note the green horizontal line). The temperature rapidly increases in the September through November period as the sun returns to the Antarctic region. This spring temperature increase is indirectly driven by the upward propagation of planetary waves [*Andrews et al.*, 1987]. These waves deposit easterly momentum in the middle stratosphere, which induces a descending circulation over the Antarctic

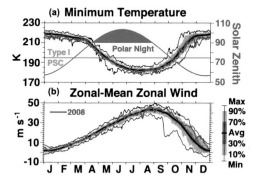

Plate 7. As in Plate 3 except for (a) minimum temperature and (b) zonal-mean zonal wind on the 50-hPa level. The temperature is the minimum value in the 50°–90°S region from each day, while the zonal-mean zonal wind is averaged in the 45°–75°S latitude band. The green line in Plate 7a indicates the temperature at which type I PSCs would form. The data are from the National Centers for Environment Prediction's (NCEP) Climate Prediction Center stratospheric analyses.

region. This descending circulation directly warms the Antarctic stratosphere. By the middle of October, the temperature is too high to support the formation of PSCs.

The polar night jet reaches its strongest point at the end of August. Plate 7b shows the annual cycle of winds around Antarctica. During the summer (December–February), winds are very weak (<10 m s^{-1}). The jet steadily accelerates from late January to late August and rapidly decelerates in the late August to middle December period. As was noted with the warming of the Antarctic stratosphere, upward propagating planetary-scale waves deposit easterly momentum in the August through November period, which causes this deceleration. These waves are key elements in the annual cycle of the Antarctic and the year-to-year variability of the Antarctic stratosphere.

The very strong jet stream acts to confine the air in the Antarctic region. This "containment" of air over Antarctica can be theoretically explained by examining potential vorticity (PV), a conserved quantity. PV is derived from the momentum, thermodynamic, and continuity equations [*Ertel*, 1942; *Andrews et al.*, 1987]. PV has two properties that make it a very useful dynamic quantity. First, PV in the lower stratosphere is conserved on a 10- to 30-day time scale, and second, if a "piece" of polar PV air is displaced northward, then this same air mass will set up a circulation pattern that acts to push it back southward, effectively producing a restoring force. The first property makes PV an excellent

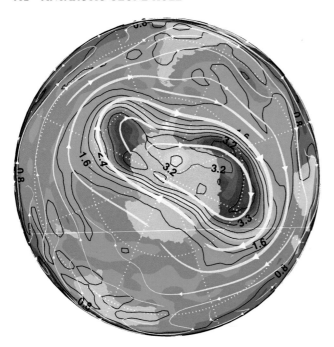

Plate 8. Potential vorticity (PV) (colors and solid black isolines) and streamlines (white lines) for 1200 UTC on 22 September 2002 on the 460-K isentropic surface (about 50 hPa or 20 km). The white streamlines show the magnitude and direction of the wind. The wind speed scales linearly with line thickness, with the maximum wind speed of 68 m s^{-1} (indicated by the thickest white line between Antarctica and Africa). The PV is estimated following *Lait* [1994] and is scaled by a factor of -10^{-5}. The data are from the NCEP/ National Center for Atmospheric Research reanalyses.

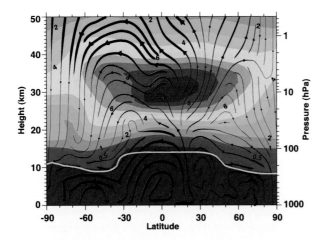

Plate 9. Ozone mixing ratios (in ppmv, colors) and the residual circulation (black streamlines) for August 2007. The ozone mixing ratios are from Aura MLS observations, while the residual circulation is derived from a GMAO meteorological analysis. The black streamlines show both the magnitude and direction of the flow. The semitransparent white line shows the tropopause.

parameter for tracing polar air. The second property acts to keep Antarctic air over Antarctica, reducing the mixing between polar latitudes and midlatitudes. In addition, this PV self-restoring force also allows the formation of waves that move about the polar vortex on 1- to 2-week time scales. Plate 8 displays PV on 22 September 2002: an extremely disturbed period for the SH stratosphere. (An undisturbed pattern would be more circular and centered on the pole.) The high values of PV (in red) constitute the Antarctic polar vortex. As is clear from the streamlines, the vortex (red region) is generally found inward of the maximum value of the jet stream. The vortex edge is found in the strong gradient region between the inner core (red colors) and the midlatitudes (blue colors). Polar vortex air has a much different chemical composition than midlatitude air. This PV and constituent relationship has a long history in stratospheric dynamics (for example, *Danielsen* [1968] analyzed the relationship of ozone, PV, and radioactive tracers near the

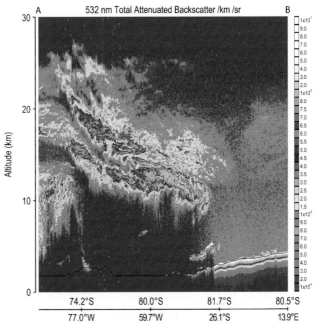

Plate 10. Polar stratospheric cloud structure observed over Antarctica by the CALIPSO satellite on 1 August 2008. The plotted quantity is total attenuated backscatter from the 532-nm laser. The red line denotes the Antarctic surface elevation. The higher elevations to the left at 74.2°S are the mountains near the base of the Antarctic Peninsula, while the steady upslope near the right shows the East Antarctic ice sheet. The detailed wavelike structure of the backscatter observations suggest that the PSCs result from waves that are forced by mountain waves, which propagate upward into the stratosphere and produce local cooling. Plate courtesy of C. Trepte (NASA/LaRC).

NH jet stream). *Newman* [1986] examined the breakdown of the polar vortex and related this to the "filling" of the Antarctic ozone hole by midlatitude air.

As was mentioned in section 1, the ozone hole is caused by chlorine and bromine released by human-produced compounds, such as CFCs, halons, and CH_3Br. How do the CFCs get from industrialized regions, predominantly in the Northern Hemisphere (NH), to the polar stratospheres? The transport circulation carries air from the troposphere into the stratosphere at the tropical tropopause. Plate 9 displays the typical "residual circulation" for late summer (August 2007) (see *Shepherd* [2007] and references therein for a complete discussion of middle atmosphere transport and the residual circulation). Air is carried into the stratosphere across the tropopause in the tropics (thick semitransparent white line). The air moves upward in the tropical stratosphere and then spreads poleward and downward. This transport is referred to as the Brewer-Dobson circulation. As the air rises in the tropics, ozone is produced by solar radiation (more in the next section), reaching values greater than 8 ppmv in the tropical middle stratosphere. This air is then carried to the Antarctic stratosphere. This Brewer-Dobson circulation is slow. The average time required for an air parcel to move from the tropical tropopause to 20 km over Antarctica is 5 to 6 years [*Hall and Plumb*, 1994]. The Brewer-Dobson circulation carries air containing the compounds to the upper stratosphere, where these compounds are photolyzed (broken down) by intense solar radiation. For example, at the tropical tropopause, the CFC-12 mixing ratio is still undiminished from the well-mixed troposphere. After a 6-year transit through the upper stratosphere to 20 km over Antarctica, a large fraction of the CFC-12 molecules have been photolyzed, and the released chlorine atoms are in the reservoir species hydrochloric acid (HCl) and chlorine nitrate ($ClONO_2$).

In this section, we have shown that the Antarctic lower stratosphere in late summer is: (1) extremely cold, (2) dominated by a strong west-east circulation called the polar night jet, (3) the polar night jet isolates and confines the air over Antarctica, (4) this air is relatively ozone rich, (5) the lower stratospheric air has taken 5 to 6 years to arrive over Antarctica, and (6) the chlorine in this air is largely in the form of HCl and $ClONO_2$.

4. CHEMISTRY OF THE OZONE HOLE

The ozone hole is caused by a confluence of factors. First, the temperature gets extremely low over Antarctica during polar night. Second, because it is so cold, chemical processes are enabled on cloud and aerosol particles over Antarctica, converting the $ClONO_2$ and HCl into molecular chlorine (Cl_2). Third, the Cl_2 is photolyzed by visible radiation as the

sun rises over Antarctica. Finally, chlorine and bromine catalytic reactions rapidly destroy ozone during the August through September period. In the previous section, we discussed temperature (see Plate 7a). In this section, we will cover the basic microphysical and chemical processes of the ozone hole.

Stratospheric particles (aerosols and PSCs) are a prime element behind the Antarctic ozone hole. Chemical heterogeneous reactions occur on the surfaces of these particles, and the particles can settle out of the stratosphere, carrying both water and nitric acid to lower altitudes. Reported occurrences of PSCs have been documented into the 18th century [*Stanford and Davis*, 1974]. PSCs are known as "mother-of-pearl" clouds because of their opalescent coloration as seen from the ground. These early reports on PSCs found that they were located at high latitudes in the lower stratosphere, were associated with very cold temperatures, occurred in winter, and that they occurred more frequently in the SH than in the NH [*Stanford and Davis*, 1974].

There are three basic types of PSCs: (1) a liquid supercooled ternary solution of water (H_2O), nitric acid (HNO_3), and sulfuric acid (H_2SO_4) (supercooled ternary solution (STS) or type Ib); (2) nitric acid trihydrate (NAT: $HNO_3 \cdot 3H_2O$) or dihydrate (NAD: $HNO_3 \cdot 2H_2O$) (type Ia); and (3) water ice particles (type II). The beautiful coloration of PSCs results from their dispersion of sunlight. This dispersion is caused by the varying size, composition, and structure of the PSCs [*Hallett and Lewis*, 1967]. The "type" classification of PSCs is derived from optical observations. Stratospheric aerosols [*Junge et al.*, 1961] are a separate class of particles, since they are optically thin (unobservable by eye except for the brilliant sunsets following large volcanic eruptions). These aerosols are also found in the extrapolar stratosphere.

PSCs generally form at temperatures below 195 K. Such temperature conditions are widespread over Antarctica in midwinter (see Plate 6). The importance of stratospheric particles is that the chlorine in reservoir species (HCl and $ClONO_2$) is converted into Cl_2, initiating the catalytic ozone loss. Plate 10 displays a snapshot of PSC structure on 1 August 2008 over Antarctica from the Cloud-Aerosol Lidar and Infrared Pathfinder Satellite Observation (CALIPSO) satellite lidar observations of total attenuated backscatter. Meteorological analyses (primarily based on satellite temperature retrievals) show that temperatures throughout this region were lower than 184 K at 50 hPa (about 18.5 km).

PSC microphysics remains a very important topic in the stratospheric community. *Noel et al.* [2008] used CALIPSO observations to show that while PSCs are ubiquitous over Antarctica during the winter, the low temperature is not a sufficient condition to explain the presence of a PSC. PSC

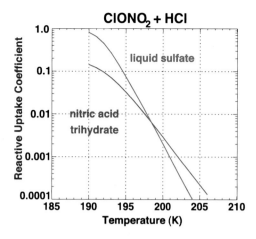

Plate 11. Reactive uptake coefficient versus temperature for a nitric acid trihydrate particle (red) and a liquid sulfate particle (blue). The two curves are taken from the result of *Kawa et al.* [1997, Figure 5], and have been computed for typical Antarctic lower stratosphere conditions.

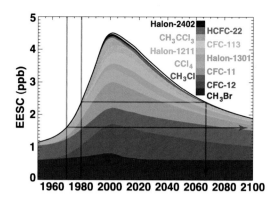

Plate 13. Sand chart of effective equivalent stratospheric chlorine (EESC) for Antarctic conditions from 1950 to 2100. The EESC is estimated using the surface observations and future estimates, empirically derived fractional release rates, an age spectrum with a 5.5-year mean age, and a bromine scaling factor of 90 (see *Newman et al.* [2007] for further details). The black line represents the total, while the contributions from various species are indicated by the color bar. Five other minor species are not shown.

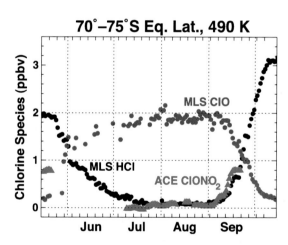

Plate 12. Daily averages of ClO (red dots) and HCl (black dots) observed by Aura MLS, and ClONO₂ (blue triangles) observed by the Fourier Transform Spectrometer from the Atmospheric Chemistry Experiment (ACE-FTS) at 490 K (about 20 km) during the 2005 Antarctic winter–spring period. All values are calculated over the 70°–75°S equivalent latitude band using the GMAO GEOS-4 temperatures and PV. Only daytime measurements are included in the averages for ClO. ClO data appear sparser because measurements in sunlight are not always available at high equivalent latitudes, especially in early winter. The sampling pattern of ACE-FTS, a solar occultation instrument, does not provide coverage of this equivalent latitude band at all times throughout the winter. Courtesy of M. Santee and adapted from *Newman and Rex* [2007, Figures 4–10].

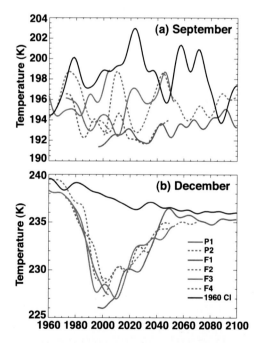

Plate 14. Temperature versus year for (a) September and (b) December at 50 hPa in the 70°–80°S latitude band. All of the model simulations are forced with the *Montzka and Fraser* [2003] Ab scenario for CFCs and halons, with the exception of the 1960 Cl simulation (black) that uses chlorine and bromine levels fixed to 1960 levels. All model simulations are forced with IPCC scenario A1b [*Houghton et al.*, 2001]. The time series have been smoothed with a Gaussian filter that has a half-amplitude point at 20 years.

formation is also dependent on the levels of H_2O and HNO_3 [*Hanson and Mauersberger*, 1988], a background of aqueous H_2SO_4 droplets, and the evolution of the air over the course of its circuit around the SH. Observations of large NAT particles by *Fahey et al.* [2001] further challenged our understanding of PSC formation. These large NAT particles were observed by an in situ instrument measuring total odd nitrogen (NO_y) and were too optically thin to be observed by normal lidar instrumentation. The formation of these large NAT particles has yet to be conclusively identified with the result that simulating the role of large NAT particles in three-dimensional (3-D) coupled chemistry-climate models is incomplete. Hence, the paradigm for PSC development has considerably evolved over the last few years, but large uncertainties remain [*Lowe and MacKenzie*, 2008]. Since PSCs do not uniformly develop in cold regions, it is critical to understand their formation and to develop parameterizations for model studies under future stratospheric climate conditions.

The primary importance of stratospheric particles is that chemical reactions can occur on the surfaces or inside these particles. While heterogeneous chemical reactions have been known for quite some time, they were not thought important in the stratosphere until the discovery of the ozone hole. *Solomon et al.* [1986] recognized that heterogeneous reactions might play an important role in the ozone hole and proposed that the reaction $ClONO_2 + HCl \xrightarrow{PSC} Cl_2 + HNO_3$ could convert reservoir chlorine species ($ClONO_2$ and HCl) into forms that could accelerate ozone depletion.

There are five principal heterogeneous reactions in the polar lower stratosphere.

$$ClONO_2 + HCl \xrightarrow{PSC} Cl_2 + HNO_3 \qquad (1)$$

$$HOCl + HCl \xrightarrow{PSC} Cl_2 + H_2O \qquad (2)$$

$$ClONO_2 + H_2O \xrightarrow{PSC} HOCl + HNO_3 \qquad (3)$$

$$N_2O_5 + HCl \xrightarrow{PSC} ClNO_2 + HNO_3 \qquad (4)$$

$$N_2O_5 + H_2O \xrightarrow{PSC} 2HNO_3 \qquad (5)$$

The reaction rates for these heterogeneous reactions are proportional to the surface area density of the particle type, the molecular thermal velocity or the square root of the temperature, and the reactive uptake coefficient (γ). The reactive uptake coefficient (sometimes referred to as the sticking coefficient) is the fraction of molecule A that reacts with molecule B after A collides with the particle surface. Plate 11 shows γ for equation (1) as adapted from *Kawa et al.* [1997]. As the Antarctic stratosphere cools during the May

and June period, the initial particles are cold liquid sulfate aerosols. These uptake coefficients are highly dependent on particle type. For example, at polar lower stratospheric temperatures, γ for $ClONO_2$ and HCl is about 0.3 for water ice (type II, $T \approx 188$ K), 0.07 for NAT (type Ia, $T \approx 193$ K), and nearly 1.0 for liquid sulfate particles at 190 K. Because the uptake coefficients of cold sulfate aerosols are so large in comparison to PSCs (as shown in Plate 11), it is now recognized that the formation of reactive chlorine in the Antarctic stratosphere is predominantly due to cold aerosols and not PSCs.

PSCs also impact ozone depletion rates by falling out of the stratosphere. This results in both dehydration (water removal) and denitrification (nitrogen removal) from the polar lower stratosphere [*Toon et al.*, 1986]. Nitrogen is a key gas for Antarctic ozone depletion because the reaction $ClO + NO \rightarrow Cl + NO_2$ short circuits the ozone depletion process by cycling Cl through nitric oxide (NO) and nitrogen dioxide (NO_2) instead of O_3. Furthermore, the reaction $ClO + NO_2 \xrightarrow{M} ClONO_2$, where M is a third body, moves reactive chlorine into the $ClONO_2$ reservoir. Both processes illustrate the key role of NO_x (NO and NO_2) in the Antarctic ozone hole. As can be seen in Plate 12, $ClONO_2$ is near zero in midwinter because of the heterogeneous conversion by equations (1)–(5). In addition, in situ observations of NO have shown near-zero values inside the Antarctic polar vortex [*Fahey et al.*, 1989]. The low NO and $ClONO_2$ results from PSC processes. From equations (1), (3), (4), and (5), the reservoir species $ClONO_2$ and dinitrogen pentoxide (N_2O_5) are converted to HNO_3 on or in PSCs. HNO_3 is not desorbed from PSCs, but can settle out of the stratosphere as PSC precipitate. *Del Negro et al.* [1997] used ER-2 aircraft observations to show that the type Ib PSC (ternary solution or STS particle) dominated the edge of the Antarctic vortex early in the winter season. These STS particles had an average size of about 1 μm (10^{-6} m). The fall speed of these liquid particles is small (about 50 m d^{-1}) [*Kasten*, 1968], meaning that these STS particles were unlikely to denitrify the stratosphere. *Santee et al.* [1995] showed definitive evidence that denitrification in the Antarctic stratosphere preceded the dehydration. *Fahey et al.* [2001] solved this denitrification mystery with their observations of large NAT particles in the midwinter Arctic stratosphere. The observed particles had sizes of 10–20 μm and, therefore, fall speeds of a few kilometers per day. Subsequent measurements of Arctic HNO_3 confirmed the denitrification.

Observations of HNO_3 from the Aura Microwave Limb Sounder (MLS) instrument show complete denitrification by midwinter [*Santee et al.*, 2008]. The absence of nitrogen in the polar lower stratosphere leads to higher ozone depletion rates. In the Antarctic, the polar lower stratosphere is almost fully denitrified. In the Arctic, the polar lower stratosphere is only partially denitrified. *Gao et al.* [2002] used observations to show that in the NH polar spring, ozone depletion greatly depended on the level of denitrification (43% denitrification resulted in a 43 ppbv d^{-1} ozone depletion, while 71% had a 63 ppbv d^{-1} depletion).

Before approaching the problem of catalyzed ozone depletion, it is important to discuss the evolution of the total amount of chlorine over Antarctica, since it is the rising levels of chlorine that have caused the ozone hole. As noted in the previous section, the Brewer-Dobson circulation carries the CFCs to altitudes in the upper stratosphere where the CFCs are photolyzed, releasing chlorine. In the mid and upper stratosphere, this chlorine is converted either into HCl (Cl + CH$_4$ → HCl + CH$_3$) or ClONO$_2$ (ClO + NO$_2$ $\xrightarrow{M}$ ClONO$_2$). These two molecules are then transported downward inside the Antarctic vortex over the course of the fall and winter seasons. Plate 12 displays satellite observations of HCl, ClONO$_2$, and chlorine monoxide (ClO) inside the Antarctic vortex from the middle of May to the middle of October at an altitude of about 20 km [*Newman and Rex*, 2007]. Prior to the appearance of PSCs, HCl and ClONO$_2$ dominate the total inorganic chlorine, respectively. As PSCs begin to appear in early June, the HCl and ClONO$_2$ are absorbed (mixing into the particle) or adsorbed (attaching to the surface of the particle), where they mainly react according to equation (1). While the HNO$_3$ from this reaction remains on or in the PSC, the Cl$_2$ is desorbed from the particle. If the particles are large enough, they can settle, carrying the HNO$_3$ out of the stratosphere (denitrification).

There are two principal catalytic cycles for ozone depletion in the polar lower stratosphere. Cycle 1 [*Molina and Molina*, 1987] is

$$ClO + ClO \xrightarrow{M} ClOOCl \tag{6a}$$

$$ClOOCl + h\nu \rightarrow ClOO + Cl \tag{6b}$$

$$ClOO \xrightarrow{M} Cl + O_2 \tag{6c}$$

$$2(Cl + O_3 \rightarrow ClO + O_2) \tag{6d}$$

$$\text{Net}: 2O_3 \rightarrow 3O_2, \tag{6e}$$

where h is Planck's constant and ν is the frequency of light. Cycle 2 [*McElroy et al.*, 1986] is

$$Cl + O_3 \rightarrow ClO + O_2 \tag{7a}$$

$$Br + O_3 \rightarrow BrO + O_2 \tag{7b}$$

$$BrO + ClO \rightarrow BrCl + O_2 \tag{7c}$$

$$BrCl + h\nu \rightarrow Br + Cl \tag{7d}$$

$$\text{Net}: 2O_3 \rightarrow 3O_2 \tag{7e}$$

The energy of photons needed to break the chemical bonds is $h\nu$, or hc/λ, where c is the speed of light, and λ is the wavelength of light. The Cl$_2$ desorbed (released) from PSCs in equations (1) and (2) is photolyzed by visible radiation ($\lambda < 470$ nm, blue light ≈ 430 nm). The Cl then reacts with ozone to form ClO and an oxygen molecule from equations (6d) and (7a). The two cycles require sunlight in equations (6b) and (7d) and sufficient chlorine and bromine. The two cycles account for 90–95% of the spring Antarctic ozone depletion, with cycle 1 being slightly more important than cycle 2 for stratospheric chlorine levels over the last decade [*Danilin et al.*, 1996].

The more typical chlorine catalytic cycle in the extrapolar stratosphere was described by *Stolarski et al.* [1974]. Cycle 3 is

$$Cl + O_3 \rightarrow ClO + O_2 \tag{8a}$$

$$ClO + O \rightarrow Cl + O_2 \tag{8b}$$

$$O_3 + h\nu \rightarrow O + O_2 \tag{8c}$$

$$\text{Net}: 2O_3 \rightarrow 3O_2. \tag{8d}$$

This catalytic reaction destroys ozone in the upper stratosphere, but it is not important in the lower stratosphere over Antarctica because the levels of oxygen atoms in equation (8b) are too low. In the polar lower stratosphere, the oxygen atom levels are low at the end of polar night because the sun is just on the horizon, and the wavelengths of light necessary to photolyze ozone in equation (8c) ($\lambda < 310$) need to penetrate a long path, resulting in a very small ozone photolysis rate.

We can estimate some of the basic processes by writing out the equations in steady state for the levels of various gases. First, we assume that there are no interference reactions (not precisely true since ClO + NO → Cl + NO$_2$, but reasonable given the large denitrification over Antarctica). Second, we neglect the bromine-chlorine reactions in equation (7). This reaction accounts for a considerable share of Antarctic ozone depletion, but it should generally follow the reactions listed in equation (8). Third, the reaction Cl + O$_3$ is very fast, so that all of the reactive chlorine ([ClO] + 2[ClOOCl], where [X] indicates the abundance of X) is partitioned between chlorine peroxide (ClOOCl) and ClO (Cl levels are small). Fourth, we assume that all of the inorganic chlorine has been converted by PSCs into reactive form (i.e., [Cl$_x$] = [Cl$_y$], see Plate 12 to note that both HCl and ClONO$_2$ are both near zero in midwinter). Fifth, we assume that the air is cold enough for the thermal decomposition of ClOOCl to be small (see Plate 7). Finally, we assume that the air is in the weak sunlight immediately after polar night. From equation (6a) (the termolecular reaction of ClO), we find

$$k[\text{ClO}]^2 = J[\text{ClOOCl}], \qquad (9)$$

where k is the reaction rate of equation (6a), and J is the photolysis rate of [ClOOCl]. By assuming that Cl levels are small, we can also write

$$[\text{Cl}_x] = [\text{ClO}] + 2[\text{ClOOCl}]. \qquad (10)$$

Solving for [ClO] from equation (9), substituting into equation (10), and then solving the quadratic equation yields

$$[\text{ClOOCl}] = \frac{\alpha}{8}\left[\left(1 + 4\frac{[\text{Cl}_x]}{\alpha}\right) - \left(1 + 8\frac{[\text{Cl}_x]}{\alpha}\right)^{1/2}\right], (11)$$

where $\alpha = J/(k[\text{M}])$ and has a typical value of about 7 ppbv in the polar lower stratosphere during the middle of September at midday conditions. From equations (6b) and (6d), we can show that the ozone depletion rate is

$$\frac{\partial \text{O}_3}{\partial t} = -2J[\text{ClOOCl}] = -2k[\text{CLO}]^2[\text{M}]. \qquad (12)$$

This peak ozone depletion rate is then computed using an estimate of Cl_y in the Antarctic lower stratosphere [*Newman et al.*, 2007]. Figure 1 displays the ozone depletion rate versus year. A comparison to Plate 1 shows that our computation captures two basic features of polar ozone depletion: (1) depletions are small in the 1970s, but not negligible (the combined chlorine and bromine over Antarctic has about 0.9 ppb that is natural) and (2) depletion accelerates in the 1980s as Cl_y rapidly increases. Importantly, while ozone depletion is proportional to [ClO]^2, the depletion is approximately linear in Cl_y (as shown in the inset) because the amount of ClO is not linearly proportional to Cl_y. As Cl_y increased from year-to-year, steadily greater amounts are partitioned into ClOOCl rather than ClO according to equation (11) [*Jiang et al.*, 1996].

From equation (12), the laboratory determined photolysis rate, J, and the reaction rate of ClO and ozone, k, are critical parameters for estimating the ozone depletions. Recommended values for J and k are taken from the work of *Sander et al.* [2003]. However, continuing laboratory work often leads to revised values. For example, *Pope et al.* [2007] remeasured J(ClOOCl) and found it to be much lower than current recommendations in the UV. If J is lower than currently assumed, then [ClOOCl] will be higher, since it is photolyzed at a slower rate. This means that ClO will be lower from

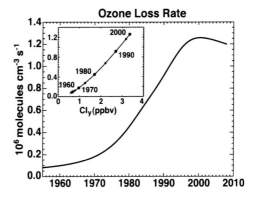

Figure 1. Ozone loss rate (10^6 molecules cm^{-3} s^{-1}) versus year over Antarctica. See the text for details about the estimate. The inset shows the same loss rates versus Cl_y. The points in the inset correspond to 5-year increments. The loss rate is estimated for midday conditions, 75°S, and 50 hPa in the middle of September. The Cl_y is estimated using the surface observations and a 6.0-year mean age of air following *Newman et al.* [2007].

equation (10), and the ozone depletion rate will also be lower from equation (12). *von Hobe et al.* [2007] and *Kawa et al.* [2009] show that the use of the *Pope et al.* [2007] J values considerably reduces our model estimates of ozone depletion, such that late September observed zero values of ozone in the lower stratosphere (see Plate 5) can no longer be simulated. Furthermore, observations of HCl and ClO also cannot be correctly matched by the modeled values. However, ongoing research suggests that the *Pope et al.* [2007] laboratory measurements may be incorrect [e.g., *Papanastasiou et al.*, 2009; *Wilmouth et al.*, 2009]. These photolysis measurement revisions highlight the importance of basic laboratory studies for our simulations of Antarctic stratospheric processes.

As can be seen in the upper panel of Plate 5, ozone levels are near zero by late September and early October. Antarctic ozone rapidly rises in the October through December period, as ozone is horizontally transported over Antarctica from midlatitudes as the polar vortex breaks up. However, two very interesting processes occur in this period prior to the breakup. First, as ozone approaches zero, there is no ozone for the Cl to react with. Further, because of denitrification, there is no nitrogen in the Antarctic stratosphere to reform ClONO$_2$. At this stage, the reaction of Cl + CH$_4$ → HCl + CH$_3$ begins to dominate the chlorine chemistry [*Prather and Jaffe*, 1990; *Douglass et al.*, 1995; *Santee et al.*, 1996], with HCl rising and ClO decreasing. Plate 12 shows this remarkable transformation of the chlorine partitioning. Since all of the chlorine has been converted to HCl, the HCl estimate in the middle of October gives a good estimate of the total inorganic chlorine in the polar lower stratosphere. Second, the complete destruction of ozone in the polar lower stratosphere alters the radiative heating. *Shine* [1986] showed that ozone destruction led to a large reduction of shortwave heating by ozone absorption of UV, with a more modest reduction of longwave heating by the upwelling infrared. The reduced heating resulted in strong polar lower stratospheric cooling [*Ramaswamy et al.*, 2001]. Hence, Antarctic ozone depletion has a long-term climate impact. The future of temperature and ozone is dealt with in the next section.

5. FUTURE OF THE OZONE HOLE

The evolution of the Antarctic ozone hole is intimately tied to the evolution of the long-lived chlorine and bromine compounds. The levels of the compounds are controlled by production regulations, emission rates, and the rate at which the compounds are transported in the troposphere-stratosphere to the Antarctic lower stratosphere in late winter. A secondary effect on the ozone hole is the alteration of the development of PSCs (necessary to activate the chlorine). As discussed in the previous section, PSC formation is tied to stratospheric water vapor, sulfuric acid, nitric acid, and temperature. Altering the levels of these compounds or temperature would impact the activation of chlorine. Finally, altering the dynamics of the SH stratosphere would impact advection of ozone (and Cl and Br) into the Antarctic lower stratosphere and change the temperature of the polar lower stratosphere.

There have been a few studies that estimate the recovery of the Antarctic ozone hole. *Hofmann et al.* [1997] empirically fit total column ozone and 12- to 20-km partial column ozone to equivalent effective stratospheric chlorine (EESC) using a 3-year shift and a bromine factor of 40, yielding a return date to 1980 conditions of about 2050. *Shindell et al.* [1998] used the GISS 3-D model with a simple chemical parameterization to predict Antarctic ozone conditions. In their model run, total ozone over Antarctica returns to 1980 levels in about 2050. *Austin et al.* [2003] used a collection of 3-D models to show a return date in about 2050, although the model estimates showed a large uncertainty in return date. These model predictions are fundamentally driven by chlorine and bromine levels over Antarctica.

The chlorine and bromine in the Antarctic stratosphere is mainly derived from CFCs and halons. During the 1970s and 1980s, the levels of these compounds were rapidly increasing in the stratosphere. In 1987, the Montreal Protocol was negotiated, limiting the production of ozone depleting substances (ODS). By 1992, production of CFCs was curtailed in the major industrial nations because of the negotiated amendments to the agreement. Plate 13 displays EESC over Antarctica. EESC represents the combined levels of chlorine and bromine over Antarctica (i.e., $Cl_y + \alpha Br_y$). EESC is computed using a semiempirical formulation that (1) accounts for the transit time from the troposphere to the Antarctic lower stratosphere using an age spectrum with a mean age of 5.5 years, (2) chlorine and bromine release rates that are estimated using ER-2 and balloon observations in the stratosphere, and (3) a scaling to account for the greater efficiency of bromine for destroying ozone than chlorine ($\alpha = 90$) [*Newman et al.*, 2007]. In 1950, the two natural species methyl chloride (CH_3Cl) and methyl bromide (CH_3Br) dominated EESC in the Antarctic stratosphere. The peak loading of EESC occurred at about the 2000–2005 period from these estimates as a result of the Montreal Protocol and the amendments to the Protocol.

The decrease of EESC is mainly controlled by the lifetime of the various species. CFC-12 has a lifetime of approximately 100 years, while methyl chloroform (CH_3CCl_3) has a lifetime of about 5 years [*Clerbaux and Cunnold*, 2007]. In Plate 13, most of the decrease in the last few years results from the decrease of methyl chloroform. The overall decrease over the 21st century is slow because of the long lifetimes of the CFCs and halons. The vertical line at 1980 indicates that EESC was approximately 2.4 ppbv in 1980 and returns to 2.4 ppbv in about 2067. The year 1980 provides a reasonable estimate for the "beginning" of the ozone hole, so we expect the ozone hole to recover to a 1980 level in this 2067 time period. However, it is clear that some ozone depletion occurred prior to 1980 (see Plate 1). If we choose 1970 as the approximate beginning of the ozone hole, then return date to a 1.6-ppbv level occurs in approximately 2129. These projections are based upon the assumption that the nations of the world will continue to abide by current Montreal Protocol production limits. However, the overall return date of the ozone hole will be quite slow because of the long lifetimes of the ODSs.

Newman et al. [2006] used a parametric model of ozone hole area that was determined from EESC in Plate 13 and observed temperature to predict a return date of about 2067. In their prediction, the effects of saturation were also modeled, and the decrease of the ozone hole area was projected to begin in about 2018. *Bodeker and Waugh* [2007] used a set of 3-D models to project the return date of the Antarctic ozone hole (see their Figures 6–14). Most of the models were run to only 2050, but their estimates of the area of the ozone hole were still considerably above zero by this time. Based upon these projections, *Bodeker and Waugh* [2007] also suggested a return date of about 2065, but with a large uncertainty.

As noted above, much of the uncertainty in model projections of ozone recovery is derived from poor representations of the evolution of Cl and Br over Antarctica. *Eyring et al.* [2007] showed that the current prediction of Cl_y levels over Antarctica is positively correlated to the return date (higher lower stratospheric Antarctic Cl_y in the model results in a later model ozone hole return date). Furthermore, most models underpredict Antarctic Cl_y, resulting in return date projections that are too early. Transport in the 3-D chemistry-climate models (CCM) is a major limitation to accurate projections of ozone hole recovery.

In addition to projecting the evolution of Cl_y, the impact of greenhouse gases on the stratosphere is a major concern for understanding the evolution of the Antarctic ozone hole. Plate 14 displays the model-simulated temperature of the Antarctic lower stratosphere during September and December from the NASA Goddard Global Modeling and Assimilation Office (GMAO) Goddard Earth Observing System Data Assimilation System GEOS 4 CCM. The model simulations have been smoothed with a 10-year half-amplitude filter to expose only the longer-term variations. September is the key period for the Antarctic ozone hole because of the activation of chlorine on PSCs. As is clear from the plot, the September temperature is dominated by long-term interannual variability with little evidence of a significant trend. Large wave events lead to warm periods (e.g., the 2020–2040 period in the F4 simulation) that mask trends. In this September period, trends are relatively small (<2 K up to 2100). However, the GEOS 4 model is only a coupled chemistry climate model and does not include an interactive ocean. Hence, the climate impact on wave driving of the stratosphere is still very uncertain.

The December temperature time series (Plate 14b) shows the interesting effects of cooling as a result of ozone depletion. In the previous section, we discussed the impact of ozone depletion on the temperature of the lower stratosphere. The model simulates this cooling (1970–2000) and also shows that temperatures increase as ODSs decline in the stratosphere (2000–2070). The model simulation with fixed 1960 chlorine levels (black line) also reveals a 3- to 4-K cooling of the polar lower stratosphere. While ozone depletions will eventually recover over Antarctica, the steady increase of greenhouse gases will impact the polar lower stratosphere.

6. SUMMARY AND ISSUES

The Antarctic ozone hole is caused by human-produced compounds, specifically, chlorine-containing CFCs and bromine-containing halons. The unique cold conditions of the Antarctic lower stratosphere in winter and spring allow for the development of PSCs. Heterogeneous reactions on these PSCs release chlorine from reservoir species into species that are easily photolyzed by visible radiation. The rising sun in the late winter over Antarctica provides the energy necessary to drive the chlorine and bromine catalytic cycles that result in massive ozone depletion during the August and September period. By early October, ozone is completely destroyed in a layer from approximately 12 to 20 km over Antarctica.

As the polar vortex breaks down in the November and December period, this ozone-depleted air is mixed across the SH, lowering ozone across the region from the subtropics to the pole. The long-term prognosis is good because ozone-depleting substances have been regulated by the Montreal Protocol and the amendments to the Protocol. The current CFC and halon scenarios that extend to the end of the century will lead to an ozone hole recovery. The models (both empirical and CCMs) that are forced with these scenarios indicate that the return date to 1980 levels will occur in approximately 2067, with an uncertainty of about 10 years.

There are still a number of problems in the understanding of the ozone hole. Some of the principal problems still revolve around basic laboratory estimates of quantities such as the photolysis of the chlorine dimer (Cl_2O_2) and other basic reaction rates. While considerable progress has been made in understanding PSCs, there are still large uncertainties, for example, reaction rates of chemicals on STS droplets [*Lowe and MacKenzie*, 2008]. In addition, most CCMs only crudely represent PSCs and the denitrification process. Aerosol formation is crucial to understanding the STS droplet PSCs. However, both changes of sulfur chemistry for the stratosphere and delivery of sulfur compounds, such as carbonyl sulfide (OCS) and sulfur dioxide (SO_2), and water into the stratosphere are poorly understood. Finally, CCMs are the best tools for understanding the interaction of ozone depletion and climate change. However, comparisons with observations reveal a number of flaws in CCM performance, in particular, the failure to properly represent levels of ozone-depleting substances over Antarctica.

REFERENCES

Allen, D. R., R. M. Bevilacqua, G. E. Nedoluha, C. E. Randall, and G. L. Manney (2003), Unusual stratospheric transport and mixing during the 2002 Antarctic winter, *Geophys. Res. Lett.*, *30*(12), 1599, doi:10.1029/2003GL017117.

Andrews, D. G., J. R. Holton, and C. B. Leovy (1987), *Middle Atmosphere Dynamics*, 489 pp., Academic, Orlando, Fla.

Austin, J., et al. (2003), Uncertainties and assessments of chemistry-climate models of the stratosphere, *Atmos. Chem. Phys.*, *3*, 1–27.

Bodeker, G. E., and D. W. Waugh (2007), The ozone layer in the 21st century, *Scientific Assessment of Ozone Depletion: 2006, Global Ozone Res. and Monitor. Proj., Rep. 50*, pp. 6.1–6.43, World Meteorol. Organ., Geneva, Switzerland.

Bowman, K. P. (1986), Interannual variability of total ozone during the breakdown of the Antarctic circumpolar vortex, *Geophys. Res. Lett.*, *13*(12), 1193–1196.

Chubachi, S. (1984), Preliminary result of ozone observations at Syowa Station from February 1982 to January 1983, *Mem. Natl. Inst. Polar Res. Spec. Issue Jpn.*, *34*, 13–19.

Clerbaux, C., and D. M. Cunnold (2007), Long-lived compounds, *Scientific Assessment of Ozone Depletion: 2006, Global Ozone Res. and Monitor. Proj., Rep. 50*, pp. 1.1–1.63, World Meteorol. Organ., Geneva, Switzerland.

Danielsen, E. F. (1968), Stratospheric-tropospheric exchange based upon radioactivity, ozone and potential vorticity, *J. Atmos. Sci.*, *25*(3), 502–518, doi:10.1175/1520-0469(1968)025<0502:STEBOR>2.0.CO;2.

Danilin, M. Y., N. Sze, M. K. W. Ko, J. M. Rodriguez, and M. J. Prather (1996), Bromine-chlorine coupling in the Antarctic ozone hole, *Geophys. Res. Lett.*, *23*(2), 153–156.

Del Negro, L. A., et al. (1997), Evaluating the role of NAT, NAD, and liquid H$_2$SO$_4$/H$_2$O/HNO$_3$ solutions in Antarctic polar stratospheric cloud aerosol: Observations and implications, *J. Geophys. Res.*, *102*(D11), 13,255–13,282.

Douglass, A. R., M. R. Schoeberl, R. S. Stolarski, J. W. Waters, J. M. Russell, III, A. E. Roche, and S. T. Massie (1995), Interhemispheric differences in springtime production of HCl and ClONO$_2$ in the polar vortices, *J. Geophys. Res.*, *100*(D7), 13,967–13,978.

Ertel, H. (1942), Ein neuer hydrodynamischer Wirbelsatz, *Meteorol. Z.*, *59*, 271–281.

Eyring, V., et al. (2007), Multimodel projections of stratospheric ozone in the 21st century, *J. Geophys. Res.*, *112*, D16303, doi:10.1029/2006JD008332.

Fahey, D. W., D. M. Murphy, K. K. Kelly, M. K. W. Ko, M. H. Proffitt, C. S. Eubank, G. V. Ferry, M. Loewenstein, and K. R. Chan (1989), Measurements of nitric oxide and total reactive nitrogen in the Antarctic stratosphere: Observations and chemical implications, *J. Geophys. Res.*, *94*(D14), 16,665–16,681.

Fahey, D. W., et al. (2001), The detection of large HNO$_3$-containing particles in the winter Arctic stratosphere, *Science*, *291*(5506), 1026–1031, doi:10.1126/science.1057265.

Farman, J. C., B. G. Gardiner, and J. D. Shanklin (1985), Large losses of total ozone in Antarctica reveal seasonal ClO$_x$/NO$_x$ interaction, *Nature*, *315*, 207–210, doi:10.1038/315207a0.

Gardiner, B. G., and J. D. Shanklin (1986), Recent measurements of Antarctic ozone depletion, *Geophys. Res. Lett.*, *13*(12), 1199–1201.

Gao, R. S., et al. (2002), Role of NO$_y$ as a diagnostic of small-scale mixing in a denitrified polar vortex, *J. Geophys. Res.*, *107*(D24), 4794, doi:10.1029/2002JD002332.

Hall, T. M., and R. A. Plumb (1994), Age as a diagnostic of stratospheric transport, *J. Geophys. Res.*, *99*(D1), 1059–1070.

Hallett, J., and R. E. J. Lewis (1967), Mother of pearl clouds, *Weather*, *22*, 56–65.

Hanson, D., and K. Mauersberger (1988), Laboratory studies of the nitric acid trihydrate: Implications for the south polar stratosphere, *Geophys. Res. Lett.*, *15*(8), 855–858.

Hofmann, D. J., J. W. Harder, S. R. Rolf, and J. M. Rosen (1987), Ballon-borne observations of the development and vertical structure of the Antarctic ozone hole in 1986, *Nature*, *326*, 59–62, doi:10.1038/326059a0.

Hofmann, D. J., S. J. Oltmans, J. M. Harris, B. J. Johnson, and J. A. Lathrop (1997), Ten years of ozonesonde measurements at the south pole: Implications for recovery of the springtime Antarctic ozone, *J. Geophys. Res.*, *102*(D7), 8931–8943.

Houghton, J. T., Y. Ding, D. J. Griggs, M. Noguer, P. J. van der Linden, X. Dai, K. Maskell, and C. A. Johnson (2001), *Climate Change 2001: The Scientific Basis*, 881 pp., Cambridge Univ. Press, Cambridge, U. K.

Jiang, Y., Y. L. Yung, and R. W. Zurek (1996), Decadel evolution of the Antarctic ozone hole, *J. Geophys. Res.*, *101*(D4), 8985–8999.

Junge, C. E., C. W. Chagnon, and J. E. Manson (1961), Stratospheric aerosols, *J. Meteorol.*, *18*(1), 81–108, doi:10.1175/1520-0469(1961)018<0081:SA>2.0.CO;2.

Kasten, F. (1968), Falling speed of aerosol particles, *J. Appl. Meteorol.*, *7*(5), 944–947, doi:10.1175/1520-0450(1968)007<0944:FSOAP>2.0.CO;2.

Kawa, S. R., et al. (1997), Activation of chlorine in sulfate aerosol as inferred from aircraft observations, *J. Geophys. Res.*, *102*(D3), 3921–3933.

Kawa, S. R., R. S. Stolarski, P. A. Newman, A. R. Douglass, M. Rex, D. J. Hofmann, M. L. Santee, and K. Frieler (2009), Sensitivity of polar stratospheric ozone loss to uncertainties in chemical reaction kinetics, *Atmos. Chem. Phys.*, *9*, 8651–8660.

Labitzke, K., and H. van Loon (1972), The stratosphere in the Southern Hemisphere, in *Meteorology of the Southern Hemisphere*, edited by C. W. Newton, *Meteor. Monogr.*, *13*, 113–138.

Lait, L. R. (1994), An alternative form for potential vorticity, *J. Atmos. Sci.*, *51*(12), 1754–1759, doi:10.1175/1520-0469(1994)051<1754:AAFFPV>2.0.CO;2.

Lowe, D., and A. R. MacKenzie (2008), Polar stratospheric cloud microphysics and chemistry, *J. Atmos. Sol. Terr. Phys.*, *70*(1), 13–40, doi:10.1016/j.jastp.2007.09.011.

McElroy, M. B., R. J. Salawitch, S. C. Wofsy, and J. A. Logan (1986), Reductions of Antarctic ozone due to synergistic interactions of chlorine and bromine, *Nature*, *321*(6072), 759–762, doi:10.1038/321759a0.

Molina, L. T., and M. J. Molina (1987), Production of chlorine oxide (Cl$_2$O$_2$) from the self-reaction of the chlorine oxide (ClO) radical, *J. Phys. Chem.*, *91*(2), 433–436, doi:10.1021/j100286a035.

Montzka, S. A., and P. J. Fraser (2003), Controlled substances and other source gases, *Scientific Assessment of Ozone Depletion: 2002, Global Ozone Res. and Monitor. Proj., Rep. 47*, pp. 1.1–1.83, World Meteorol. Organ., Geneva, Switzerland.

Müller, R. (2009), A brief history of stratospheric ozone research, *Meteorol. Z.*, *18*(1), 3–24, doi:10.1127/0941-2948/2009/352.

Newman, P. A. (1986), The final warming and polar vortex disappearance during the Southern Hemisphere spring, *Geophys. Res. Lett.*, *13*(12), 1228–1231.

Newman, P. A., and E. R. Nash (2005), The unusual Southern Hemisphere stratospheric winter of 2002, *J. Atmos. Sci.*, *62*(3), 614–628, doi:10.1175/JAS-3323.1.

Newman, P. A., and M. Rex (2007), Polar ozone: Past and Present, *Scientific Assessment of Ozone Depletion: 2006, Global Ozone*

Res. and Monitor. Proj., Rep. 50, pp. 4.1–4.48, World Meteorol. Organ., Geneva, Switzerland.

Newman, P. A., E. R. Nash, S. R. Kawa, S. A. Montzka, and S. M. Schauffler (2006), When will the Antarctic ozone hole recover?, *Geophys. Res. Lett.*, *33*, L12814, doi:10.1029/2005GL025232.

Newman, P. A., J. S. Daniel, D. W. Waugh, and E. R. Nash (2007), A new formulation of equivalent effective stratospheric chlorine (EESC), *Atmos. Chem. Phys.*, *7*, 4537–4552.

Noel, V., A. Hertzog, H. Chepfer, and D. M. Winker (2008), Polar stratospheric clouds over Antarctica from the CALIPSO spaceborne lidar, *J. Geophys. Res.*, *113*, D02205, doi:10.1029/2007JD008616.

Papanastasiou, D. K., V. C. Papadimitriou, D. W. Fahey, and J. B. Burkholder (2009), UV absorption spectrum of the ClO dimer (Cl_2O_2) between 200 and 420 nm, *J. Phys. Chem. A*, *113*(49), 13,711–13,726, doi:10.1021/jp9065345.

Pope, F. D., J. C. Hansen, K. D. Bayes, R. R. Friedl, and S. P. Sander (2007), Ultraviolet absorption spectrum of chlorine peroxide, ClOOCl, *J. Phys. Chem. A*, *111*(20), 4322–4332, doi:10.1021/jp067660w.

Prather, M., and A. H. Jaffe (1990), Global impact of the Antarctic ozone hole: Chemical propagation, *J. Geophys. Res.*, *95*(D4), 3473–3492.

Ramaswamy, V., et al. (2001), Stratospheric temperature trends: Observations and model simulations, *Rev. Geophys.*, *39*(1), 71–122, doi:10.1029/1999RG000065.

Randel, W. J., and P. A. Newman (1998), The stratosphere in the Southern Hemisphere, in *Meteorology of the Southern Hemisphere*, edited by D. J. Karoly and D. G. Vincent, *Meteorol. Monogr.*, *27*, 243–282.

Sander, S. P., et al. (2003), Chemical kinetics and photochemical data for use in atmospheric studies, Evaluation 14, *Jet Propulsion Lab. Pub. 02-25*, California Institute of Technology, Pasadena, Calif.

Santee, M. L., W. G. Read, J. W. Waters, L. Froidevaux, G. L. Manney, D. A. Flower, R. F. Jarnot, R. S. Harwood, and G. E. Peckham (1995), Interhemispheric differences in polar stratospheric HNO_3, H_2O, ClO, and O_3, *Science*, *267*(5199), 849–852, doi:10.1126/science.267.5199.849.

Santee, M. L., L. Froidevaux, G. L. Manney, W. G. Read, J. W. Waters, M. P. Chipperfield, A. E. Roche, J. P. Kumer, J. L. Mergenthaler, and J. M. Russell, III (1996), Chlorine deactivation in the lower stratospheric polar regions during late winter: Results from UARS, *J. Geophys. Res.*, *101*(D13), 18,835–18,859.

Santee, M. L., I. A. MacKenzie, G. L. Manney, M. P. Chipperfield, P. F. Bernath, K. A. Walker, C. D. Boone, L. Froidevaux, N. J. Livesey, and J. W. Waters (2008), A study of stratospheric chlorine partitioning based on new satellite measurements and modeling, *J. Geophys. Res.*, *113*, D12307, doi:10.1029/2007JD009057.

Schoeberl, M. R., and D. L. Hartmann (1991), The dynamics of the stratospheric polar vortex and its relation to springtime ozone depletions, *Science*, *251*(4989), 45–62, doi:10.1126/science.251.4989.46.

Schoeberl, M. R., and J. M. Rodriguez (2009), The rise and fall of dynamical theories of the Ozone Hole, in *Twenty Years of Ozone Decline*, edited by C. Zerefos et al., pp. 263–272, doi:10.1007/978-90-481-2469-5_19, Springer, Athens.

Shepherd, T. G. (2007), Transport in the middle atmosphere, *J. Meteorol. Soc. Jpn.*, *85B*, 165–191, doi:10.2151/jmsj.85B.165.

Shepherd, T. G., R. A. Plumb, and S. C. Wofsy (2005), Preface, *J. Atmos. Sci.*, *62*(3), 565–566, doi:10.1175/JAS-9999.1.

Shindell, D. T., D. Rind, and P. Lonergan (1998), Increased polar stratospheric ozone losses and delayed eventual recovery owing to increasing greenhouse-gas concentrations, *Nature*, *392*(6676), 589–592, doi:10.1038/33385.

Shine, K. P. (1986), On the modeled thermal response of the Antarctic stratosphere to a depletion of ozone, *Geophys. Res. Lett.*, *13*(12), 1331–1334.

Solomon, S., R. R. Garcia, F. S. Rowland, and D. J. Wuebbles (1986), On the depletion of Antarctic ozone, *Nature*, *321*(6072), 755–758, doi:10.1038/321755a0.

Solomon, S., R. W. Portmann, T. Sasaki, D. J. Hofmann, and D. W. J. Thompson (2005), Four decades of ozonesonde measurements over Antarctica, *J. Geophys. Res.*, *110*, D21311, doi:10.1029/2005JD005917.

Stanford, J. L., and J. S. Davis (1974), A century of stratospheric cloud Reports: 1870–1972, *Bull. Am. Meteorol. Soc.*, *55*(3), 213–219, doi:10.1175/1520-0477(1974)055<0213:ACOSCR>2.0.CO;2.

Stolarski, R. S., and R. J. Cicerone (1974), Stratospheric chlorine: A possible sink for ozone, *Can. J. Chem.*, *52*(8), 1610–1615, doi:10.1139/v74-233.

Stolarski, R. S., A. J. Kreuger, M. R. Schoeberl, R. D. McPeters, P. A. Newman, and J. C. Alpert (1986), Nimbus-7 satellite measurements of the springtime Antarctic ozone decrease, *Nature*, *322*(6082), 808–811, doi:10.1038/322808a0.

Toon, O. B., P. Hamill, R. P. Turco, and J. Pinto (1986), Condensation of HNO_3 and HCl in the winter polar stratospheres, *Geophys. Res. Lett.*, *13*(12), 1284–1287.

von Hobe, M., R. J. Salawitch, T. Canty, H. Keller-Rudek, G. K. Moortgat, J.-U. Grooß, R. Müller, and F. Stroh (2007), Understanding the kinetics of the ClO dimer cycle, *Atmos. Chem. Phys.*, *7*, 3055–3069.

Waugh, D. W., and L. M. Polvani (2010), Stratospheric polar vortices, in *The Stratosphere: Dynamics, Transport, and Chemistry, Geophys. Monogr. Ser.*, doi: 10.1029/2009GM000887, this volume.

Wilmouth, D. M., T. F. Hanisco, R. M. Stimpfle, and J. G. Anderson (2009), Chlorine-catalyzed ozone destruction: Cl atom production from ClOOCl photolysis, *J. Phys. Chem. A*, *113*(51), 14,099–14,108, doi:10.1021/jp9053204.

P. A. Newman, NASA GSFC, Code 613.3, Greenbelt, MD 20771, USA. (Paul.A.Newman@nasa.gov)

Solar Variability and the Stratosphere

Joanna D. Haigh

Department of Physics, The Blackett Laboratory, Imperial College London, London, UK

The absorption of solar radiation in the atmosphere is considered, focusing on its spectral composition as it travels downward through the stratosphere. The evidence for variations in solar radiative output is reviewed, and how these impact heating rates and ozone concentrations is examined. The signal of solar variability in the stratosphere, as found in observational data and model simulations, is described, and the mechanisms whereby these signals are produced are discussed. Finally, the potential for a solar impact on the stratosphere to influence tropospheric climate through dynamical coupling is considered.

1. INTRODUCTION

Absorption of solar radiation is fundamental in determining the thermodynamic structure and the composition of the middle atmosphere. Solar UV radiation is involved in both the creation and destruction of stratospheric ozone, and the related diabatic processes result in the increase with height of temperature in the stratosphere and thus the existence of the stratopause. Seasonal variations in solar irradiance (solar irradiance, or flux, is the radiative energy crossing unit area per unit time) create strong latitudinal gradients in heating rates, which are partially cancelled by fluxes of heat radiation and also by adiabatic effects associated with atmospheric motions.

As solar radiation is crucial in establishing the climatology of the middle atmosphere, it is pertinent to consider the extent to which variations in the Sun's activity might contribute to its variability. This chapter will describe the interaction of solar radiation with the middle atmosphere and its influence on the temperature structure and chemical composition. It will discuss what is known about variations in solar irradiance and how these affect radiative fluxes and heating rates in the stratosphere. It will consider the observational evidence for solar signals in temperature and ozone and the extent to which atmospheric models are able to represent them. It will go on to investigate the dynamical, as well as the radiative, mechanisms involved in determining these signals and briefly consider how stratosphere-troposphere coupling may produce signals of solar activity in the lower atmosphere.

2. SOLAR RADIATION AND THE STRATOSPHERE

2.1. Absorption of Solar Spectral Radiation by the Atmosphere

The spectrum of solar radiation incident at Earth is approximately that of a black body at 5750 K, the temperature of the Sun's photosphere, peaking at visible wavelengths. The top-of-atmosphere spectrum is shown in Figure 1a as well, by the dashed line, as that at the Earth's surface. The absorption of radiation by the atmosphere is largely due to molecular oxygen and ozone in the UV and visible regions with water vapor and carbon dioxide being more important in the near-IR.

Figure 2a shows how the solar spectral irradiance (spectral irradiance is the irradiance per unit wavelength interval) progresses down through the atmosphere in the 200- to 700-nm wavelength region, which is of prime importance for the stratosphere. An alternative perspective is provided by Figure 1c, which shows the altitude at which the flux of radiation is reduced by a factor *e*. Figure 2b indicates how the

The Stratosphere: Dynamics, Transport, and Chemistry
Geophysical Monograph Series 190

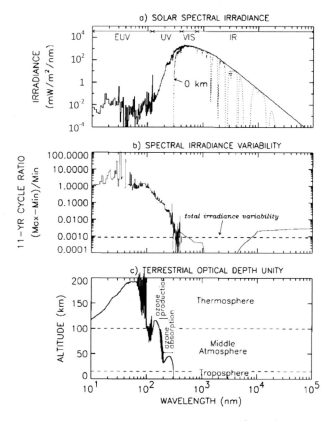

Figure 1. (a) Solar spectral irradiance (mW m^{-2} nm^{-1}): top of atmosphere (solid line) and at surface of Earth (dashed line). (b) Fractional variation over a typical 11-year solar cycle. (c) Altitude within the Earth's atmosphere at which irradiance is reduced to $1/e$ of its incident magnitude. From *Lean and Rind* [1998]. Copyright American Meteorological Society.

absorption of radiation translates into atmospheric heating rates. In the 200- to 242-nm region, absorption by molecular oxygen produces the oxygen atoms important in the production of ozone and heats the stratopause region. Between 200 and 350 nm, the radiation is responsible for the photodissociation of ozone and for strong radiative heating in the upper stratosphere and lower mesosphere. The ozone absorption bands, 440–800 nm, are much weaker but, because they absorb broadly across the peak of the solar spectrum, their energy deposition into the lower stratosphere is significant.

Plate 1 (top) presents vertical profiles of spectrally integrated solar fluxes and heating rates (The heating rate is the rate of temperature increase resulting from the absorption of radiation; it is derived from the vertical divergence of the irradiance, scaled by the heat capacity.), showing the contribution of each of the UV, visible, and near-IR wavelength regions. This vertical structure in the absorption

of solar radiation is intrinsic in determining the profile of atmospheric temperatures and plays an important role in atmospheric chemistry and thus composition.

2.2. Variations in Total Solar Irradiance

Observations of solar activity, in terms of the numbers of sunspots on the solar surface, have been made for many centuries, but attempts using ground-based instruments to measure coherent variations in the strength of the solar energetic output were confounded by uncertainties and fluctuations in atmospheric absorption. The advent of the satellite era in the late 1970s enabled high-precision direct measurements of total solar irradiance (TSI) (Total solar irradiance is the flux integrated across the entire spectrum. The value at the top of the atmosphere, approximately 1370 W m^{-2}, is often referred to as the solar constant.) to be made outside the Earth's atmosphere and, hence, accurate determination of its temporal variation.

Plate 2a presents all available satellite measurements of TSI, and it is clear that significant uncertainties remain related to its absolute value. For example, data from the newest instrument, the Total Irradiance Monitor on the SORCE satellite, show values 4–5 W m^{-2} lower than other contemporaneous instruments, which themselves show a range of 1–2 W m^{-2} (i.e., a few parts per thousand). There is a related uncertainty in the existence of any underlying trend in TSI over the past 2 cycles (peaking around 1980 and 1990 and numbered 21 and 22, respectively). Plate 2b presents one attempt to composite the measurements into a best estimate. It shows essentially no difference in the value of TSI between the solar "11-year" cycle minima occurring in 1986 and 1996 but significantly lower values during the current solar minimum. Another such composite, Plate 2c, suggests higher values in 1986 than in 1996, while in a third, Plate 2d, shows similar values in the last two solar minima but higher values in the present minimum. The discrepancies hinge on assumptions made concerning the degradations over time of the different radiometers contributing to the record, making their intercalibration very difficult. The result is a serious uncertainty in the multidecadal variation in solar output, as well as an uncertainty of a few W m^{-2} in the value of the solar "constant." Nevertheless, the precision of the individual instruments means that the solar cycle variation in TSI is known to greater accuracy, showing approximately 0.08% (~1.1 W m^{-2}) peak to peak.

2.3. Variations in Solar Spectral Irradiance

The response of the middle atmosphere to solar variability depends critically on the spectral composition of the changes

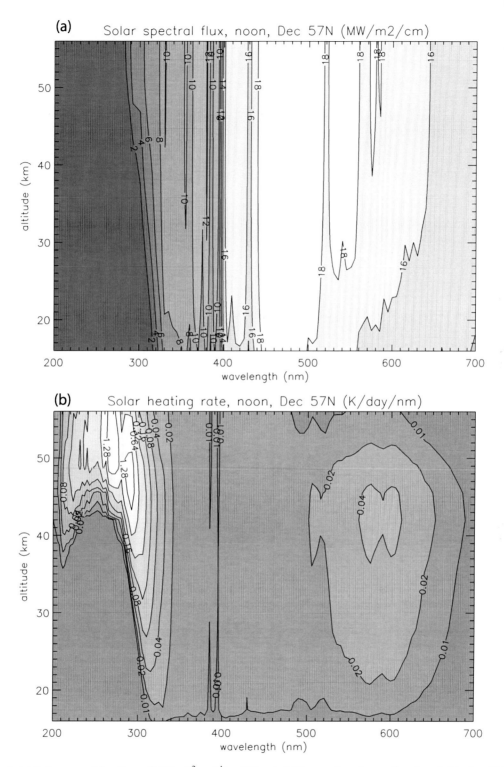

Figure 2. (a) Solar spectral irradiance (MW m^{-2} cm^{-1}) at UV and visible wavelengths as a function of altitude within the Earth's atmosphere, calculated for latitude 57°N, 21 December, noon, using spectral irradiance data at top of atmosphere for year 2000. (b) Solar spectral heating rate (K d^{-1} nm^{-1}).

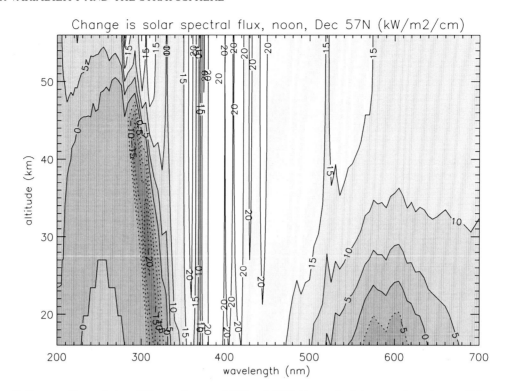

Figure 3. Same as Figure 2a but difference between 2007 and 2000 (dates of minimum and maximum of the last solar activity cycle).

in solar irradiance. Direct observation of this started with the SME and UARS satellites, which provided measurements of the solar UV spectrum during the 1980s and 1990s. The spectral range was extended into the visible and near IR through the instruments on ERS-2, 1994–2003 (http://www.iup.uni-bremen.de/gome/), EnviSat, 2002–present (http://www.iup.uni-bremen.de/sciamachy/) and SORCE, 2003–present (http://lasp.colorado.edu/sorce/). Otherwise, estimates have been based on models of solar output based, semiempirically, on proxy measures of solar activity such as sunspot numbers. Figure 1b, based mainly on such a model [*Lean*, 2000], shows changes in spectral irradiance deduced for an 11-year solar cycle corresponding to a peak-to-peak TSI variation of 1.2 W m^{-2} (similar to solar cycles 21 and 22). It is apparent that much larger fractional changes take place at shorter wavelengths than the ~0.1% variation in TSI, shown in Plate 2, which effectively represents the changes in the visible portion of the spectrum. This means that the direct impact of solar irradiance variability is much larger in the middle and upper atmosphere than it is at lower altitudes.

2.4. Variability of Fluxes and Heating Rates

Figure 3 shows the field of spectral irradiance as in Figure 2a but for the difference between solar cycle maximum and minimum conditions, based on spectral variability of *Lean* [2000]. In these plots, which were produced using a coupled radiative-chemical-dynamical 2-D model [*Haigh*, 1994], the effects of changes in ozone concentration resulting from the enhanced solar irradiance are included. This means that the vertical penetration of the enhanced irradiance is not spectrally uniform, depending on the ozone perturbation and the ozone absorption spectrum. Much of the increased irradiance penetrates to the surface, but the increases in ozone result in lower values of irradiance throughout the stratosphere at wavelengths less than 320 nm, despite the increased irradiance above, and also in the lower stratosphere at visible wavelengths. This effect is more marked at high solar zenith angles, as illustrated by the strong latitudinal gradients in the difference in spectrally integrated irradiance in winter midlatitudes shown in Figure 4. This implies that changes in solar heating rates in response to solar variability are highly nonlinear and very sensitive to details of the photochemical response in ozone.

Plate 1 (bottom) presents an example of differences between solar maximum and minimum in the downward solar radiative fluxes and heating rates. In this region of the atmosphere, the direct radiative effects are small, with the largest change in heating rate about 0.3 K d^{-1} near the stratopause and of an order of a few hundredths of a K d^{-1} in

Shortwave Fluxes

Shortwave Heating Rates

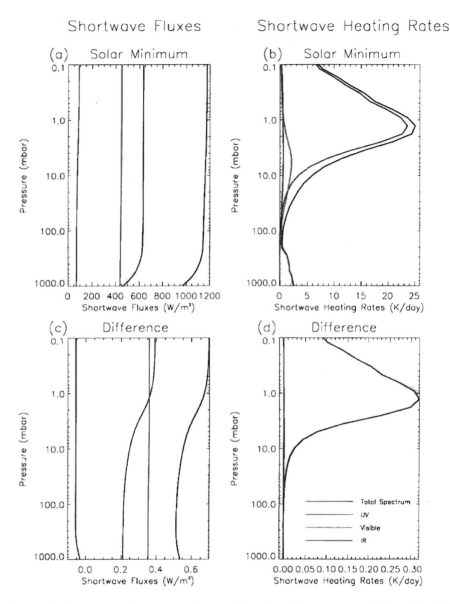

Plate 1. Profiles of (left) solar flux and (right) heating rate, integrated in three spectral bands: UV (220–320 nm) in blue, visible (320–690) in green, and near IR (690–1000 nm) in red for an overhead Sun. (top) Values for solar minimum. (bottom) Difference between solar maximum and minimum. From *Larkin* [2000]. See also *Larkin et al.* [2000].

the lower stratosphere. In the near IR, the changes are negative, reflecting the decrease in irradiance in this spectral region shown in Figure 1b.

New data from the SIM instrument on the SORCE satellite suggest that solar UV radiation varies by a much larger factor than shown in Figure 1b, while radiation at visible wavelengths is also lower when the Sun is more active [*Harder et al.*, 2009]. These preliminary results, from 3 years of data when the Sun was descending into its current minimum state, would significantly affect estimates of the changes in heating rates, such as shown in Figures 3 and 4. If such a spectral variability were shown to hold over the 11-year solar cycle or, indeed, on longer timescales, it would raise questions concerning current understanding of the response to solar variability of the temperature and composition of the middle atmosphere (see sections 3 and 4) and, potentially, also the troposphere (see section 5) [*Haigh et al.*, 2010].

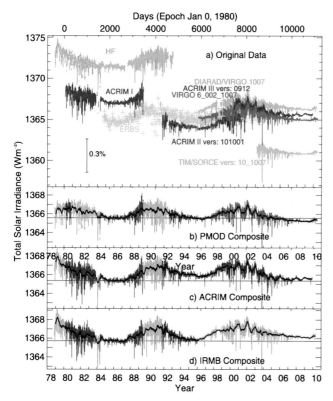

Plate 2. (a) Daily averaged total solar irradiance from radiometers on different space platforms November 1978 to March 2009. (b)–(d) Composites by three different authors. The colors indicate the source of the data. Image courtesy of Physikalisch-Meteorologisches Observatorium Davos, Davos, Switzerland (www.pmodwrc.ch).

3. SOLAR CYCLE VARIABILITY IN TEMPERATURE AND WIND

3.1. Observations

Numerous studies have demonstrated a solar cycle variation in stratospheric temperatures. Early work using radiosonde data focused on correlations between geopotential height and a solar activity index; Figure 5, for example, shows the annual mean of the geopotential height of the 30 hPa pressure surface (a measure of the mean temperature of the atmosphere below about 24 km altitude) at a location near Hawaii in the subtropical Pacific Ocean. It varies in phase with the solar 10.7 cm radio wave flux (The solar radiation at radio wavelengths is essentially unaffected by the atmosphere and has been measured at the Earth's surface since the late 1950s.) over three and a half solar cycles with an amplitude suggesting that the lower atmosphere is 0.5–1.0 K warmer at solar maximum than at solar minimum. This is a large response, but from this figure alone, it is not clear

whether it might apply to other locations or how the temperature anomaly is distributed in the vertical. Furthermore, the existence of such a correlation does not necessarily exclude the potential influence of factors other than the Sun in producing the variations in temperature.

Attempts to isolate the solar influence have been carried out by various authors using multiple linear regression analysis. An example is given in Figure 6a, which presents the solar cycle signal in zonal mean, annual mean temperature from an analysis of data from the ERA-40 Reanalysis dataset [*Uppala et al.*, 2005] in which the effects of other factors (a linear trend for climate change, quasi-biennial oscillation (QBO), El Niño–Southern Oscillation (ENSO), and stratospheric aerosol) have been simultaneously extracted. In the tropics, it shows largest warming, of over 1.5 K, in the upper stratosphere near 1 hPa, a minimum response around 5–30 hPa, and lobes of warming in the subtropical lower stratosphere. In the troposphere, maximum warming does not appear in the tropics but in midlatitudes, with vertical bands of temperature increase around 0.4 K. This confirms an earlier study [*Haigh*, 2003], using National Centers for Environmental Prediction (NCEP) Reanalysis data, (http://www.cdc.noaa.gov/). Direct analyses of stratospheric temperatures from satellite data, however, shows a somewhat different pattern in the stratosphere without the double peak profile in the tropics and with a broad maximum of over 1 K in the upper stratosphere (see Figure 7). It has been suggested that the discrepancy might be due to inadequate separation in the regression studies of the solar from other signals (e.g., volcanic eruptions or the QBO) [*Lee and Smith*, 2003]. An alternative explanation is that the broad vertical weighting functions of the satellite instruments are not able to resolve the vertical structure [*Gray et al.*, 2009]. Modeling studies show that the temperature signal is closely related to that in ozone (see next section).

At the winter pole, lower stratospheric temperatures depend on the phase of the QBO. Figure 8 shows a positive correlation between solar activity and 30 hPa winter temperatures above the North Pole when the QBO is westerly but a negative, though weaker, correlation when it is easterly. Warmer winter polar temperatures are associated with a higher frequency of sudden stratospheric warming events, and it has been observed previously [*Holton and Tan*, 1980] that the pole tends to be more disturbed when the QBO is easterly. The results of *Labitzke et al.* [2006] suggest that this applies only when the Sun is less active, and that the relationship is reversed when the Sun is more active. Other work, however, finds no clear QBO relationship at solar maximum [*Camp and Tung*, 2007]. The conflicting interpretations are explained by *Roy and Haigh* [2010] as due to the use of different pressure levels in the definition of

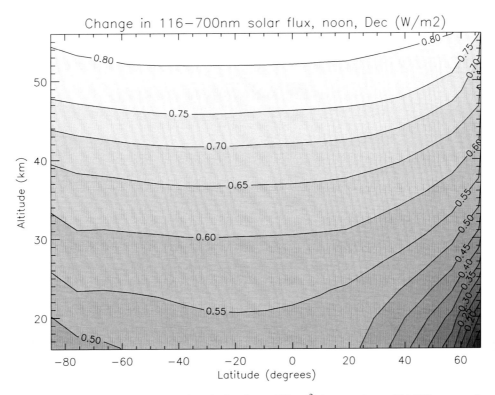

Figure 4. Difference between 2007 and 2000 in solar irradiance (W m^{-2}) integrated over 116–700 nm as a function of latitude and altitude for noon, December.

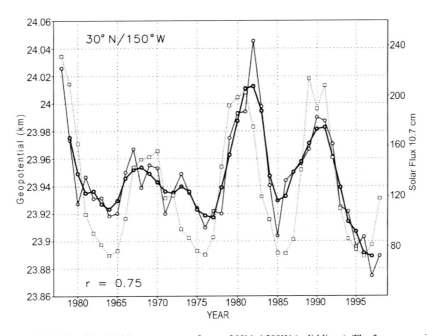

Figure 5. Geopotential height of the 30-hPa pressure surface at 30°N, 150°W (solid lines). The 3-year running mean (bold line). Solar activity as measured by 10.7 cm radio flux (dashed line). From *Labitzke* [2001]. Reproduced by permission of the publisher (http://borntraeger-cramer.de).

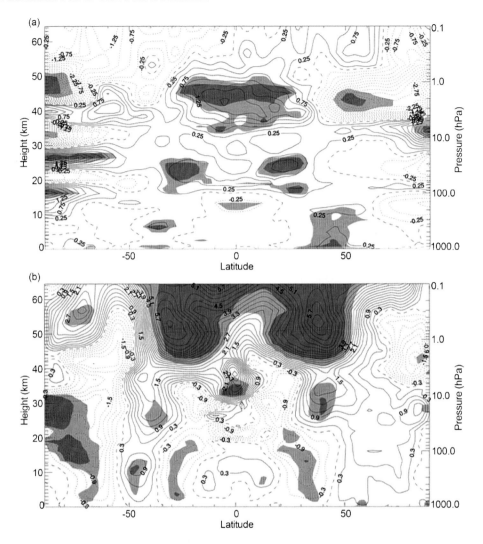

Figure 6. Difference between solar cycle minimum and maximum in (a) annual mean zonal mean temperature (K) and (b) DJF zonal mean zonal wind (m s^{-1}) from multiple linear regression analysis of ERA-40 data. Shaded regions are deemed statistically significant at the 5% and 1% levels. From *Crooks and Gray* [2005]. Copyright American Meteorological Society.

the phase of the QBO. Such a sensitivity indicates the caution which should be applied in making interpretations of the observational analyses.

Figure 6b presents some results from a multiple regression analysis of zonal mean zonal winds for the boreal winter season. When the Sun is more active, there is a strong positive zonal wind response in the winter hemisphere subtropical lower mesosphere and upper stratosphere, as previously noted in analyses of rocketsonde and NCEP data [*Hood et al.*, 1993; *Kodera and Yamazaki*, 1990]. The zonal wind anomaly is observed to propagate downward with time over the course of the winter [*Kodera and Kuroda*, 2002]. In the troposphere, the wind anomalies indicate that the midlatitude jets are weaker and positioned farther poleward when the Sun is more

active, as found by *Haigh et al.* [2005]. This has implications for the positions of the storm tracks and provides evidence for a solar signal in midlatitude climate.

3.2. Dynamical Response to Changes in Irradiance

While absorption of solar radiation, as illustrated by Figure 3, can explain the observed warming of the upper stratospheric when the Sun is more active, it is clear that radiative processes alone cannot account for the entire solar signal in zonal mean temperature. This is particularly apparent in the latitudinal gradient of temperature change near the stratopause and in much of the solar signal in the lower stratosphere and troposphere.

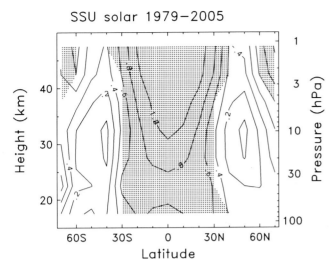

SSU solar 1979-2005

Figure 7. Solar cycle signal in zonal mean temperature (K) as a function of latitude and altitude from multiple linear regression analysis of SSU and MSU channel 4 data. Shading denotes that the solar fit is significant at the two-sigma level. From *Randel et al.* [2009].

That variations in solar heating might influence the dynamical structure of the middle atmosphere was first appreciated during early days of research in middle atmosphere dynamics. *Hines* [1974] proposed that changes in meridional temperature gradients would influence the zonal wind structure and, thus, the propagation of planetary waves. Further developments of this idea [*Bates*, 1981; *Geller and Alpert*, 1980; *Kodera*, 1995] proposed that the momentum deposited by the planetary waves influences the strength of the mean overturning of the stratosphere (Brewer-Dobson circulation) and, thus, the temperature and wind structure. The changed wind structure then has a further effect on wave propagation. With greater solar heating near the tropical stratopause, the stratospheric jets are stronger, the polar vortices less disturbed, and the Brewer-Dobson circulation weaker [*Kodera and Kuroda*, 2002]. This produces cooling in the polar lower stratosphere due to weaker descent and warming at low latitudes through weaker ascent; a cartoon representing these processes is reproduced in Figure 9. Thus, in the lower stratosphere, the tropics are warmer than would result from radiative processes alone, and the poles are cooler.

An additional perspective [*Gray et al.*, 2001] demonstrates with rocketsonde and satellite data that temperature anomalies

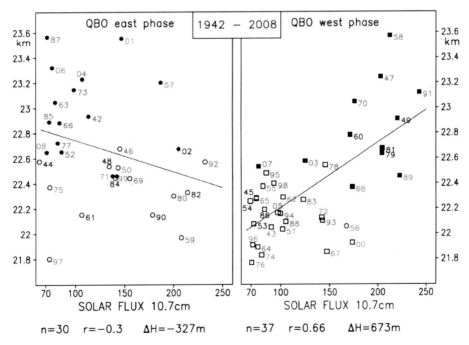

Figure 8. The 30-hPa geopotential height (km) over the North Pole in February for each year from 1942 to 2006 plotted against solar radio flux, sorted into two groups according to the phase of the quasi-biennial oscillation: 1958–1986 (plus signs); 1987–2006 (solid squares); 1948–1957 (solid circles); and 1942–1947 (diamonds). Definitions are n, number of years; r, correlation coefficient; and H, the mean difference of the heights (m) between solar maxima and minima. From *Labitzke et al.* [2006]. Reproduced by permission of the publisher (http://borntraeger-cramer.de).

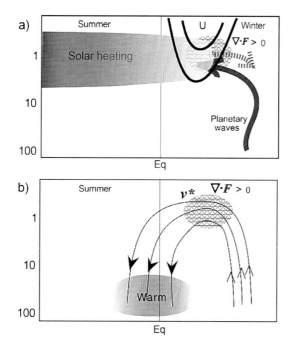

Figure 9. Schematic indicating mechanisms whereby variations in solar heating of the upper stratosphere may influence the atmosphere below through dynamical coupling. (a) At solar maximum, an increase in the latitudinal gradients of solar heating strengthens the stratospheric winter polar jet and, thus, affects the upward propagation of planetary-scale waves. (b) Increased divergence of wave momentum weakens the mean meridional circulation of the middle atmosphere, thus warming the tropical lower stratosphere. From *Kodera and Kuroda* [2002].

in the tropical upper stratosphere (potentially related to solar activity) are correlated with subsequent temperature anomalies in the polar lower stratosphere. Results from a mechanistic model suggest that zonal wind anomalies in the subtropical upper stratosphere can influence the timing and amplitude of sudden stratospheric warmings. Such a relationship between equatorial winds in the upper stratosphere and the timing of sudden stratospheric warmings could help to explain the interaction between the solar cycle and QBO influences on polar temperatures [*Gray et al.*, 2004].

4. SOLAR CYCLE VARIABILITY IN STRATOSPHERIC OZONE

4.1. Observations

As outlined in section 2.1 stratospheric ozone is produced by short wavelength solar UV radiation and destroyed by radiation at longer UV and visible wavelengths. Because the amplitude of solar cycle variability is greater in the far UV (see

section 2.3), ozone production is more strongly modulated by solar activity than its destruction, and this leads to a higher net production of stratospheric ozone during periods of higher solar activity. Multiple regression analyses of ozone measurements from ground- and space-based instruments suggest that in low to midlatitudes, the vertically integrated ozone column varies by 1%–3% in phase with the 11-year solar cycle, with the largest signal in the subtropics [*Randel and Wu*, 2007].

The vertical distribution of the solar signal in ozone is more difficult to establish because of the short length of individual observational records and problems of intercalibration of the various instruments. Figure 10, [*Randel and Wu*, 2007; see also *Chipperfield et al.*, 2007; *Soukharev and Hood*, 2006], shows the latitude-height solar signal from a multiple regression analysis of a composite ozone data set for 1979–2005. Solar cycle variations in tropical upper stratospheric ozone are around 2% to 4% peak-peak, and positive responses are also present at middle and higher latitudes in the middle stratosphere and in the tropics below the 20-hPa level. A smaller, statistically insignificant, response is found in the tropical middle stratosphere. The lower stratospheric ozone response is reflected in the ozone column signal.

4.2. UV Photochemistry

Attempts to predict the response of the stratosphere to solar UV variability were first carried out with two-dimensional chemistry transport models. These predicted a solar cycle response with a peak warming of around 1 K near the stratopause and peak increases in ozone of around 2% at altitudes around 40 km, with perturbations in both temperature and ozone monotonically decreasing toward the tropopause [*Fleming et al.*, 1995; *Garcia et al.*, 1984; *Haigh*, 1994; *Huang and Brasseur*, 1993]. They did not reproduce the more complex latitudinal and vertical gradients demonstrated in Figures 6a and 10, which, in particular, show a secondary peak around 20–25 km in temperature and ozone change, respectively. This indicates that the ozone response, at least in the middle and lower stratosphere, is influenced by modifications to its transport brought about by solar-induced changes in atmospheric circulation. Furthermore, as the structure in the temperature signal is fundamentally related to the ozone response, it is unlikely that any simulation will satisfactorily reproduce the one without the other.

One recent study, which used a fixed dynamical heating model, suggested that the solar signal in temperature is strongly determined by the ozone distribution [*Gray et al.*, 2009]. Earlier studies using general circulation models in which ozone changes were prescribed did not produce the secondary peak in temperature, but none of these used an ozone

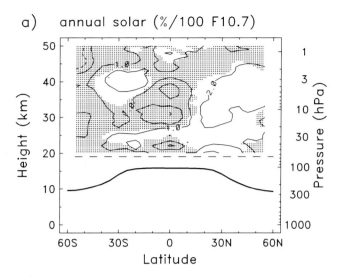

a) annual solar (%/100 F10.7)

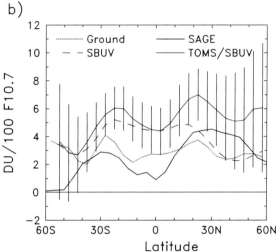

b)

Figure 10. Ozone response to solar variability from multiple linear regression analysis (a) as a function of latitude and altitude from zonal mean Stratospheric Aerosol and Gas Experiment (SAGE) I and II data, 1979–2005. Shading denotes that the fit is not statistically significant. (b) Latitudinal profile of the solar cycle variations in column ozone, derived from vertically integrated SAGE I and II data (over 20–50 km), and three column ozone data sets (ground-based, solar backscatter ultraviolet (SBUV), and merged Total Ozone Mapping Spectrometer (TOMS)/SBUV data). Error bars on the TOMS/SBUV curve denote two-sigma uncertainty in the fit. The results are presented as the change for 100 units of solar radio flux; over a typical solar cycle, this varies by about 130 units. From *Randel and Wu* [2007].

field with the similar structure [*Balachandran and Rind*, 1995; *Haigh*, 1996, 1999; *Matthes et al.*, 2004; *Shindell et al.*, 1999].

More recently, models with fully interactive chemistry have been employed so that the imposed irradiance variations affect both the radiative heating and the ozone

photolysis rates, allowing feedback between ozone concentrations, diabatic heating, and transport (see the review by *Austin et al.* [2008]). These models are now simulating an improved vertical structure of the annual mean ozone signal in the tropics, with some including the lower stratospheric maximum (see Plate 3) although there is no clear picture as to what factor is responsible for the apparent improvement; candidates include time-varying sea surface temperatures [*Austin et al.*, 2007], transient solar input [*Garcia et al.*, 2007], high vertical resolution or the ability of the model to produce an internally generated QBO [*Schmidt et al.*, 2010], or inadvertent aliasing with the signal of ENSO [*Marsh and Garcia*, 2007].

4.3. Solar Energetic Particles

Solar activity is manifest not only in variations of the Sun's emission of electromagnetic radiation but also in a range of other parameters. One of these is the occurrence and severity of coronal mass ejections, which result in the emission of energetic particles, some of which reach the Earth. Precipitating electrons and solar protons follow the Earth's magnetic field lines and so have greatest initial impact at high latitudes. They dissociate and ionize background constituents leading to increases in the budgets of nitrogen oxides (NO_y) and hydrogen oxides (HO_x). The lifetime of the HO_x is only hours to days, but the lifetime of NO_y is much longer, and decreases of ozone lasting several months have been observed in the polar mesosphere and upper stratosphere following large solar proton events (SPEs) [*Jackman et al.*, 2005]. The impact on ozone can last longer and penetrate to lower altitudes in winter polar regions, but the impact of temperature and circulation is very small at these altitudes [*Jackman et al.*, 2009; *Lu et al.*, 2008]. The influence of less energetic SPEs higher in the atmosphere may also be felt in the winter polar upper stratosphere due to the downward transport of the NO_y produced [*Baumgaertner et al.*, 2009; *Randall et al.*, 2007].

It is interesting to note that the effect of energetic particle events on ozone is in the opposite sense to that of enhanced UV irradiance. As particle events tend to occur more often when the Sun is in a declining phase of the solar cycle, the combined effect on ozone may be complex in its geographical, altitudinal, and temporal distribution and may confuse regression analyses (such as shown in Figure 10), which use only radiative indicators of solar activity, especially at high latitudes.

5. COUPLING BETWEEN THE STRATOSPHERE AND THE TROPOSPHERE

While evidence of a solar signal in the stratosphere is unequivocal, it is far more subtle in the troposphere. Nevertheless,

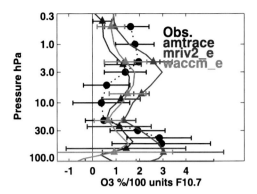

Plate 3. Solar cycle signal in tropical ozone profile from satellite data (black) and three coupled chemistry-climate model ensembles of simulations (colors). Horizontal bars indicate 95% confidence intervals. From *Austin et al.* [2008].

there is a growing body of evidence for solar cycle signals in tropospheric temperature [*Haigh*, 2003; *van Loon and Labitzke*, 1998; *van Loon and Shea*, 2000], zonal wind [*Haigh et al.*, 2005], and circulation [*Gleisner and Thejll*, 2003]. Solar radiation reaching these levels is predominantly at visible wavelengths, which penetrate to the Earth's surface and which show only a very small response to solar variability (see Plate 1, bottom), so a direct radiative response seems unlikely. Furthermore, modeling studies of the impact of varying UV radiation, in which sea surface temperatures have been fixed, are at least qualitatively successful in simulating the tropospheric patterns of response to solar variability [*Haigh*, 1996, 1999; *Larkin et al.*, 2000; *Matthes et al.*, 2006; *Shindell et al.*, 1999] indicating a dynamical influence of changes in the stratosphere on the troposphere rather than a direct radiative effect.

More generally, a number of different studies have indicated that such a downward influence does take place. Observational analyses suggest a downward propagation of polar circulation anomalies in the northern [*Hansen et al.*, 2002; *Kennaugh et al.*, 1997] and southern [*Haigh and Roscoe*, 2009; *Shindell et al.*, 2004] hemispheres. Coherent variations have also been observed between temperatures in the polar lower stratosphere, the tropical tropopause, and tropospheric temperatures that are consistent with possible changes in the Hadley circulation, tropical convection, and latent heat release [*Salby and Callaghan*, 2005]. These signals, however, do not necessarily imply causality from above [*Plumb and Semeniuk*, 2003].

Model studies have also demonstrated a downward influence from Antarctic stratospheric ozone depletion on the Southern Annular Mode [*Gillett and Thompson*, 2003] and from stratospheric temperature trends on the NAO [*Scaife et al.*, 2005]. While these studies do not specifically address the

impact of solar variability on climate, they do suggest that the troposphere responds to (any) perturbation imposed in the stratosphere.

There are many proposed mechanisms for a dynamical influence from the lower stratosphere into the troposphere (see reviews by *Haynes* [2005] and *Shepherd* [2002]). These include response of the mean meridional circulation to angular momentum forcing above ("downward control,") [*Haynes et al.*, 1991] quasi-instantaneous geostrophic adjustment within the troposphere to changes in the potential vorticity structure of the tropopause region [*Black*, 2002; *Hartley et al.*, 1998], modification of the refraction [*Hartmann et al.*, 2000] or reflection [*Perlwitz and Harnik*, 2003] of upward propagating planetary scale waves, and feedbacks between changes in the mean-flow and tropospheric baroclinic eddies [*Kushner and Polvani*, 2004; *Song and Robinson*, 2004].

Coupling between the Hadley circulation and midlatitude eddies may also play a key part. Studies with a simple climate model, in which an anomalous adiabatic heating was imposed in the tropical lower stratosphere, found a zonal mean tropospheric response qualitatively similar to that observed in response to the solar cycle [*Haigh et al.*, 2005]. Consistent with this, the Hadley circulation response in coupled chemistry simulations have been linked to the additional heating introduced by solar-induced ozone into the tropical upper troposphere and lower stratosphere [*Shindell et al.*, 2006].

Other experiments with a simple GCM have involved ensembles of model spin-ups to analyze the time development of the response to an applied stratospheric perturbation in order to investigate the chain of causality involved in converting the stratospheric thermal forcing to a tropospheric climate signal [*Haigh and Blackburn*, 2006]. It was found that the initial effect of the change in static stability at the tropopause is to reduce the eddy momentum flux convergence in this region. This is followed by a vertical transfer of the momentum forcing anomaly by an anomalous mean circulation to the surface, where it is partly balanced by surface stress anomalies. The unbalanced part drives the evolution of the vertically integrated zonal flow. It was concluded that solar heating of the stratosphere may produce changes in the circulation of the troposphere even without any direct forcing below the tropopause and that the impact of the stratospheric changes on wave propagation is key to this effect. This idea has been developed further by *Simpson et al.* [2009], who identified a feedback mechanism between the eddy propagation and the wind anomalies, which serves to maintain the initial perturbations introduced by a thermal (solar) impact on the tropopause region. They found that changes to the thermal structure of the lower stratosphere influenced the propagation of synoptic scale waves, creating anomalies in eddy heat and momentum fluxes, which drove changes in zonal wind and meridional circulation throughout the troposphere. These tropospheric changes then influenced the subsequent propagation of waves, so as to reinforce the initial perturbations.

Although details of the mechanisms involved are still not fully established, it is becoming increasingly clear that solar variability may influence the climate of the troposphere through processes whereby UV heating of the stratosphere indirectly influences the troposphere through dynamical coupling.

6. SUMMARY

Measurements from satellite instruments have established that the radiation emitted by the Sun varies in time alongside other indicators of solar activity, such as the number of sunspots. Total (spectrally integrated) irradiance varies by only a small fraction, around 0.08% of the total, over the 11-year sunspot cycle, but this conceals a very different behavior across the spectrum. Models of solar activity suggest that the variations are of much greater amplitude at shorter wavelengths (e.g., a few percent around 200 nm), and available satellite data confirm this, although the observational data show a wide range of values for the actual magnitude.

Solar UV radiation is absorbed in the stratosphere, where it influences temperatures and ozone concentrations. The response in the tropical upper stratosphere over the 11-year cycle is of the order 1 K in temperature and 2%–3% in ozone, which can largely be explained by direct absorption of solar UV radiation and associated photochemical effects. In the lower stratosphere, the temperature signal, of the order 0.5–1 K cannot be explained by direct radiative heating, indicating some dynamical response to the Sun, and similarly, the 2%–3% variation in lower stratospheric in ozone is also most likely produced by transport processes. Coupled chemistry-climate models are able to reproduce these signals, within the wide bounds of uncertainty, although interpretation of the processes involved is not complete. At winter higher latitudes, the solar signal is modulated by the phase of the quasi-biennial oscillation; this may be associated with a modulation of the transmission of planetary waves through the middle atmosphere in a mean state altered by the changes in solar irradiance.

A growing body of evidence that the state of the stratosphere influences the troposphere by dynamical coupling suggests that any solar influence in the middle atmosphere may produce an impact on tropospheric climate manifest in changes in phase of polar modes of variability and also the mean circulation in the tropics.

REFERENCES

Austin, J., L. L. Hood, and B. E. Soukharev (2007), Solar cycle variations of stratospheric ozone and temperature in simulations of a coupled chemistry-climate model, *Atmos. Chem. Phys.*, 7(6), 1693–1706.

Austin, J., et al. (2008), Coupled chemistry climate model simulations of the solar cycle in ozone and temperature, *J. Geophys. Res.*, 113, D11306, doi:10.1029/2007JD009391.

Balachandran, N. K., and D. Rind (1995), Modeling the effects of UV variability and the QBO on the troposphere-stratosphere system .1. The middle atmosphere, *J. Clim.*, 8(8), 2058–2079.

Bates, J. R. (1981), A dynamical mechanism through which variations in solar ultraviolet-radiation can influence tropospheric climate, *Sol. Phys.*, 74(2), 399–415.

Baumgaertner, A. J. G., P. Jockel, and C. Bruhl (2009), Energetic particle precipitation in ECHAM5/MESSy1-Part 1: Downward transport of upper atmospheric NOx produced by low energy electrons, *Atmos. Chem. Phys.*, 9(8), 2729–2740.

Black, R. X. (2002), Stratospheric forcing of surface climate in the Arctic oscillation, *J. Clim.*, 15(3), 268–277.

Camp, C. D., and K. K. Tung (2007), The influence of the solar cycle and QBO on the late-winter stratospheric polar vortex, *J. Atmos. Sci.*, 64(4), 1267–1283.

Chipperfield, M. P., et al. (2007), Global ozone: Past and present, *Rep. 50*, World Meteorol. Organ. Geneva, Switzerland.

Crooks, S. A., and L. J. Gray (2005), Characterization of the 11-year solar signal using a multiple regression analysis of the ERA-40 dataset, *J. Clim.*, 18(7), 996–1015.

Fleming, E. L., S. Chandra, C. H. Jackman, D. B. Considine, and A. R. Douglass (1995), The middle atmospheric response to short and long-term solar UV variations: Analysis of observations and 2D model results, *J. Atmos. Terr. Phys.*, 57(4), 333–365.

Garcia, R. R., S. Solomon, R. G. Roble, and D. W. Rusch (1984), A numerical response of the middle atmosphere to the 11-year solar-cycle, *Planet. Space Sci.*, 32(4), 411–423.

Garcia, R. R., D. R. Marsh, D. E. Kinnison, B. A. Boville, and F. Sassi (2007), Simulation of secular trends in the middle atmosphere, 1950–2003, *J. Geophys. Res.*, 112, D09301, doi:10.1029/2006JD007485.

Geller, M. A., and J. C. Alpert (1980), Planetary wave coupling between the troposphere and the middle atmosphere as a possible sun-weather mechanism, *J. Atmos. Sci.*, 37(6), 1197–1215.

Gillett, N. P., and D. W. J. Thompson (2003), Simulation of recent Southern Hemisphere climate change, *Science*, 302(5643), 273–275.

Gleisner, H., and P. Thejll (2003), Patterns of tropospheric response to solar variability, *Geophys. Res. Lett.*, 30(13), 1711, doi:10.1029/2003GL017129.

Gray, L. J., S. J. Phipps, T. J. Dunkerton, M. P. Baldwin, E. F. Drysdale, and M. R. Allen (2001), A data study of the influence of the equatorial upper stratosphere on northern-hemisphere stratospheric sudden warmings, *Q. J. R. Meteorol. Soc.*, *127*(576), 1985–2003.

Gray, L. J., S. Crooks, C. Pascoe, S. Sparrow, and M. Palmer (2004), Solar and QBO influences on the timing of stratospheric sudden warmings, *J. Atmos. Sci.*, *61*(23), 2777–2796.

Gray, L. J., S. T. Rumbold, and K. P. Shine (2009), Stratospheric temperature and radiative forcing response to 11-year solar cycle changes in irradiance and ozone, *J. Atmos. Sci.*, *66*(8), 2402–2417.

Haigh, J. D. (1994), The role of stratospheric ozone in modulating the solar radiative forcing of climate, *Nature*, *370*(6490), 544–546.

Haigh, J. D. (1996), The impact of solar variability on climate, *Science*, *272*(5264), 981–984.

Haigh, J. D. (1999), A GCM study of climate change in response to the 11-year solar cycle, *Q. J. R. Meteorol. Soc.*, *125*(555), 871–892.

Haigh, J. D. (2003), The effects of solar variability on the Earth's climate, *Philos. Trans. R. Soc. London, Ser. A*, *361*(1802), 95–111.

Haigh, J. D., and M. Blackburn (2006), Solar influences on dynamical coupling between the stratosphere and troposphere, *Space Sci. Rev.*, *125*(1–4), 331–344.

Haigh, J. D., and H. K. Roscoe (2009), The final warming date of the Antarctic polar vortex and influences on its interannual variability, *J. Clim.*, *22*(22), 5809–5819.

Haigh, J. D., M. Blackburn, and R. Day (2005), The response of tropospheric circulation to perturbations in lower-stratospheric temperature, *J. Clim.*, *18*(17), 3672–3685.

Haigh, J. D., A. R. Winning, R. Toumi, and J. W. Harder (2010), An influence of solar spectral variations on radiative forcing of climate, *Nature*, in press.

Hansen, J., et al. (2002), Climate forcings in Goddard Institute for Space Studies SI2000 simulations, *J. Geophys. Res.*, *107*(D18), 4347, doi:10.1029/2001JD001143.

Harder, J. W., J. M. Fontenla, P. Pilewskie, E. C. Richard, and T. N. Woods (2009), Trends in solar spectral irradiance variability in the visible and infrared, *Geophys. Res. Lett.*, *36*, L07801, doi:10.1029/2008GL036797.

Hartley, D. E., J. T. Villarin, R. X. Black, and C. A. Davis (1998), A new perspective on the dynamical link between the stratosphere and troposphere, *Nature*, *391*(6666), 471–474.

Hartmann, D. L., J. M. Wallace, V. Limpasuvan, D. W. J. Thompson, and J. R. Holton (2000), Can ozone depletion and global warming interact to produce rapid climate change?, *Proc. Natl. Acad. Sci. U. S. A.*, *97*(4), 1412–1417.

Haynes, P. (2005), Stratospheric dynamics, *Annu. Rev. Fluid Mech.*, *37*, 263–293.

Haynes, P. H., C. J. Marks, M. E. McIntyre, T. G. Shepherd, and K. P. Shine (1991), On the downward control of extratropical diabatic circulations by eddy-induced mean zonal forces, *J. Atmos. Sci.*, *48*(4), 651–679.

Hines, C. O. (1974), Possible mechanism for production of sun-weather correlations, *J. Atmos. Sci.*, *31*(2), 589–591.

Holton, J. R., and H. C. Tan (1980), The influence of the equatorial quasi-biennial oscillation on the global circulation at 50 mb, *J. Atmos. Sci.*, *37*(10), 2200–2208.

Hood, L. L., J. L. Jirikowic, and J. P. McCormack (1993), Quasi-decadal variability of the stratosphere - influence of long-term solar ultraviolet variations, *J. Atmos. Sci.*, *50*(24), 3941–3958.

Huang, T. Y. W., and G. P. Brasseur (1993), Effect of long-term solar variability in a 2-dimensional interactive model of the middle atmosphere, *J. Geophys. Res.*, *98*(D11), 20,413–20,427.

Jackman, C. H., M. T. DeLand, G. J. Labow, E. L. Fleming, D. K. Weisenstein, M. K. W. Ko, M. Sinnhuber, J. Anderson, and J. M. Russell (2005), The influence of the several very large solar proton events in years 2000-2003 on the neutral middle atmosphere, in *Influence of the Sun's Radiation and Particles on the Earth's Atmosphere and Climate*, edited by J. Lastovicka, pp. 445–450, Elsevier, Orlando, Fla.

Jackman, C. H., D. R. Marsh, F. M. Vitt, R. R. Garcia, C. E. Randall, E. L. Fleming, and S. M. Frith (2009), Long-term middle atmospheric influence of very large solar proton events, *J. Geophys. Res.*, *114*, D11304, doi:10.1029/2008JD011415.

Kennaugh, R., S. Ruth, and L. J. Gray (1997), Modeling quasi-biennial variability in the semiannual double peak, *J. Geophys. Res.*, *102*(D13), 16,169–16,187.

Kodera, K. (1995), On the origin and nature of the interannual variability of the winter stratospheric circulation in the northern-hemisphere, *J. Geophys. Res.*, *100*(D7), 14,077–14,087.

Kodera, K., and Y. Kuroda (2002), Dynamical response to the solar cycle, *J. Geophys. Res.*, *107*(D24), 4749, doi:10.1029/2002JD002224.

Kodera, K., and K. Yamazaki (1990), Long-term variation of upper stratospheric circulation in the northern-hemisphere in December, *J. Meteorol. Soc. Jpn.*, *68*(1), 101–105.

Kushner, P. J., and L. M. Polvani (2004), Stratosphere-troposphere coupling in a relatively simple AGCM: The role of eddies, *J. Clim.*, *17*(3), 629–639.

Labitzke, K. (2001), The global signal of the 11-year sunspot cycle in the stratosphere: Differences between solar maxima and minima, *Meteorol. Z.*, *10*(2), 83–90.

Labitzke, K., M. Kunzel, and S. Broonnimann (2006), Sunspots, the QBO and the stratosphere in the North Polar Region–20 years later, *Meteorol. Z.*, *15*(3), 355–363.

Larkin, A. (2000), An investigation of the effects of solar variability on climate using atmospheric models of the troposphere and stratosphere, Ph.D. thesis, Univ. of London, London, U. K.

Larkin, A., J. D. Haigh, and S. Djavidnia (2000), The effect of solar UV irradiance variations on the Earth's atmosphere, *Space Sci. Rev.*, *94*(1–2), 199–214.

Lean, J. L. (2000), Evolution of the sun's spectral irradiance since the Maunder Minimum, *Geophys. Res. Lett.*, *27*(16), 2425–2428.

Lean, J. L., and D. Rind (1998), Climate forcing by changing solar radiation, *J. Clim.*, *11*(12), 3069–3094.

Lee, H., and A. K. Smith (2003), Simulation of the combined effects of solar cycle, quasi-biennial oscillation, and volcanic forcing on stratospheric ozone changes in recent decades, *J. Geophys. Res.*, *108*(D2), 4049, doi:10.1029/2001JD001503.

Lu, H., M. A. Clilverd, A. Seppala, and L. L. Hood (2008), Geomagnetic perturbations on stratospheric circulation in late winter and spring, *J. Geophys. Res.*, *113*, D16106, doi:10.1029/2007JD008915.

Marsh, D. R., and R. R. Garcia (2007), Attribution of decadal variability in lower-stratospheric tropical ozone, *Geophys. Res. Lett.*, *34*, L21807, doi:10.1029/2007GL030935.

Matthes, K., U. Langematz, L. L. Gray, K. Kodera, and K. Labitzke (2004), Improved 11-year solar signal in the Freie Universität Berlin Climate Middle Atmosphere Model (FUB-CMAM), *J. Geophys. Res.*, *109*, D06101, doi:10.1029/2003JD004012.

Matthes, K., Y. Kuroda, K. Kodera, and U. Langematz (2006), Transfer of the solar signal from the stratosphere to the troposphere: Northern winter, *J. Geophys. Res.*, *111*, D06108, doi:10.1029/2005JD006283.

Perlwitz, J., and N. Harnik (2003), Observational evidence of a stratospheric influence on the troposphere by planetary wave reflection, *J. Clim.*, *16*(18), 3011–3026.

Plumb, R. A., and K. Semeniuk (2003), Downward migration of extratropical zonal wind anomalies, *J. Geophys. Res.*, *108*(D7), 4223, doi:10.1029/2002JD002773.

Randall, C. E., V. L. Harvey, C. S. Singleton, S. M. Bailey, P. F. Bernath, M. Codrescu, H. Nakajima, and J. M. Russell (2007), Energetic particle precipitation effects on the Southern Hemisphere stratosphere in 1992–2005, *J. Geophys. Res.*, *112*, D08308, doi:10.1029/2006JD007696.

Randel, W. J., and F. Wu (2007), A stratospheric ozone profile data set for 1979–2005: Variability, trends, and comparisons with column ozone data, *J. Geophys. Res.*, *112*, D06313, doi:10.1029/2006JD007339.

Randel, W. J., et al. (2009), An update of observed stratospheric temperature trends, *J. Geophys. Res.*, *114*, D02107, doi:10.1029/2008JD010421.

Roy, I., and I. D. Haigh (2010), Solar cycle signals in sea level pressure and sea surface temperature, *Atmos. Chem. Phys.*, *10*(6), 3147–3153.

Salby, M. L., and P. F. Callaghan (2005), Interaction between the Brewer-Dobson circulation and the Hadley circulation, *J. Clim.*, *18*(20), 4303–4316.

Scaife, A. A., J. R. Knight, G. K. Vallis, and C. K. Folland (2005), A stratospheric influence on the winter NAO and North Atlantic surface climate, *Geophys. Res. Lett.*, *32*, L18715, doi:10.1029/2005GL023226.

Schmidt, H., G. P. Brasseur, and M. A. Giorgetta (2010), Solar cycle signal in a general circulation and chemistry model with internally generated quasi-biennial oscillation, *J. Geophys. Res.*, *115*, D00I14, doi:10.1029/2009JD012542.

Shepherd, T. G. (2002), Issues in stratosphere–troposphere coupling, *J. Meteorol. Soc. Jpn.*, *80*(4B), 769–792.

Shindell, D. T., D. Rind, N. Balachandran, J. L. Lean, and P. Lonergan (1999), Solar cycle variability, ozone, and climate, *Science*, *284*(5412), 305–308.

Shindell, D. T., B. P. Walter, and G. Faluvegi (2004), Impacts of climate change on methane emissions from wetlands, *Geophys. Res. Lett.*, *31*, L21202, doi:10.1029/2004GL021009.

Shindell, D. T., G. Faluvegi, R. L. Miller, G. A. Schmidt, J. E. Hansen, and S. Sun (2006), Solar and anthropogenic forcing of tropical hydrology, *Geophys. Res. Lett.*, *33*, L24706, doi:10.1029/2006GL027468.

Simpson, I. R., M. Blackburn, and J. D. Haigh (2009), The role of eddies in driving the tropospheric response to stratospheric heating perturbations, *J. Atmos. Sci.*, *66*(5), 1347–1365.

Song, Y. C., and W. A. Robinson (2004), Dynamical mechanisms for stratospheric influences on the troposphere, *J. Atmos. Sci.*, *61*(14), 1711–1725.

Soukharev, B. E., and L. L. Hood (2006), Solar cycle variation of stratospheric ozone: Multiple regression analysis of long-term satellite data sets and comparisons with models, *J. Geophys. Res.*, *111*, D20314, doi:10.1029/2006JD007107.

Uppala, S. M., et al. (2005), The ERA-40 re-analysis, *Q. J. R. Meteorol. Soc.*, *131*(612), 2961–3012.

van Loon, H., and K. Labitzke (1998), The global range of the stratospheric decadal wave. Part I: Its association with the sunspot cycle in summer and in the annual mean, and with the troposphere, *J. Clim.*, *11*(7), 1529–1537.

van Loon, H., and D. J. Shea (2000), The global 11-year solar signal in July–August, *Geophys. Res. Lett.*, *27*(18), 2965–2968.

J. D. Haigh, Department of Physics, The Blackett Laboratory, Imperial College London, London SW7 2AZ, UK. (j.haigh@imperial.ac.uk)

AGU Category Index

Acoustic-gravity waves, 109

Atmospheric sciences, vii

Climate change and variability, 59, 123

General circulation, 23, 59, 93, 109

General or miscellaneous, 137

Global climate models, 157

Instruments and techniques, 137

Middle atmosphere dynamics, vii, xiii, 1, 5, 23, 43, 59, 93, 109, 123, 157

Middle atmosphere: composition and chemistry, xiii, 1, 5, 43, 93, 123, 137, 157, 173

Middle atmosphere: constituent transport and chemistry, vii, xiii, 1, 5, 43, 137, 157

Radiative processes, 173

Solar and stellar variability, xiii, 1

Solar variability, 173

Stratosphere/troposphere interactions, 5, 23, 43, 59, 93, 123, 173

Theoretical modeling, vii

Tides and planetary waves, 23

Index

Note: Page numbers with italicized *f* and *t* refer to figures and tables

A

absolute vorticity, 45
aerosols, 19
age of air, 150–151
Agung, 125
Airborne Antarctic Ozone Experiment, 152
Airborne Arctic Stratosphere Expedition, 152
Aleutian Islands, 46
angular momentum, 29, 31
annular modes, 51, 59–87, 65. *See also* polar vortices
  calculation of, 64
  climate responses, 85–86
  dynamics, 69–75
  empirical orthogonal function analysis, 60–62, 64–65
  in general circulation models, 75–85
    climate responses, 79–82
    climate trends, 79–82
    fluctuation dissipation theory, 82–85
    ozone-depletion effect, 82
    simulation of intraseasonal variability, 76–79
    stratospheric representation, 85–86
  hemispherically symmetric, 65
  Holton-Mass model, 74–75
  index, 77f
  quasigeotrophic dynamics, 68f
  regression in zonal wind, 77f
  research papers, 60
  stratospheric, 74
  structure, 60–69
Antarctic circulation, 159–163
Antarctic ozone hole, 157–169
  Antarctic circulation, 159–163
  chemistry, 163–168
  evolution of, 168–169
  observations, 158–159
  overview, 157
  recovery of, 168
Antarctic temperature, 45
Antarctic vortex, 44, 46, 52, 52f
Arctic oscillation, 105
Arctic temperatures, 45

Arctic vortex, 46, 52, 52f
atmospheric tides, 17–18
Aura Microwave Limb Sounder, 129–130, 138, 139f, 165

B

balloon measurements, 5
boreal winter maps, 115f
Brewer-Dobson circulation, 6
  chlorofluorocarbons and, 163, 166
  methane transport and, 104
  planetary waves and, 98, 181
  quasi-biennial oscillation, 97
  solar cycle and, 103
  trace gas transport and, 138
  wave driving of, 110
bromine, 157–158, 166
Brunt-Väisälä frequency, 6, 8

C

Canadian Middle Atmosphere Model (CMAM), 78f
carbon dioxide, climate response to, 113–114
Charney-Drazin theorems, 8, 25, 34
Chemical Model of the Stratosphere (CLaMS), 146
chemistry-climate models (CCMs), 52–53, 111, 114f
chemistry transport models, 152–153
chlorine, 166
chlorine monoxide, 166
chlorine peroxide, 166–167
chlorofluorocarbons, 19, 163, 166, 168
cirrus clouds, 113
climate change, 113–114
climate responses, 79–82
climate trends, 79–82
Cloud-Aerosol Lidar and Infrared Pathfinder Satellite Observation (CALIPSO), 163
contour advection, 146
convection, 117–118
Coriolis parameter, 6, 25
Coriolis torque, 98

Courant-Friedrich-Lewy (CFL) condition, 152
critical layer, 28–29
critical level, 8–12

D

diabatic heating, 29–30
dinitrogen pentoxide, 165
diurnal tide, 17–18
Dobson units, 158
downward control principle, 144

E

eddy forcing, 29
eddy momentum flux, 71
eddy transports, 19–20
effective diffusion, 149–150
El Chichon volcano, 104, 125, 126*f*
Eliassen and Palm's first theorem, 7
Eliassen-Palm (EP) flux, 28, 32–33, 34–36, 68*f*,
        70–71, 110
Eliassen response, 30
El Niño–Southern Oscillation (ENSO), 63–64,
        125–126, 178
empirical orthogonal function (EOF), 60, 64–65
equatorial dynamics, 93–106
equatorial stratosphere, 93–106. *See also* stratosphere
    extratropics, 99–102
    impact on troposphere, 105–106
    meridional structure, 98–99
    quasi-biennial oscillation, 95–98
    semiannual oscillation, 95–98
    solar cycle and, 102–103
    trace gas transport, 103–105
    zonal winds, 93–95
equivalent effective stratospheric chlorine (EESC),
        168
equivalent latitude, 149–150
ERA-40, 178
ERS-2, 176
Ertel potential vorticity, 143–144
extratropical winter stratosphere, 23–39
extratropics, 99–102
extreme events, 49–52

F

Fickian diffusion, 20
final warming, 51
fluctuation dissipation theory, 82–85
fluid density, 45

G

general circulation models (GCMs), 106
    annular modes, 79–86
        climate responses, 79–82
        climate trends, 79–82
        fluctuation dissipation theory, 82–85
        ozone-depletion effect, 82
        stratospheric representation, 85–86
    atmosphere, 73
    simplified, 73, 76
Goddard Earth Observing System (GEOS), 169
gravity waves, 8–12, 109–119
    climate change and, 113–114
    convection-generated, 117–118
    effects on stratosphere, 110–114
    energy flux, 9
    extratropical effects, 110–112
        cold-pole problem, 111
        ozone chemistry, 111–112
        polar stratospheric clouds, 112
        transition to summer easterlies, 111
        wave driving of Brewer-Dobson circulation, 110
    group velocity, 8–9
    internal, 8–9, 8*f*
    jet sources, 118
    kinetic energy per unit volume, 9
    momentum fluxes, 114–116
    momentum forcing, 109–110
    mountain wave parameterization effects, 113–114
    phase velocity, 7, 8–9
    Rossby-gravity waves, 97
    satellite observations, 116
    steady, 7
    thermal forcing, 109–110
    tropical effects, 112–113
        quasi-biennial oscillation, 112
        semiannual oscillation, 112–113
        tropical cirrus, 113
greenhouse gases (GHGs), 52–53
Green's function, 150–151
group velocity, 8–9

H

Hadley circulation, 184
Halley Bay station, 157, 159*f*
Halogen Occultation Experiment (HALOE), 103*f*, 129
Holton-Mass model, 34, 74, 75*f*
Holton-Tan anticorrelative behavior, 101
Hooke, Robert, 5
Hough functions, 17

I

inertia-gravity waves, 97
internal gravity waves, 8–9, 8*f*

J

jet sources, 118

K

Kelvin cat's eyes, 28
Kelvin wave, 12*t*, 97

L

Lagrangian methods, 145–146
Laplace's tidal equation, 17
Limb Infrared Monitor of the Stratosphere, 138
linear theory, 33–37
Lyapunov coefficient, 146

M

meridional, 98–99
meridional circulation, 30
meridional heat flux, 71
meridional momentum, 144
meridional wave number, 6, 26
mesospheric winds, 10–12
methyl bromide (CH_3Br), 168
methyl chloride (CH_3Cl), 168
methyl chloroform (CH_3CCl_3), 168
Microwave Limb Sounder (MLS), 129, 138, 139*f*, 165
Microwave Sounding Unit (MSU), 124–125
middle atmosphere research, 5–20
    atmospheric tides, 17–18
    critical level, 8–12
    gravity waves, 8–12
    history, 5–6
    planetary waves, 14–17
    quasi-biennial oscillation, 12–14
    stratospheric ozone chemistry, 18–19
    stratospheric sudden warming, 14–17
    stratospheric transport, 19–20
    wave breaking, 8–12
    wave-flow mean interactions, 6–7
modified potential vorticity, 144
momentum fluxes, 114–116
    zonal and meridional components of, 115*f*
momentum forcing, 109–110
Moon, gravitation forces of, 17

mountain wave parameterization, 113–114
Mount Pinatubo, 125, 126*f*
"moving flame" phenomenon, 6
Murgatroyd, R.J., 5

N

National Centers for Environmental Prediction (NCEP),
        178
National Oceanographic and Atmospheric Administration
        (NOAA), 158
Newton, Isaac, 5
nitric acid, 163
nitric acid dihydrate (NAD), 163
nitric acid trihydrate (NAT), 163
nitric oxide, 165
nitrogen, 165
nitrogen oxide, 19, 165
Northern Annular Mode (NAM), 51, 65–66. *See also* annular
        modes; Southern Annular Mode (SAM)
    Arctic oscillation, 105
    climate trends, 79–82
    Eliassen-Palm flux, 68*f*
    index, 49–50
    seasonal cycle of time scales, 67*f*
Northern Hemisphere (NH), 163
    chemistry-climate model, 114*f*
    jet, 65–66
    sudden warmings, 50
    troposphere, 64
nuclear weapons, testing of, 19–20
numerical transport schemes, 151–153. *See also* trace gas
        transport
    chemistry transport models, 152–153
    development of, 152
    requirements, 151–152

O

OH radicals, 19
ozone hole, 157–169
    Antarctic circulation, 159–163
    area, 160*f*
    chemistry, 163–168
    evolution of, 168–169
    observations, 158–159
    overview, 157
    recovery of, 168
Ozone Monitoring Instrument (OMI), 158
ozone/ozone depletion, 18–19
    catalytic cycles for, 166

ozone/ozone depletion (*continued*)
  chemistry of, 163–168
  cold-pole problem and, 111–112
  polar stratospheric clouds and, 112, 165

P

photochemistry, 182–185
photons, 166
Planck's constant, 166
planetary vorticity, 143–144
planetary waves, 14–17, 28–29, 99–101
  amplifications, 37–38
  barrier to propagation, 14
  breaking, 28–29, 47, 49
  Brewer-Dobson circulation and, 98, 181
  critical layer, 28–29
  dispersion relation, 7
  effective refraction index, 6
  in equatorial lower stratosphere, 12*t*
  impact on stratospheric structure, 29–31
  polar vortices, 46
  propagation, 14, 24–28, 44, 99–101
  pump, 30
  quasi-geostrophic, 15–16
  stationary, 7, 14–15
  vertical propagation, 7
Plumb, R. Alan, vii–xi, 5
polar cap potential vorticity, 70
polar cap wave guide, 25
polar night jet, 83*f*, 159–161
polar stratospheric clouds, 157–158. *See also* stratosphere
  Antarctic circulation, 160–161
  formation of, 45
  heterogenous reactions, 165
  ozone chemistry, 112
  ozone depletion rates, 165
  types of, 163
  volume, 52
polar temperatures, 45
polar vortices, 43–54
  climatological structure, 44–45
  extreme events, 49–52
  formation of, 44
  hemispheric differences, 44–45
  potential vorticity, 45–49
  simulations, 48*f*
  stratosphere-troposphere coupling, 51–52
  trends, 52–53
potential vorticity, 45–49
  coordinates, 149

effective diffusivity, 149–150
Eliassen-Palm flux, 70–71
equivalent latitude, 149–150
inversion, 46, 69
maps, 46
quasi-geostrophic, 8
as tracer, 149
tracer transport and, 143–144
zonal mean potential vorticity anomaly, 69
pseudo potential vorticity, 86–87

Q

quasi-biennial oscillation (QBO), 12–14. *See also* semiannual
    oscillation (SAO)
  amplitude, 96*f*
  easterly phase, 100–101
  equatorial temperature anomalies, 99*f*
  equatorial winds, 101
  gravity wave forcing, 112
  Holton-Tan anticorrelative behavior, 101
  impact on extratropics, 99–102
  impact on troposphere, 105–106
  mechanisms, 95–98
  meridional structure, 98–99
  momentum forcing, 12
  numerical modeling of, 106
  overview, 93
  positive/negative phase speed waves, 14
  propagating waves, 97
  radiative forcing, 12
  schematic representation, 13*f*
  solar cycle and, 102–103
  solar cycle variability and, 178
  stages of half cycle, 97*f*
  tracer transport, 103–105
  westerly phase, 100
  zonal winds, 93–95
  zonal winds for westerly and easterly years, 100*f*
quasi-geostrophic dynamics, 69
quasi-geostrophic potential vorticity, 8, 25, 26–27, 69, 86–87
quasi-resonances, 37–38

R

Radiosonde Atmospheric Temperature Products for Assessing
    Climate (RATPAC), 124
Radiosonde Innovation Composite Homogenization (RICH),
    124
reactive uptake coefficient, 165
reconstruction, 149

refractive index, 24–26
residual continuity, 144
reverse domain fill, 146
Richardson number, 9, 146
Rossby critical velocity, 25, 28
Rossby gravity wave, 12*t*, 97
Rossby number, 8
Rossby radius of deformation, 7
Rossby waves. *See* planetary waves

S

satellite observations
    Antarctic ozone hole, 158–159, 162*f*
    gravity waves, 116
    solar radiation, 176–177
    trace gas transport, 138–143
sea level pressure (SLP), 81
sea surface temperatures (SSTs), 73
second-order moment (SOM), 152
semiannual oscillation (SAO), 93. *See also* quasi-biennial
        oscillation (QBO)
    amplitude, 96*f*
    gravity wave forcing, 112–113
    impact on extratropics, 99–102
    impact on troposphere, 105–106
    mechanisms, 95–98
    meridional structure, 98–99
    propagating waves, 97–98
    tracer transport, 103–105
semidiurnal tide, 17–18
solar cycle, 102–103
    dynamical response to changes in irradiance, 180–181
    observations, 178–180
    temperature variability, 178–182
    variability in stratospheric ozone, 182–184
        observations, 182
        solar energetic particles, 183
        UV photochemistry, 182–183
    wind variability, 178–182
    zonal mean temperature and, 181*f*
solar energetic particles, 182–183
solar fluxes, 176–177
solar heating, 17–18
solar radiation, 173–177
    absorption by atmosphere, 173–174
    heating rates, 176–177
    variability of fluxes, 176–177
    variations in solar spectral irradiance, 174–176
    variations in total solar irradiance, 174
solar spectral irradiance, 174–176

SORCE, 176, 177
SORCE satellite, 176–177
Southern Annular Mode (SAM), 51, 65–66. *See also* annular
        modes; Northern Annular Mode (NAM)
    climate trends, 79–82
    Eliassen-Palm flux, 68*f*
    seasonal cycle of time scales, 67*f*
Southern Hemisphere (SH)
    jet, 65–66
    troposphere, 64
stationary mountain waves, 7
sticking coefficient, 165
stratopause, 112–113
stratosphere
    annular modes, 59–87
    equatorial dynamics, 93–106
        extratropics, 99–102
        impact on troposphere, 105–106
        meridional structure, 98–99
        quasi-biennial oscillation, 95–98
        semiannual oscillation, 95–98
        solar cycle and, 102–103
        trace gas transport, 103–105
    gravity waves in, 109–119
        convection-generated, 117–118
        extratropical effects, 110–112
        from jet sources, 118
        model responses to climate changes, 113–114
        momentum fluxes, 114–116
        tropical effects, 112–113
    planetary waves and, 29–31
    solar cycle and, 182–184
        observations, 182
        solar energetic particles, 183
        UV photochemistry, 182–183
    solar radiation and
        absorption of solar spectral radiation, 173–174
        fluxes, 176–177
        heating rates, 176–177
        variations in solar spectral irradiance, 174–176
        variations in total solar irradiance, 174
    temperature, 123–128
        anomalies, 127*f*, 128*f*
        data, 124–125
        observations, 125–128
    trace gas transport, 137–153
        diagnostics, 145–151
        numerical transport schemes, 151–153
        observation, 138–143
    troposphere and, 183–185
    water vapor, 123–131

stratosphere-troposphere coupling, 31–33, 51–52
stratospheric meteorology, 24–28
stratospheric ozone chemistry, 18–19
Stratospheric Sounding Unit (SSU), 124–125
stratospheric sudden warmings (SSWs), 14–17
    barotropic, 48
    chemistry-climate models, 53
    extreme events and, 49–51
    frequency of, 52
    model, 16f, 53
    polar vortex and, 46
    potential vorticity and, 46
    quasi-biennial oscillation and, 99
    wave amplification and, 37–38
stratospheric transport, 19–20
summer easterlies, 111
supercooled ternary solution (STS), 163
surf zone, 15, 29, 47
Syowa ozone, 160f

T

Teiserenc De Bort, Léon, 5
temperatures
    anomalies, 125–126
    Antarctic, 45
    Arctic, 45
    polar, 45
    solar cycle, 176–177
    stratospheric, 123–128
        data, 124–125
        observations, 125–128
thermal forcing, 109–110
Total Ozone Mapping Spectrometer, 157
total solar irradiance (TSI), 174
trace gas transport, 137–153, 145–146. *See also* stratosphere
    diagnostics, 145–151, 149
        age of air, 150–151
        contour advection, 146
        potential vorticity coordinates, 149
        potential vorticity as tracer, 149
        reverse domain fill, 146
        tracer-tracer correlations, 146–149
        trajectory methods, 145–146
    numerical transport schemes, 151–153
        chemistry transport models, 152–153
        development of, 152
    observation, 138–143
    theory, 143–145
trajectory methods, 145–146
transformed Eulerian mean theory, 24
tropical cirrus, 113

troposphere
    annular modes, 59–87
    quasi-biennial oscillation and, 105–106
    semiannual oscillation and, 105–106
    stratosphere and, 183–185
tropospheric jet, 83f

U

ultraviolet (UV) radiation, 182–183
Upper Atmosphere Research Satellite (UARS), 138

V

Venus, 6
vertical wave propagation, 7
vortex displacement events, 50
vortex events, 49–52
vortex split events, 50
vortex stripping, 47
vorticity flux, 71

W

warm vortex events, 67f
water vapor, 128–131
wave amplification, 37–38
wave breaking, 8–12, 9f, 28–29, 47
wave-mean flow interactions, 6–8
wave momentum flux, 9
wave propagation, linear theory of, 24–28, 33–37
Wentzel-Kramers-Brillouin (WKB) theory, 26
Whipple, Fred Lawrence, 5
winds, 77f, 93–95
    over equatorial situations, 10f, 11f
    quasi-biennial oscillation, 93–95
    semiannual oscillation, 93–95
    solar cycle variations in, 178–182
    time-height section, 10f, 11f
    winter and summer patterns, 13f

Z

zonal mean geopotential anomaly, 61f, 62f
zonal mean potential vorticity anomaly, 69
zonal wave number, 6
zonal winds, 77f, 93–95
    over equatorial situations, 10f, 11f
    quasi-biennial oscillation, 93–95
    semiannual oscillation, 93–95
    solar cycle variations in, 178–182
    time-height section, 10f, 11f
    winter and summer patterns, 13f